Homogeneous Hydrogenation with Non-Precious Catalysts

Homogeneous Hydrogenation with Non-Precious Catalysts

Edited by
Johannes F. Teichert

WILEY-VCH

Editor

Prof. Johannes F. Teichert
Technische Universität Berlin
Institut für Chemie
Straße des 17. Juni 115
10623 Berlin
Germany

Cover
Inspiration for the cover picture by Birte M. Zimmermann, Lea T. Brechmann, Benyapa Kaewmee and Trung Tran Ngoc.

Cover Images
(Balloons) © Anastasiia_M/Getty Images, (Cylinder) © golostt/Getty Images, © Martin McCarthy/Getty Images, (Background) © Jaemsri praikajee/Shutterstock

Library of Congress Card No.:
applied for

British Library Cataloguing-in-Publication Data
A catalogue record for this book is available from the British Library.

Bibliographic information published by the Deutsche Nationalbibliothek
The Deutsche Nationalbibliothek lists this publication in the Deutsche Nationalbibliografie; detailed bibliographic data are available on the Internet at <http://dnb.d-nb.de>.

Print ISBN: 978-3-527-34439-0
ePDF ISBN: 978-3-527-81422-0
ePub ISBN: 978-3-527-81420-6
oBook ISBN: 978-3-527-81423-7

Typesetting SPi Global, Chennai, India
Printing and Binding Markono Print Media Pte Ltd, Singapore

Printed on acid-free paper

10 9 8 7 6 5 4 3 2 1

Contents

Preface *ix*

Prelude – A Critical Assessment from an Industrial Point of View *1*
Hans-Ulrich Blaser
1 Some Introductory Remarks *1*
1.1 What Is the Motivation for Developing Non-Noble Metal Catalysts? *2*
1.2 Crucial Parameters for Process Development *3*
1.2.1 The Catalyst: Metal Complex, (Chiral) Ligand *4*
1.2.2 The Process *5*
2 Comparison with the Established Catalysts *5*
2.1 Chemoselective Hydrogenations: Alternatives to Heterogeneous Catalysts *6*
2.1.1 Hydrogenation of Esters *6*
2.1.2 E-Selective Hydrogenation of Alkynes *7*
2.1.3 Selective C=O Reduction of Unsaturated Carbonyl Groups *7*
2.2 Enantioselective Hydrogenations *8*
2.3 A Few Remarkable Transformations Using Cu and Co Complexes *9*
3 Epilogue *10*
References *13*

1 Iron-Catalyzed Homogeneous Hydrogenation Reactions *15*
Thomas Zell and Robert Langer
1.1 Introduction *15*
1.2 Fundamental Differences Between Noble and 3d Metal Complexes *17*
1.3 Mechanistic Scenarios and the Role of Substrates *22*
1.3.1 Nonpolar Substrates *22*
1.3.2 Polar Substrates *24*
1.3.3 Exceptions *25*
1.4 Iron-Catalyzed Hydrogenation of C—C Multiple Bonds *26*
1.4.1 Hydrogenation of Olefins *26*
1.4.2 Hydrogenation of Alkynes *27*
1.5 Iron-Catalyzed Hydrogenation of C—O Multiple Bonds *28*
1.5.1 Hydrogenation of Aldehydes and Ketones *29*

1.5.2 Hydrogenation of Esters *31*
1.5.3 Hydrogenation of Amides *32*
1.6 Iron-Catalyzed Hydrogenation of C—N Multiple Bonds *32*
1.7 Conclusion *34*
Abbreviations *35*
References *35*

2 Cobalt-Catalyzed Hydrogenations *39*
Felicia Weber and Gerhard Hilt
2.1 Introduction *39*
2.2 Hydrogenation Reactions *39*
2.2.1 Activation of Molecular Hydrogen–Dihydrogen Complexes vs. Dihydride Complexes *39*
2.2.2 Hydrogenation of CO_2, Carboxylic Acids, Carboxylic Esters, and Nitriles *40*
2.2.3 Hydrogenation of C=O, C=N, C=C, C≡C, and (Hetero)arenes *46*
2.3 Conclusion *57*
References *57*

3 Homogeneous Nickel-Catalyzed Hydrogenations *63*
Marlene Böldl and Ivana Fleischer
3.1 Introduction *63*
3.2 Hydrogenation of Alkenes *65*
3.2.1 Hydrogenation of Alkyl- and Aryl-Substituted Alkenes *65*
3.2.2 Hydrogenation of Electron-Deficient Alkenes *76*
3.3 Hydrogenation of Alkynes *78*
3.4 Hydrogenation of Carbonyl Groups *79*
3.4.1 Hydrogenation of Ketones *79*
3.4.2 Hydrogenation of Carbon Dioxide *81*
3.5 Conclusions *83*
References *83*

4 Homogeneous Hydrogenation with Copper Catalysts *87*
Niklas O. Thiel, Felix Pape, and Johannes F. Teichert
4.1 Introduction *87*
4.1.1 Early Studies on Copper-Catalyzed Hydrogenations *87*
4.2 Hydrogenation of (α,β-Unsaturated) Carbonyl and Carboxyl Compounds *88*
4.2.1 Conjugate Reduction *88*
4.2.2 1,2-Hydrogenation of α,β-Unsaturated Ketones and Aldehydes *91*
4.2.3 Asymmetric 1,2-Hydrogenation of Simple (Nonconjugated) Ketones and Aldehydes *92*
4.3 CO_2 Reduction to Formate *93*
4.4 Allylic Substitutions with a Hydride Nucleophile Generated from H_2 *95*
4.5 *Z*-Selective Alkyne Semihydrogenation *98*

4.6 Alkyne Transfer Semihydrogenation and Transfer Conjugate Reduction with Ammonia Borane *104*
4.7 Dihydrogen-Mediated Cross-Coupling of Internal Alkynes and Aryl iodides *105*
4.8 Conclusions and Perspectives *106*
References *107*

5 Hydrogenation Reactions Using Group III to Group VII Transition Metals *111*
Matthew L. Clarke and Magnus B. Widegren
5.1 Introduction *111*
5.2 Group III Metals: Scandium and Yttrium *111*
5.3 Group IV Metals: Titanium, Zirconium, and Hafnium *112*
5.3.1 Asymmetric Hydrogenation Using Titanium and Zirconium Catalysts *115*
5.4 Group V Metals: Vanadium, Niobium, and Tantalum *119*
5.5 Group VI Metals: Chromium, Molybdenum, and Tungsten *121*
5.6 Group VII Metals: Manganese and Rhenium *122*
5.7 Summary and Conclusions *137*
References *137*

6 Early Main Group Metal Catalyzed Hydrogenation *141*
Heiko Bauer and Sjoerd Harder
6.1 Introduction *141*
6.2 Hydrogenation of C=C Double Bonds *144*
6.3 Hydrogenation of C=N Double Bonds *153*
6.4 Hydrogenation of C=O Double Bonds *157*
6.5 Summary and Perspectives *160*
References *163*

7 Frustrated Lewis Pair-Catalyzed Reductions Using Molecular Hydrogen *167*
Jan Paradies and Sebastian Tussing
7.1 Introduction *167*
7.2 Mechanistic Considerations *168*
7.3 Influence of the Lewis Acid and Lewis Base on Hydrogenation Reactivity *171*
7.3.1 Choice of Lewis Acid *172*
7.4 Balance Between Lewis Acidity and Lewis Basicity *173*
7.4.1 Hydrogenation of Olefins *178*
7.4.2 Dehydrogenative Coupling *179*
7.4.3 Acceptorless Dehydrogenation *180*
7.4.4 Intramolecular Frustrated Lewis Pairs *181*
7.4.5 Air-Stable FLPs *184*
7.5 Application of Frustrated Lewis Pairs in Hydrogenations *187*
7.5.1 Hydrogenation of Aldimines and Ketimines *187*

7.5.2 Hydrogenation of Enamines and Silylenol Ethers *193*
7.5.3 Hydrogenation of Ketones *198*
7.5.4 Reductive Deoxygenations *198*
7.6 Hydrogenation of Heterocycles *201*
7.7 Hydrogenation of Enones, Alkylidene Malonates, and Nitroolefins *207*
7.8 Hydrogenation of Unpolarized Olefins and Polycyclic Aromatic Hydrocarbons *211*
7.9 Electrophilic Phosphonium Cations (EPCs) *215*
7.10 Summary *217*
Abbreviations *217*
References *218*

8 Recent Advances in Selective Biocatalytic (Hydrogen Transfer) Reductions *227*
Gonzalo de Gonzalo and Iván Lavandera
8.1 Introduction *227*
8.2 Ketoreductases *228*
8.2.1 Alcohol Dehydrogenases in "Nonconventional" Media *229*
8.2.2 Dynamic Processes Employing Ketoreductases *230*
8.2.3 Alcohol Dehydrogenases in Multicatalytic Processes *232*
8.2.4 Application of Ketoreductases to the Synthesis of Valuable Compounds *236*
8.3 Ene-Reductases *241*
8.3.1 Substrate Scope of Ene-Reductases *242*
8.3.2 Ene-Reductases in Multicatalytic Processes *244*
8.4 Imine Reductases *248*
8.5 Carboxylic Acid Reductases *249*
8.6 Emerging Enzymes: Nitrile Reductases and Nitroreductases *250*
8.7 Summary and Outlook *251*
Abbreviations *252*
References *253*

9 Organocatalytic Transfer Hydrogenation *261*
Jie Wang and Yong-Gui Zhou
9.1 Introduction *261*
9.2 Asymmetric Transfer Hydrogenation of C=C Bonds *262*
9.3 Asymmetric Transfer Hydrogenation of C=N Bonds *266*
9.4 Asymmetric Transfer Hydrogenation of C=O Bonds *273*
9.5 Asymmetric Transfer Hydrogenation of Heteroaromatics *274*
9.6 Summary and Outlook *280*
Abbreviations *280*
References *281*

Index *285*

Preface

Catalytic homogeneous hydrogenations are among the most valuable catalytic transformations in synthetic chemistry due to the wide variety of substrates that can be converted with high atom economy, often in a stereoselective fashion. After Calvin reported the first homogeneous hydrogenations based on copper catalysts in 1938, however, the field of homogeneous hydrogenation was taken over by the much more efficient noble metal catalysts based on rhodium, iridium, and ruthenium. These catalysts dominate the field to this day, with many catalytic homogeneous hydrogenations being run on industrial scale. However, in the last two decades, a resurrection of interest in the first row transition metals as catalysts for homogeneous hydrogenations has taken place. In this sense, there is quite a parallel with the ligand development in asymmetric catalysis, in which the first chiral ligands had been monodentate, followed by a long-term rush on bidentate ligands, only to lead to chiral monodentate ligands to be rediscovered much later. The 3d metals tend to be more readily available, which might be an important factor for large-scale processes. The chapter on iron catalysts gives a good comparison on 3d metals vs. noble metals. Many of the processes studied so far are still quite limited in terms of catalyst efficiency, which hampers their applicability for industrial applications. But maybe even more importantly, though, catalysts based on 3d metals and other less frequently studied elements can offer new reactivity through reaction pathways less studied so far. This might be the most important aspect of these catalysts. In any case, an overview of the present methods so far with non-noble catalysts seems timely, and this book tries to offer a collection of the various catalysts that have been studied, mostly in an academic setting. Nevertheless, from an industrial vantage point, other factors might be of importance. Therefore, the stage is set by a specialist view on the challenges of homogeneous hydrogenation in an industrial and process setting. This lays out the main challenges for the following chapters (and the catalysts included in them) to be tackled. Certainly no easy tasks!

Next to the transition metal catalysts and main group elements covered in this book, also other approaches to homogeneous hydrogenations in the broader sense are presented. New ways of activating dihydrogen by frustrated Lewis pairs (FLPs) have recently been undergoing a gold rush. A chapter devoted to this research area is included here to counterbalance well the transition metals with main group catalysts. Even though transfer hydrogenations are, strictly speaking, not "real" hydrogenations, two approaches have been included in this book to give the reader a good overview of other entries to reductive transformations. Consequently, the book ends with a perspective on transfer hydrogenations based on enzymes and organocatalysts.

Bringing together these various entryways to homogeneous hydrogenations, I hope that this book offers a thorough overview on alternative non-noble catalysts to activate dihydrogen. I am indebted to all the authors that have contributed to this book and would like to express my gratitude to their efforts.

Technische Universität Berlin
Berlin
April 2015

Johannes F. Teichert

Prelude – A Critical Assessment from an Industrial Point of View

Hans-Ulrich Blaser

1 Some Introductory Remarks

This preface has been written from the personal viewpoint of the author who has been active in industrial research and the development of catalytic methodologies for the production of fine chemicals, agrochemicals, and pharmaceuticals but who also has a strong interest in mechanistic aspects of catalysis [1].

While catalysis is THE key technology for the production of base and bulk chemicals, this is not (yet) the case for the fine chemical and pharmaceutical industry where classical organic synthesis dominates. Nevertheless, catalysis is able to play a valuable role in the endeavor to minimize waste production in this industry by:

- Replacing reagents which produce wastes with detrimental effects to the environment (e.g. using hydrogen instead of other reducing agents).
- Offering access to shorter, more efficient synthetic routes (e.g. direct synthesis avoiding the use of protective groups).
- Combining several different catalytic and non-catalytic steps to a "one-pot reaction," thereby avoiding workup procedures.
- Taking advantage of unique selectivity properties of a catalytic system (e.g. the possibility to introduce stereogenic centers by enantioselective transformation of prochiral substrates).

Among the different catalytic methods, catalytic hydrogenation is arguably the most valuable catalytic transformation with the highest practical impact. The versatility of this method allows for the selective conversion of an amazing number of functional groups, even in highly functionalized molecules with high yield under often relatively mild conditions. Up to now, heterogeneous catalysts such as Pd, Pt, Rh, Ru, and Ni on various supports have dominated the field of chemoselective hydrogenation [2], while soluble Rh, Ru, and Ir complexes with chiral ligands are applied predominantly for stereoselective hydrogenations [3]. These well-established catalytic systems will be the standard reference when comparing and assessing the various alternative catalysts presented in this monograph [4–12].

Homogeneous Hydrogenation with Non-Precious Catalysts, First Edition. Edited by Johannes F. Teichert.

The dominance of classical catalysts can in part be explained by the historical development of the mechanistic understanding of the activation and transfer of dihydrogen. For decades, the prevailing opinion was that ensembles of metal atoms, i.e. heterogeneous catalysts, were necessary for the dissociative activation of dihydrogen and its transfer to an unsaturated function (also adsorbed and thereby activated on the same metal surface). It was only in the 1960s (Halpern, Wilkinson, and others) that it was shown that properly ligated transition (mostly noble) metal complexes could actually perform the same reactions. Still, the presence of a transition metal was considered indispensable. This hypothesis was challenged by the finding that certain main group metal complexes display hydrogenation activity and even more by the discovery by Stephan in 2006 that certain frustrated Lewis pairs (FLPs) are able to activate dihydrogen.

Considering the rapid development of "nontraditional" hydrogenation catalysts, the topic of this monograph is timely, allowing a closer look at the development of alternative homogeneous catalytic systems both in respect of the mechanistic understanding as well as how their performance compares to the established competition described [13]. In the following, I undertake this comparison from the point of view of their synthetic application and their potential for eventual use on a technical scale.

1.1 What Is the Motivation for Developing Non-Noble Metal Catalysts?

Besides the interest in mechanistic aspects such as new modes of hydrogen activation, the following practical aspects and expectations are most often mentioned in the various chapters:

- *Cost of the metal* (e.g. Pd is about 2000 times more expensive than Ni). However, this relation has to be qualified, since, very often, similar ligands are used and especially chiral ligands are often just as (sometimes even more) expensive as the noble metal. Further, much of the higher costs of the noble metal complexes are often compensated for by their much higher catalyst activities and productivities (and that is what counts for the process chemist).
- *Abundance of the metal.* There is no doubt that noble metals are scarce and their supply is limited. The often high catalytic activity, the typically relatively low production volume of fine chemicals, and a usually very high recovery rate (often >90% for Pd or Rh) make their application more acceptable. Nevertheless, there is a green chemistry aspect in avoiding scarce metals (and maybe even all metals) which deserves consideration.
- *Metal toxicity.* Due to the relatively high toxicity of (noble) metals, there are strict regulations in place for the removal of metals to the low parts per million level from active pharmaceutical intermediates [14]. Of the several classes of toxicities defined, Au, Pd, Pt, Ir, Os, Rh, Ag, Ru, as well as **Co**, Mo, and V belong to class 2 (appreciable toxicity); Sb, **Ba**, **Li**, Cr, **Cu**, Sn, and **Ni** to class 3 (relatively low toxicity); and **Al**, **B**, **Fe**, **Zn**, **K**, **Ca**, **Na**, **Mn**, **Mg**, and W to class 4 (low inherent toxicity). However, it has to be stressed that the toxicity of a metal strongly depends on the nature of the metal complex as well as on how it is administered [15].

And, just a reminder: While organocatalytic methods obviously do not apply metals, many of the (chiral) catalysts used are actually quite toxic and have to be carefully removed from the product as well. Furthermore, the Lewis acid component in FLPs are mostly heavily fluorinated B compounds which cannot be considered as being particularly green.

- *Catalyst separation and trace metal removal.* In any case, metallic catalysts have to be removed from the product mixture. In order to separate the bulk of a heterogeneous catalyst, filtration is generally unproblematic. For soluble catalysts, this is usually more of a problem. Classical solutions are distillation for volatile (which also removes trace metals) and crystallization for solid products. In rare cases, the catalyst can be precipitated and removed via filtration. There are several strategies to remove trace metals to get acceptable rest concentrations; recrystallization and adsorption by scavengers which are then either extracted or filtered off are the most useful ones [16].
- *Different reactivity of non-noble metals.* Obviously, the various metal complexes have quite different chemical properties compared to the established noble metal complexes. Often, this will be a disadvantage for their hydrogenation properties which has to be compensated for, e.g. with suitable ligands or higher reaction temperatures and/or hydrogen pressures. However, there are also contrasting examples as, for example, the higher reactivity of low-valent Ni complexes which, compared to Pd catalysts, often react faster with unreactive organic molecules such as aryl chlorides.
- *No need for high-pressure equipment for transfer hydrogenations (THs).* This argument invoked for organo- and biocatalytic reductions might be valid when no such equipment is available, a case rather seldom in modern fine chemicals development and production. With hydrogen donors such as isopropanol, there might be a problem with incomplete conversion due to equilibria. Hantzsch esters are relatively expensive hydrogen donors and the resulting pyridines have to be separated; and in the case of the cofactors used in biocatalysis, sophisticated regeneration strategies are required. This means that such catalysts will be used only when their catalytic performance is superior to alternative solutions, which for selected biocatalytic reductions is indeed the case.

1.2 Crucial Parameters for Process Development

In process development, there is usually a hierarchy of goals (or criteria) to be met. It is simply not possible to reach all the requirements for a technically useful process in one step. As depicted in Table 1, the catalyst selectivity (combined, of course, with an acceptable activity) is the first criterion – just as in academic research. However, when a reasonable selectivity has been obtained, other criteria will become important: catalyst activity, productivity and stability, catalyst separation (and, maybe, recycling). Then, questions like the cost and availability of the (chiral) catalyst and other materials have to be addressed. The final process is usually a compromise since often not all of these requirements can be fulfilled maximally. It is useful to divide the development of a manufacturing process into different phases; however, it is rarely possible to proceed in a linear manner and

Table 1 Catalyst choice: Criteria and requirements during different process development phases.

Development phase	Activities	Performance criteria	Milestones	Catalyst sourcing aspects
Feasibility phase of development	Screening	Selectivity; (activity)	Chemical feasibility	Availability in screening amounts (c. 100 mg)
Bench scale phase, demonstration of technical feasibility	Optimization, scale-up, catalyst handling quality risk analysis	Selectivity, activity, productivity, robustness	Technical feasibility	Catalyst supply <1 kg typically, quality, lead times
Pilot and production process	Scale-up, process adaptation to infrastructure, first production on scale	Selectivity, activity, productivity, recycling/ refining, metal removal, cost	Commercial feasibility	>1 kg quantities of ligand and metal precursor; lead times, quality, metal refining

very often one has to go back to an earlier phase in order to answer additional questions before it is possible to go on.

1.2.1 The Catalyst: Metal Complex, (Chiral) Ligand

The first and foremost requirement for choosing a particular catalyst is its *selectivity* (chemo-, regio-, stereo, or enantioselectivity) for reducing the desired unsaturated function while tolerating other functional groups. Since enantioselective catalysts are especially quite substrate specific, catalysts that can easily be adapted, for example, by optimizing the structure of the chiral ligand will be preferred.

The second criterion for technical application will be *catalyst productivity* (turnover number, TON) and *activity* (turnover frequency, TOF), which determine catalyst cost. What catalyst costs can be tolerated is determined by the added value of the transformation – usually defined by the price of the product. This means that even inexpensive catalysts must have sufficient activity to be competitive.

A further criterion is the *stability* of the metal complex and whether its preparation and handling needs special equipment (glove boxes are not standard equipment in organic laboratories). This might be, for example, a problem for the rather sensitive early transition metal complexes. Last but not least, either the catalyst complexes or the appropriate metal precursors and the ligands should be commercially available in order to get a chance to be evaluated. At the moment, most ligands described in the monograph are probably not yet available commercially, certainly not in larger quantities.

Finding and developing efficient biocatalysts entails a high effort, especially when the starting material is not a very close analog to the natural substrate. A variety of techniques have been developed to genetically engineer enzymes via directed evolution in order to adapt them to catalyze a specific reaction [17]. Through repeated cycles of gene mutagenesis, expression, screening for catalytic performance, and selection, an enzyme has to be developed which has the required performance concerning activity, enantioselectivity, stability, or lower product inhibition. This requires special knowhow and equipment, only available in large companies or specialized technology companies such as Codexis, which also offer enzyme screening kits selected for specific transformations such as C=O and C=N reductions [18].

1.2.2 The Process

Several factors determine whether a process is industrially feasible. As can be seen in the various chapters, several catalysts give best results in chlorinated solvents, especially dichloromethane. The very low emission limits, which in many countries apply to dichloromethane (DCM), lead to high costs or even make industrial production impossible. In order to be commercially attractive, TOF values as well as space–time yield (how much can be produced in a given volume and time) are important. This could be problematic for enzymatic reactions, which often have to be operated at low concentrations (product inhibition, solubility problems). For hydrogenation reactions, pressures above 20–50 bar might be a problem for many companies. Finally, product isolation and purification as well as the problem of catalyst separation (and maybe recycling) must be solved.

2 Comparison with the Established Catalysts

As stated in the introduction, heterogeneous catalysts dominate the field of chemoselective hydrogenation, while homogeneous noble metal complexes with chiral ligands are predominantly applied for enantioselective hydrogenations. These well-established catalytic systems are the standard reference when comparing and assessing the performance of the alternative catalysts presented in this monograph. The comparison is not meant to disqualify these alternative catalyst systems. It should be noted that the application of noble metal catalysts started in the 1980s and has been perfected over decades, while significant research efforts to develop alternative catalysts and catalytic reactions started only about 10–15 years ago.

As summarized in Table 2 in [2], heterogeneous Pd, Pt, Rh, Ru, and Ni catalysts are quite efficient and selective for the selective hydrogenation of reducible functions such as C=X (X = C, N, O), RC≡X (X = C, N), RN_3, $ArNO_2$, and (hetero)aromatic compounds, as well as the hydrogenolysis of R–X (X = halogen, OBn, NBn) under moderate dihydrogen pressures and temperatures. In contrast, the hydrogenation of esters and amides requires hydrogen pressures of 100–300 bar and temperatures of 100–200 °C. There exists a very large body of knowhow for the chemoselective hydrogenation of various combinations of reducible functions. Some of the most frequent selectivity problems and how heterogeneous catalysts perform are summarized in Table 2. More details can be found in [2] (Tables 4–6) and several classical monographs [19–22].

Table 2 Selected chemoselectivity problems (0 unselective, 1 partially selective, 2 selective).

		To be reduced			
		C=C	C≡C	C=O	$ArNO_2$
To be retained	C=C	–	2	0–1	2
	C≡C	0	–	0	1–2
	C=O	1–2	1	–	2
	$ArNO_2$	0–1	1	0	–

As a general rule, it can be stated that when an effective heterogeneous catalyst is available, homogeneous catalysts in general and homogeneous non-noble catalysts in particular will not be a realistic option. Therefore, their application should be focused on cases where heterogeneous catalysts do not perform well, and this is illustrated for the reduction of acid derivatives, the *E*-selective hydrogenation of alkynes, and the selective C=O reduction of unsaturated ketones.

The same applies to the case of enantioselective hydrogenations: For many enantioselective reactions, established chiral noble metal catalysts show not only good enantioselectivity but also high catalytic activity. A comparison is made of the performance of established noble metals with that of the alternative catalysts described in the monograph for the enantioselective hydrogenation of C=C, C=O, C=N functions and of selected heteroaromatic compounds.

2.1 Chemoselective Hydrogenations: Alternatives to Heterogeneous Catalysts

2.1.1 Hydrogenation of Esters

The hydrogenation of esters and amides using supported Cu, Ru catalysts often requires hydrogen pressures of 100–300 bar and temperatures of 100–200 °C. Several examples described in the monograph show that homogeneous Mn, Fe, and Co complexes might present a viable alternative (see Table 3). Of particular interest are the high activity of a Co complex with a tridentate P^P^P ligand (TON up to 8000) and the high functional group tolerance of Fe/P^N^P complex **[Fe-I]**, which is able to selectively reduce an ester function of the polypeptide depicted in

Table 3 $RCOOR' + H_2 \rightarrow RCH_2OH + R'OH$.

Catalyst	*p* (bar)	*T* (°C)	TON	TOF	*y* (%)	References, comment
Mn/P^N^N	50	50	100	~7	60–96	[8], Tables 1–3
Fe/P^N^P	50	120	10–100			[4], Figure 7
Co/P^P^P	80	100	8 000	400	50–90	[4], Scheme 6
Co/P^N^P	50–55	120	20–50	2–3	50–99	[4], Schemes 7, 9

X^Y^Z tridentate ligand with X, Y and Z coordinating atom.

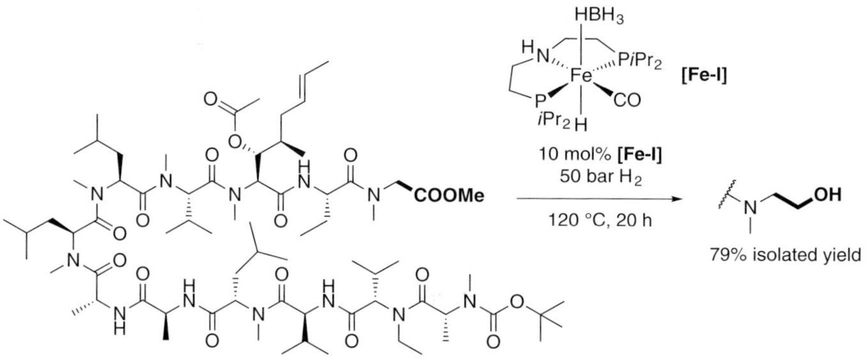

Figure 1 Selective hydrogenation of a polypeptide.

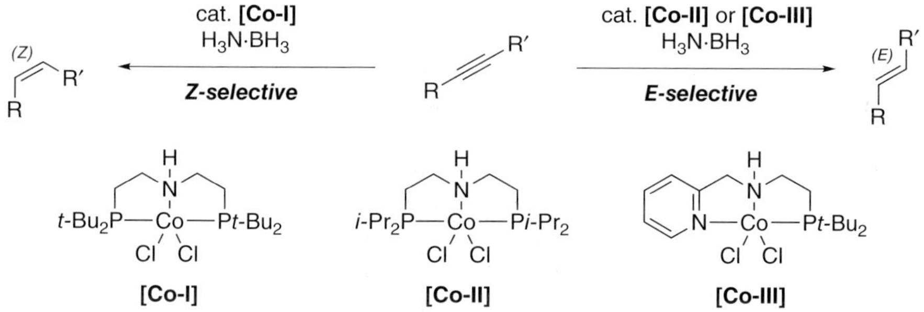

Figure 2 Stereoselective reduction of alkynes.

Figure 1 tolerating C=C, OAc and without epimerization of stereogenic centers [23].

2.1.2 E-Selective Hydrogenation of Alkynes

As a general rule, heterogeneous catalysts add dihydrogen in *cis* manner to the unsaturated function, leading to *cis*-alkenes in the hydrogenation of alkynes (Figure 2). Several examples described in the monograph show that homogeneous Cu [7] and Co [5] complexes are able to produce *trans*-alkenes. Of special interest is the Co-catalyzed transfer semihydrogenation of alkynes (Figure 3), where stereoselectivity is controlled by the nature of the ligand. Furthermore, a wide range of functional groups is tolerated and catalyst loadings as low as 0.2 mol% can be applied [24].

2.1.3 Selective C=O Reduction of Unsaturated Carbonyl Groups

While some modified heterogeneous catalysts are able to selectively reduce unsaturated aldehydes to give the corresponding allylic alcohol, this is not possible for analogous unsaturated ketones or tolerating a C≡C function. The results reported for Cu [7] and Mn [8] catalysts depicted in Figures 3 and 4 are therefore of high interest.

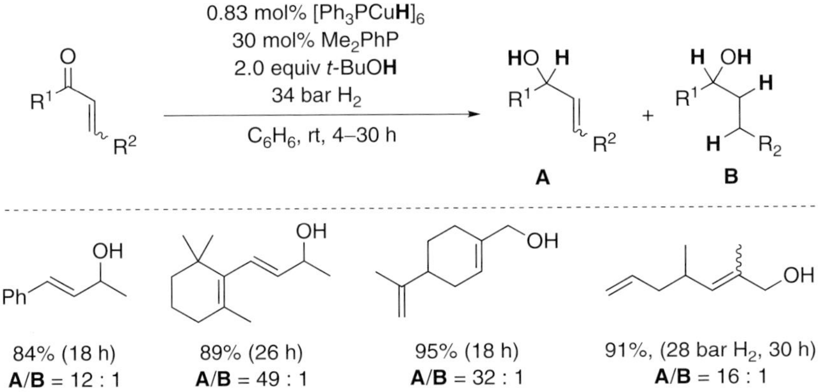

Figure 3 Cu-catalyzed hydrogenation of α,β-unsaturated ketones and aldehydes [7].

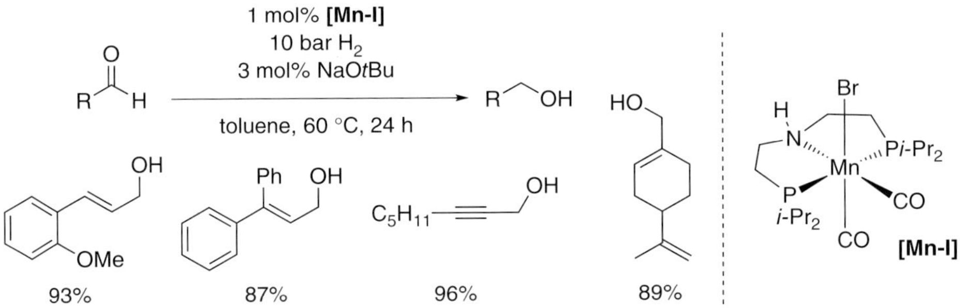

Figure 4 Mn-catalyzed hydrogenation of unsaturated aldehydes [8].

2.2 Enantioselective Hydrogenations

The reference for the enantioselective hydrogenation of olefins, ketones, and imines are Rh, Ru, and Ir complexes, usually with chiral diphosphine ligands. After decades of research, a rich repertoire of effective catalyst systems has accumulated [3]. Listed here are selected results for the hydrogenation of olefins (Table 4), ketones, and imines (Table 5) as well as for heteroaromatic compounds (Table 6, Figures 6 and 7). In general, comparable enantioselectivities, as for the classical noble metal catalysts, are achieved for all substrate classes. However, with a few exceptions, both activities and productivities are much lower for the alternative catalysts and generally not (yet) in a range useful for larger scale applications. Results for a Cu/P^P catalyst for ketone reductions, a Fe/P^N^P catalyst for imine reductions, and a Co/P^P catalyst for the hydrogenation of dehydro amino acid derivatives [25] look promising and show that higher activities are possible with non-noble metal complexes. Furthermore, some organocatalysts were able to reduce nitro-substituted olefins and quinolines with high chemoselectivity, which would be very difficult with metal catalysts.

That non-noble catalysts are able to perform the difficult hydrogenation of non-functionalized olefins has been demonstrated for Co/N^N^N complex

Table 4 Selected examples for the enantioselective hydrogenation of C=C functions.

Substrates	*ee* (%)	TON	TOF (h^{-1})	Catalyst types, references, comment
Enamides, enol acetates, itaconates	*90–99*	*1 000–50 000*	*200–5 000*	*Rh/P^P, Ru/P^P,* [3][a)]
	up to 99	20–150	3–50	Co/P^P, [5], Scheme 15
α,β-Unsaturated acids or esters	*85–95*	*2 000–20 000*	*500–3 000*	*Rh/P^P, Ru/P^P,* [3][a)]
	70–96	20–100	1–5	Ni/P^P, [6], Scheme 14
Cyclic enones	98–99	200	10	Cu/P^P, [7], Scheme 4
	70–98	5–20	<1	TH, organocat., [12], Scheme 4
Nitroolefins	90–97	5	<1	TH, organocat., [12], Scheme 5

a) Typical range for suitable substrates and optimized catalysts.

Table 5 Selected examples for the enantioselective hydrogenation of ketones and imines.

Substrates	*ee* (%)	TON	TOF (h^{-1})	Catalyst types, references, comment
ArCOR	*90–95*	*5 000–10 000*	*500–10 000*	*Ru/P^P + N^N,* [3][a)]
	90–96	50–30 000	2–2 000	Cu/P^P, [7], Scheme 7
	50–74	50	1	Co/P^N^N^P, [5], Scheme 16
	20–91	100	5–10	Mn/P^N^N, [8], Scheme 8
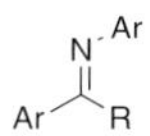	*80–90*	*500–1 000*	*5–100*	*Ir/P^P, Ir/P^N,* [3][a)]
	77–99	20–50	<1	TH, chiral phosphoric acid, [12], Figure 2
	70–99	20–50	<1	Chiral FLP, [10], Figures 32, 33
	up to 85	up to 5 000	up to 2 000	Fe/P^N^P, [4], Figure 3, ref. 35
Cyclic imines, enamines	90–98	50–1 000	1–50	cpTi catalysts, [3, 8], Scheme 5

a) Typical range for suitable substrates and optimized catalysts.

[Co-IV] with very good enantioselectivity and yield, albeit with high catalyst concentration (Figure 5).

2.3 A Few Remarkable Transformations Using Cu and Co Complexes

A Cu carbene catalyst reduces a variety of alkyl-substituted *Z*-allylic chlorides selectively to the corresponding terminal alkenes ($\gamma/\alpha > 20:1$). Deuteration

Table 6 Enantioselective hydrogenation of selected heteroaromatic compounds (Figure 6).

Catalyst types, references, comment	*ee* (%)	TON	TOF (h^{-1})
Ir/P^P, [3][a]	*70–99*	*100–>20 000*	*up to 1 000*
TH, chiral phosphoric acid [12], Figure 3	70–99	10–50	<1
Chiral FLP, [10], Figures 59–61	60–99	10–20	<1

The first lines in italic are always the reference results for the noble metal catalysts.
a) Typical range for suitable substrates and optimized catalysts.

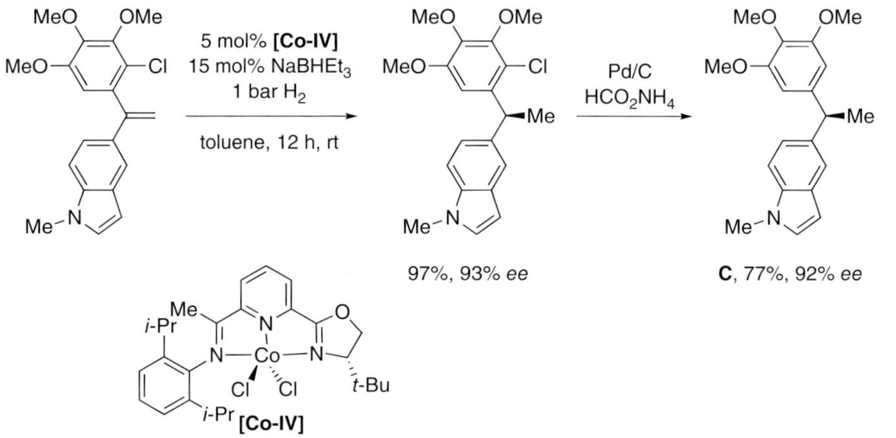

Figure 5 Co-catalyzed hydrogenation as a key step for the synthesis of the anticancer agent **C** (Scheme 18, [5]).

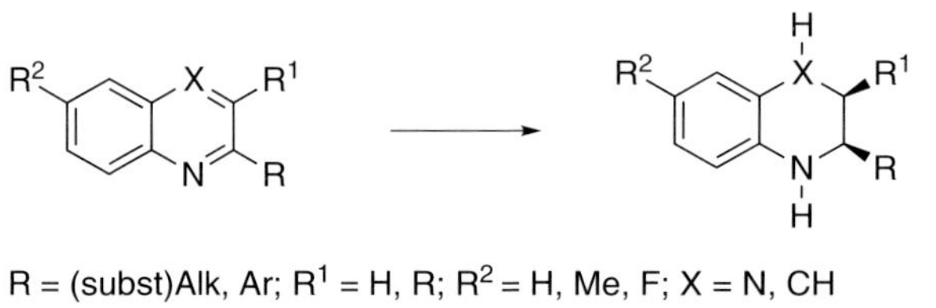

Figure 6 Hydrogenation of substituted quinolines and quinoxalines.

experiments show that the hydride is added selectively to the γ position (Figure 8). Another unexpected transformation is the reductive cross-coupling of internal alkynes with aryl iodides by Pd/Cu cooperative catalysis to give preferentially E olefins (Figure 9). Another interesting result is the Co-catalyzed hydroboration/hydrogenation sequence of internal alkynes to generate chiral benzylic boronic esters in a one-pot reaction (Figure 10).

3 Epilogue

Most catalysts described in this monograph are not (yet) competitive for synthetic, and much less for industrial, applications. In some cases, inherent chemical properties might prevent giving useful hydrogenation catalysts, but for many

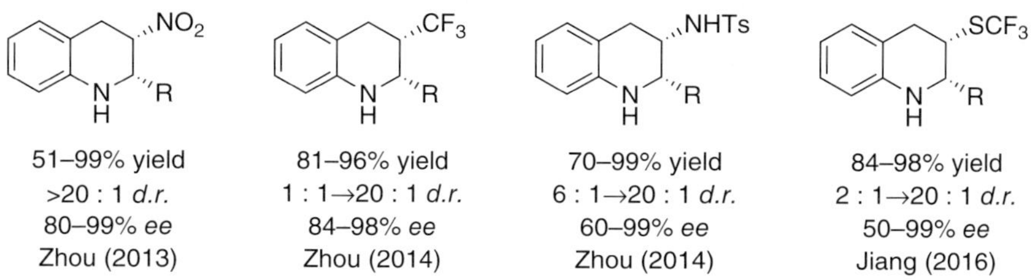

Figure 7 Selected examples of hydrogenated quinolines (TH with binaphthol-based phosphoric acids, Hantzsch ester) (see [12], Figure 3).

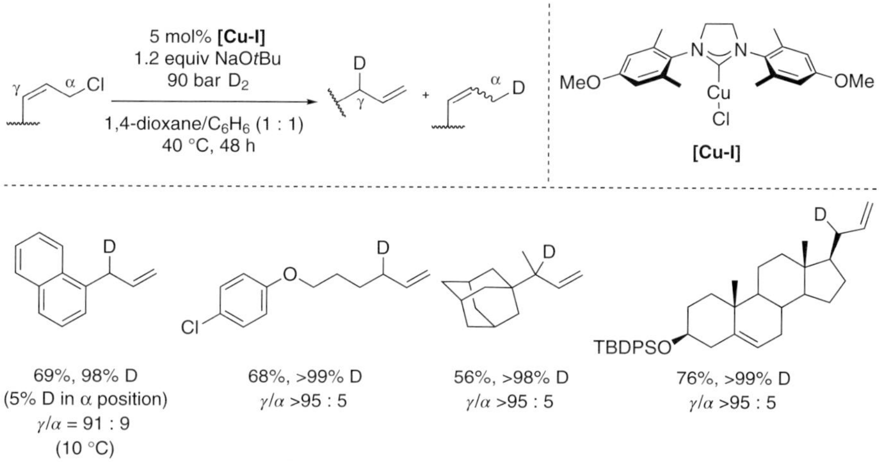

Figure 8 Cu-catalyzed deuteride transfer from D_2 to the γ position of an allylic chloride. (Scheme 1, [7]).

of the other examples described here, there is ample room for improved performance. One should always keep in mind that it is very hard to predict what new insights diligent investigations can give. For example, not very long ago gold was considered a "non-catalytic" metal, or one should remember how the novel concept of FLPs developed from "unimaginable" to a viable catalytic concept within less than a decade! If we look at which factors have contributed most to the success of non-noble metal catalysis, it is probably the design of new ligands. It is well known in homogeneous catalysis that once the metal is chosen, besides oxidation state and anion the nature of the organic ligands plays the decisive role in its catalytic performance. While for Rh, Ru, and Ir bidentate ligands, mostly of the P^P and P^N type dominate, these are obviously not ideal for non-noble metals where often tri- and sometimes even tetradentate give better results. The potential of carbene-type ligands, maybe in combination with P or N functions, has also not yet been fully exploited. So it is certainly worthwhile to continue research in the area of homogeneous non-noble hydrogenation catalysts.

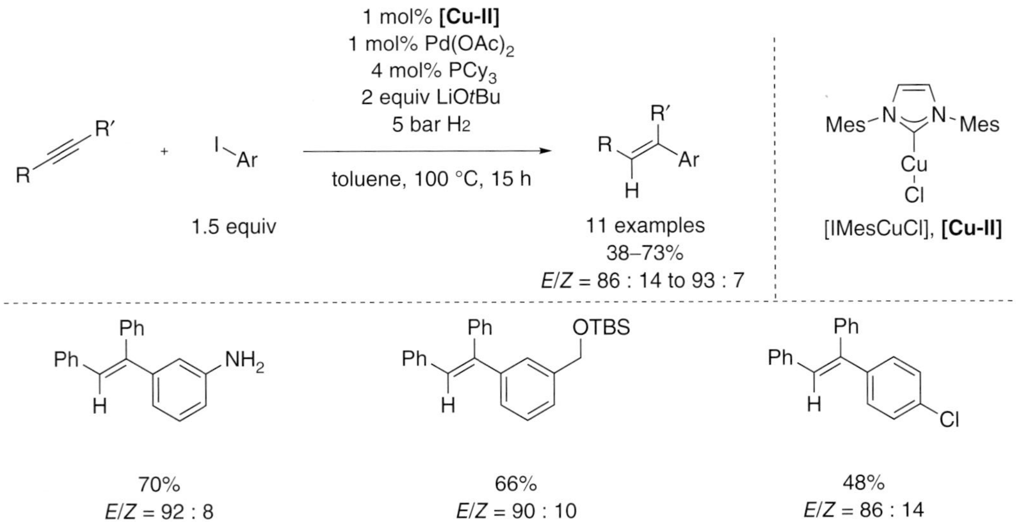

Figure 9 Hydrogenative cross-coupling of internal alkynes and aryl iodides (Scheme 21, [7]).

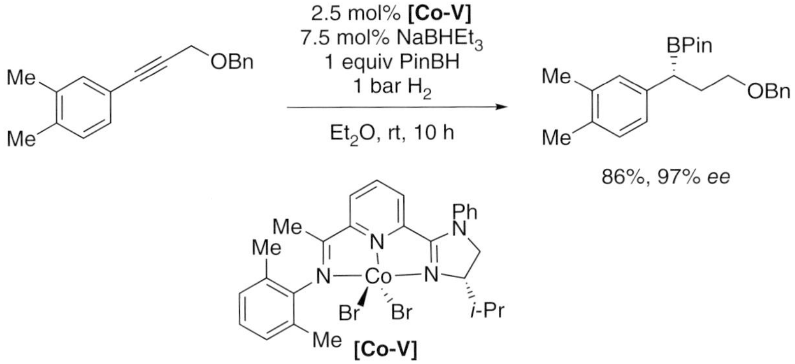

Figure 10 Co-catalyzed hydroboration/hydrogenation sequence of alkynes (Scheme 24, [5]).

References

1 H.U. Blaser, Looking back on 35 years of industrial catalysis. *Chimia* 2015, 69, 393–406.

2 For a concise overview on heterogeneous hydrogenation see H.U. Blaser, A. Schnyder, H. Steiner, F. Roessler, P. Baumeister, Selective hydrogenation of functionalized hydrocarbons. In: *Handbook of Heterogeneous Catalysis*, 2e (eds. G. Ertl, H. Knözinger, F. Schüth and J. Weitkamp) Weinheim: Wiley-VCH Verlag, 2008, 3284–3308.

3 For a recent overview on enantioselective hydrogenation H.U. Blaser, B. Pugin, F. Spindler, Industrial application of asymmetric hydrogenation. In: *Applied Homogeneous Catalysis with Organometallic Compounds* (eds. B. Cornils, W.A. Hermann, M. Beller and R. Paciello) Wiley-VCH, 2017, 621–645.

4 T. Zell, R. Langer, Iron-catalyzed homogeneous hydrogenation reactions. In: *Homogeneous Hydrogenation with Non-Precious Catalysts*, Chapter 1 (ed. J.F. Teichert) Wiley-VCH, 2020, 15–38.

5 F. Weber, G. Hilt, Cobalt-catalyzed hydrogenations. In: *Homogeneous Hydrogenation with Non-Precious Catalysts*, Chapter 2 (ed. J.F. Teichert) Wiley-VCH, 2020, 39–62.

6 M. Böldl, I. Fleischer, Homogeneous nickel catalyzed hydrogenations. In: *Homogeneous Hydrogenation with Non-Precious Catalysts*, Chapter 3 (ed. J.F. Teichert) Wiley-VCH, 2020, 63–86.

7 N.O. Thiel, F. Pape, J.F. Teichert, Homogeneous hydrogenation with copper catalysts. In: *Homogeneous Hydrogenation with Non-Precious Catalysts*, Chapter 4 (ed. J.F. Teichert) Wiley-VCH, 2020, 87–110.

8 M.L. Clarke, M.B. Widegren, Hydrogenation reactions using Group III to Group VII transition metals. In: *Homogeneous Hydrogenation with Non-Precious Catalysts*, Chapter 5 (ed. J.F. Teichert) Wiley-VCH, 2020, 111–140.

9 H. Bauer, S. Harder, Early main group metal catalyzed hydrogenation. In: *Homogeneous Hydrogenation with Non-Precious Catalysts*, Chapter 6 (ed. J.F. Teichert) Wiley-VCH, 2020, 141–166.
10 J. Paradies, S. Tussing, Frustrated Lewis pair-catalyzed reductions using molecular hydrogen. In: *Homogeneous Hydrogenation with Non-Precious Catalysts*, Chapter 7 (ed. J.F. Teichert) Wiley-VCH, 2020, 167–224.
11 G. de Gonzalo, I. Lavandera, Recent advances in selective biocatalytic (hydrogen transfer) reductions. In: *Homogeneous Hydrogenation with Non-Precious Catalysts*, Chapter 8 (ed. J.F. Teichert) Wiley-VCH, 2020, 225–256.
12 J. Wang, Y.-G. Zhou, Organocatalytic transfer hydrogenation. In: *Homogeneous Hydrogenation with Non-Precious Catalysts*, Chapter 9 (ed. J.F. Teichert) Wiley-VCH, 2020, 257–280.
13 For another recent review see also G.A. Filonenko, R. van Putten, E.J.M. Hensen, E.A. Pidko, Catalytic (de)hydrogenation promoted by non-precious metals – Co, Fe and Mn: recent advances in an emerging field. *Chem. Soc. Rev.* 2018, 47, 1459–1483.
14 See Guideline for elemental impurities by ICH Expert Working Group https://www.ich.org/fileadmin/Public_Web_Site/ICH_Products/Guidelines/Quality/Q3D/Q3D_Step_4.pdf
15 For a recent overview see K.S. Egorova, V.P. Ananikov, Toxicity of metal compounds: knowledge and myths. *Organometallics* 2017, 36, 4071–4090.
16 For a recent overview see L.C. Forfar, P.M. Murray, Meeting metal limits in pharmaceutical processes. *Top. Organomet. Chem.* 2018, 13, 1–36.
17 U.T. Bornscheuer, B. Hauer, K.E. Jaeger, U. Schwaneberg, Directed evolution empowered redesign of natural proteins for the sustainable production of chemicals and pharmaceuticals. *Angew. Chem. Int. Ed.* 2019, 54, 36–40.
18 https://www.codexis-estore.com/; accessed December 2018.
19 P.N. Rylander, *Hydrogenation Methods*. New York: Academic Press, 1985.
20 Houben-Weyl, *Methoden der organischen Chemie, Vierte Auflage, Reduktionen I, Band IV/1c*. Stuttgart: Georg Thieme Verlag, 1980.
21 S. Nishimura, *Handbook of Heterogeneous Catalytic Hydrogenation for Organic Synthesis*. New York: Wiley, 2001.
22 R.L. Augustine, *Heterogeneous Catalysis for the Synthetic Chemist*. New York, Basel and Hong Kong: Marcel Dekker, Inc., 1996.
23 S. Werkmeister, K. Junge, B. Wendt, E. Alberico, H. Jiao, W. Baumann, H. Junge, F. Gallou, M. Beller, Hydrogenation of esters to alcohols with a well-defined iron complex. *Angew. Chem. Int. Ed.* 2014, 53, 8722–8726.
24 S. Fu, N.Y. Chen, X. Liu, Z. Shao, S.P. Luo, Q. Liu, Ligand-controlled cobalt-catalyzed transfer hydrogenation of alkynes: stereodivergent synthesis of *Z*- and *E*-alkenes. *J. Am. Chem. Soc.* 2016, 138, 8588–8594.
25 M.R. Friedfeld, H. Zhong, R.T. Ruck, M. Shevlin, P.-J. Chirik, *Science* 2018, 360, 888–893.

1

Iron-Catalyzed Homogeneous Hydrogenation Reactions

Thomas Zell and Robert Langer

Philipps-University Marburg, Department of Chemistry, Hans-Meerwein-Straße, 35043 Marburg, Germany

1.1 Introduction

During the past decade, increasing attention has been drawn to iron-based catalysts as a substitute for ruthenium and other noble metal catalysts in homogeneous hydrogenation reactions. This is in part due to the fact that iron is more abundant, much less expensive, and it is also believed to be less toxic than are noble metals. Most importantly, nature uses 3d metals such as iron in highly active metalloenzymes like hydrogenases. However, nowadays, most reports, claiming to provide inexpensive and environmentally benign catalyst alternatives based on iron, often obviate the cost of the employed ligand(s) and usually provide no evidence for a reduced toxicity or the environmental sustainability of the reported iron complexes.

Moreover, iron based catalysts are currently, in most cases, less active (lower TOF[1]) and exhibit lower productivities (lower TON[2]) than, for example, their ruthenium-based counterparts. While in the chemical industry, by scale hydrogenation reactions represent one of the biggest homogeneously catalyzed processes for the production of bulk chemicals, none of the current processes uses an iron-based catalyst. From an economic point of view, iron catalysts are obviously not yet sufficiently attractive to replace noble metal catalysts, indicating that still substantial improvements in catalyst development have to be made to achieve the goal of providing ***real*** alternatives to noble metal based catalyst systems for hydrogenation reactions. Nonetheless, the extremely high activity of hydrogenase enzymes as well as the activity of certain iron catalysts in different reactions involving hydrogen indicate the enormous potential of iron-based hydrogenation catalysts and may give rise to the assumption that this field of research will be a "hot topic" in the upcoming decades, too. This chapter intends to illustrate fundamental differences between iron and noble metal complexes that have to be considered in catalyst design, including fundamental

1 TOF = turnover frequency = $\frac{\text{TON}}{t}$.
2 TON = turnover number = $\frac{\text{mole of product}}{\text{mole of catalyst}}$.

Homogeneous Hydrogenation with Non-Precious Catalysts, First Edition. Edited by Johannes F. Teichert.

coordination chemistry aspects as well as mechanistic considerations. Finally, for various substrates, state-of-the-art iron catalyst systems are compared.

At this point, it may be noted that the field of Fe-based hydrogenation catalysis is rather young, when compared to classic and well-established noble metal counterparts. Throughout the past decade, there was a significant synthetic progress in the development of synthetically valuable protocols for Fe-based hydrogenation reactions. In contrast, it turns out that often substantial mechanistic information is rather scarce. The fact that, in many cases, comprehensive analytic data is missing can in part be attributed to restrictions of spectroscopic methods and complications in the analysis of pressurized reactions. In addition, quantum chemical simulations are often more complicated than those for noble metal catalysis, simply due to the fact that more electronic configurations are energetically accessible, which is a result of lower ligand field splitting. Hence, for the calculations, high accuracy methods are required in order to assign the energetically most favored reaction pathways and possible decomposition pathways of active catalyst species.

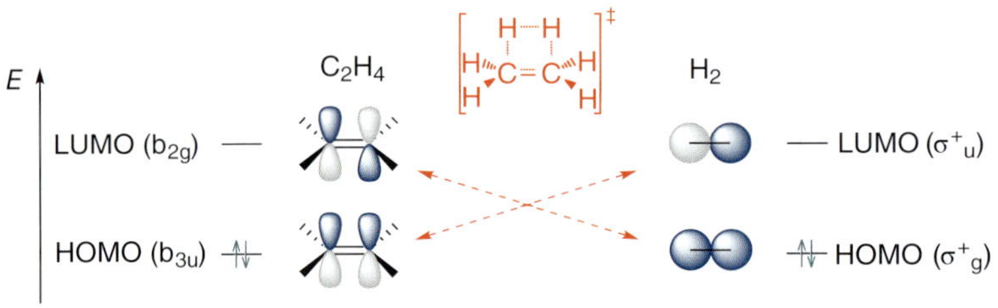

Figure 1.1 Schematic representation of LCAO-derived frontier orbitals of ethylene and H_2, illustrating that direct orbital interactions are forbidden by symmetry.

The direct concerted addition of H_2 to an unsaturated organic molecule is usually not favorable and exhibits too high reaction barriers, due to the lack of suitable frontier orbital interactions. Using the simplest example of ethylene and dihydrogen, Figure 1.1 visualizes that the two possible HOMO–LUMO combinations are forbidden by symmetry. Although the reaction of ethylene with dihydrogen to ethane is kinetically hindered by a high energetic barrier, from a thermodynamic point of view it is highly favorable; it is exergonic by $\Delta G = 101$ kJ/mol. For this reason, a catalyst is required to facilitate this thermodynamically favorable reaction by lowering the activation energies (Scheme 1.1).

The role of such a homogeneous hydrogenation catalyst can be reduced to two simple reaction steps (Scheme 1.1): (i) the activation of H_2 by heterolytic cleavage

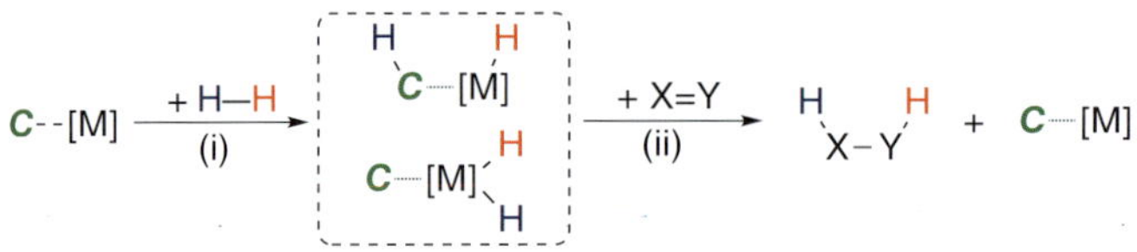

Scheme 1.1 Basic roles of homogeneous hydrogenation catalysts: H_2 activation (i) and transfer of the activated H_2 (ii). C indicates a functional ligand that can act as cooperative site).

or oxidative addition and (ii) the transfer of the activated H_2 to the substrate. While step (i) mainly depends on the type of ligand, the central metal atom and its formal oxidation state, the most viable mechanistic pathway of step (ii) strongly depends on the substrate and the polarity of the multiple bond that gets hydrogenated.[3]

1.2 Fundamental Differences Between Noble and 3d Metal Complexes

Common approaches in catalyst design with 3d metals either involve mimicking the reactivity of noble metal based analogs or circumventing undesired reactivity patterns, such as one-electron redox steps, by utilization of special ligands, such as *redox-active* ligands. For both strategies, it is of utmost importance to understand basic differences in physical and chemical properties and reactivity between 3d metals and noble metals. In the following sections, we focus on comparing properties of iron complexes with those of ruthenium complexes and in part with other noble metals (Table 1.1).

As geometric constraints of the utilized ligands strongly influence the catalyst activity, the difference in ionic radii deserves some attention. Comparing Shannon radii for the formal oxidation state +III, it becomes evident that low-spin iron(III) is significantly smaller than ruthenium(III), which exclusively appears in a low-spin configuration [1]. However, the Shannon radius of high-spin iron(III) is only slightly smaller than that of ruthenium(III). When aiming to mimic the reactivity of ruthenium complexes with iron catalysts, it is necessary to design systems which operate in low-spin configurations throughout the catalytic cycle. A significant difference of ionic radii should be taken into account, although

Table 1.1 Selected properties of iron, ruthenium, and their complexes.

	Fe	Ru
Ionic radii(pm)	55 (Fe^{III}, ls)[a) 64.5 (Fe^{III}, hs)[a)	68 (Ru^{III}, ls)[a)
$E°(M^{2+}/M^0)$ (V)	−0.44	+0.45
Occurrence of coordination numbers[b)	$6 > 5 \approx 4$	$6 \gg 5$
pK_a of $[L_nM(H_2)]$[c)	11.5	14.1

a) ls, low spin; hs, high spin.
b) Based on the number of entries in the CCDC database, reported in order of decreasing occurrence.
c) The pK_a values are reported for $[(dppe)M(H)(H_2)]^+$ (M = Fe, Ru; dppe = 1,2-bis(diphenylphosphino)ethane).

---

3 It should be noted that, in principle, all the discussed mechanistic pathways can be realized for a certain substrate, but the most efficient catalysts usually operate via certain mechanisms, depending on the substrate and its polarity.

common polydentate ligands are usually sufficiently flexible to give stable complexes with both elements, Ru and Fe, in the same oxidation state.

As hydrogenation reactions occur naturally in reducing environments, a closer look at standard redox potentials reveals a second important difference. While ruthenium with a standard potential of +0.45 V for the Ru^{2+}/Ru^0 redox couple is a typical noble metal, the standard potential of iron for the Fe^{2+}/Fe^0 couple is only −0.44 V. In consequence, the reduction to ruthenium(0) would be expected to be a dominating deactivation pathway in catalytic hydrogenation reactions. This is, however, usually not observed. Based on the limited available mechanistic information, iron complexes rather get deactivated by reduction under hydrogenation conditions, demonstrating that the actual situation is more complex and influenced by a number of factors.

In contrast to noble metals, which preferably react via two-electron redox steps, iron shows a distinct preference for one-electron steps. Against common beliefs, these redox steps are, in most cases, of secondary importance for iron-catalyzed hydrogenation reactions. With the reducing environment of hydrogenation reactions, oxidation is usually not a conceivable activation pathway and reduction often simply leads to inactive iron(0) complexes or catalyst decomposition, e.g. by loss of ligands. Furthermore, for catalysts operating via a concerted mechanism, the formal oxidation state does not change within the catalytic cycle. In general, kinetic phenomena seem to outweigh thermodynamic properties in these reactions. A closer look at typical reaction steps involved in the catalytic hydrogenation of carbonyl compounds reveals that the change of spin multiplicity in reaction steps can be problematic and result in additional energetic barriers (Scheme 1.2).

The square planar and diamagnetic Vaska's complex with a d^8 electron count (Scheme 1.2a) shows a reactivity pattern, typical for noble metals. Coordination of H_2 results in a *penta*-coordinated diamagnetic intermediate that undergoes subsequent oxidative addition to a diamagnetic, octahedral dihydride complex. While no change of spin multiplicity is observed in this sequence, the situation changes for iron. From gas-phase studies it is known that $[Fe(CO)_4]$ (d^8 electron count) exhibits a distorted tetrahedral coordination geometry with a triplet ground state ($S = 1$), suggesting that the ligand field splitting of carbonyl ligands is too low for iron(0) to affect spin pairing and the stabilization of a square planar complex (Scheme 1.2b) [2]. Distorted tetrahedral coordination geometries are observed with phosphine ligands too, which are an important class of ligands in hydrogenation reactions for both iron and ruthenium. However, the coordination of a fifth carbonyl ligand to $[Fe(CO)_4]$ under formation of the *penta*-carbonyl complex $[Fe(CO)_5]$ or the oxidative addition of H_2 to give $[Fe(CO)_4(H)_2]$ results, in both cases, in diamagnetic complexes with a singlet ground state ($S = 0$). This change of spin multiplicity results in additional (spin-induced) barriers, which, in the current case, are observed for CO and H_2 coordination to $[Fe(CO)_4]$. This is in accordance with the observation of inverse kinetic isotope effects.

Ruthenium(II) complexes containing a cooperative ligand site are among the most active catalysts for the hydrogenation of carbonyl compounds. Similarly, the corresponding iron(II) complexes show impressive catalyst performances, too. In most cases, these catalysts share the same type of intermediates (Scheme 1.2c):

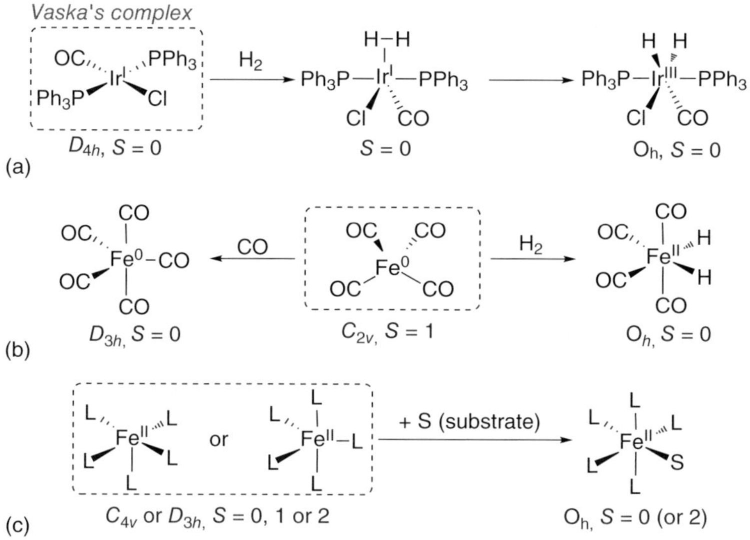

Scheme 1.2 (a) Oxidative addition without change of spin multiplicity, typical for noble metals. (b) Oxidative addition and ligand coordination to $[Fe(CO)_4]$, which is associated with a change of spin multiplicity. (c) Possible spin multiplicities for the coordination of substrate (e.g. H_2) to a *penta*-coordinated d^6 metal fragment.

a *penta*-coordinated intermediate that allows for H_2 coordination prior to heterolytic H—H bond cleavage. The resulting hydride complex facilitates the proton–hydride transfer, regenerating the *penta*-coordinated intermediate. The latter species can exhibit square pyramidal (C_{4v}) or a trigonal bipyramidal geometry (D_{3h}). Depending on the metal and the utilized ligands, singlet ($S = 0$), triplet ($S = 1$), and quintet ($S = 2$) ground states can by realized. The additional hydrido ligand usually gives rise to a very strong ligand field in the corresponding octahedral complexes, so that a singlet ground state is commonly observed. In consequence, spin-induced barriers are avoided only if the *penta*-coordinated intermediates are diamagnetic, too. For ruthenium(II), these *penta*-coordinated intermediates are exclusively diamagnetic, but for iron, this strongly depends on the ligand environment. An alternative scenario involves solvent coordination and formation of an octahedral, diamagnetic intermediate, which, in principle, can react via an interchange mechanism with H_2 to give the octahedral hydride complex.

For example, the square pyramidal complex **1** is diamagnetic and one of the most active catalysts for the transfer hydrogenation of ketones to alcohols using isopropanol as hydrogen source (Figure 1.2) [3]. Its reaction with isopropanol yields the corresponding octahedral and diamagnetic hydrido complex, which is the active reducing species. The square pyramidal amido iron complex **2** is diamagnetic as well and the most active iron catalyst known to date for the hydrogenation of carbonyl and carboxyl compounds to alcohols [4]. The coordination of H_2 and its subsequent heterolytic cleavage across the Fe—N bond yields the active dihydride species. The diamagnetic and trigonal bipyramidal complex **3**

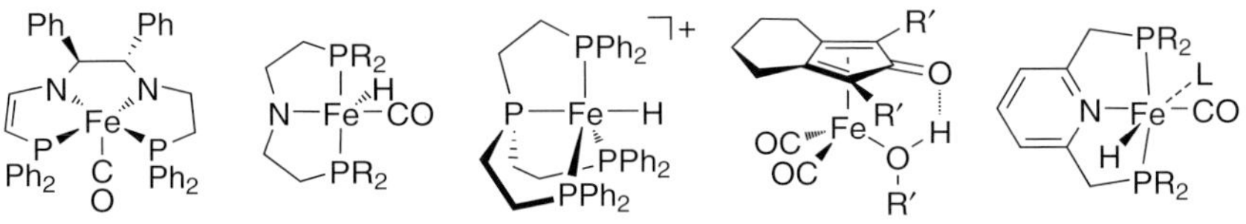

Figure 1.2 Detectable or isolable, *penta*-coordinated or solvent coordinated intermediates with a singlet ground state (R = *i*-Pr; R′ = $SiMe_3$).

is an active catalyst for the hydrogenation of carbon dioxide [5]. The coordination of CO_2 in *cis* position to the hydrido ligand leads as well to an octahedral intermediate with $S = 0$.

The hypothesis that highly active iron catalysts for the hydrogenation of carbonyl compounds avoid a change of spin state within the catalytic cycle is provided by the observation of different solvent coordinated complexes, such as **4** [6] and **5** [7], which therewith avoid the formation of potentially paramagnetic, *penta*-coordinated intermediates by associative or interchange substitution mechanisms. However, for olefins and other nonpolar substrates, high-spin complexes are frequently employed as catalysts, and they also were identified in many cases as intermediates in catalytic reactions. The participation of different spin surfaces has been demonstrated for elementary steps of iron complexes in the gas phase and is currently discussed as a viable mechanistic scenario for these catalysts [8]. The next paragraphs discusses the aspect that this violation of the spin conservation paradigm is related to a stepwise mechanism.

Further important points, when it comes to catalyst design, are the coordination number and the number of donor groups in the employed ligand (denticity). An analysis of the occurrence of certain coordination numbers for iron and ruthenium in the Cambridge Crystallographic Data Centre (CCDC) revealed that both metals have a preference for the coordination number 6, while 5 is common for both metals as well.[4] Most interestingly, tetra-coordinated iron complexes are very common as well, but in the case of ruthenium, the coordination number 4 is rare and can only be observed when bulky polydentate ligands are employed. It is well known that in a weak ligand field, iron(II) easily forms tetrahedral complexes, which are usually catalytically inactive. These findings have important consequences for the type of ligand used in iron catalysts: while mono- and bidentate spectator ligands allow for the formation of tetrahedral deactivation products (e.g. by dissociation of one ligand), these pathways can partially be suppressed when using tri- and tetradentate ligands.

The catalytic hydrogenation of carbonyl compounds involves the transfer of a hydrido ligand to the carbonyl carbon atom. The hydride donor ability of metal hydrido complexes can be quantified by the so-called hydricity, which represents the Gibbs enthalpy ($\Delta G^{\circ}_{H^-}$) for Eq. (1.1), the heterolytic cleavage of the M—H bond to a hydride and a corresponding metal fragment. Although experimental data on hydricity of iron(II) and ruthenium(II) complexes is limited, it becomes

4 The discussed results are based on a CCDC database search for iron and ruthenium complexes with different coordination numbers in August 2016.

evident that the hydride donor ability of iron complexes is significantly lower than that of ruthenium complexes [9].

$$[L_nM(H)] \rightarrow [L_nM]^+ + H^- \ (\Delta G^\circ_{H^-} > 0) \qquad (1.1)$$

For catalysts that activate H_2 by heterolytic cleavage, the acidity of the corresponding dihydrogen complexes is an essential property. These heterolytic cleavage processes can, in other words, also be regarded as deprotonation reactions of the dihydrogen complexes with an internal or external base. Available experimental data that allows a direct comparison between iron and ruthenium dihydrogen complexes is limited, but the pK_a values of complexes with the general formulae $[(dppe)_2M(H)(H_2)]^+$ (M = Fe, Ru; dppe = 1,2-bis(diphenylphosphino)ethan) indicate that the iron dihydrogen complex is even more acidic (pK_a = 11.5) than the corresponding ruthenium complex (pK_a = 14.1) [10]. However, using an increment system developed by Morris, the estimated pK_a values for different relevant dihydrogen complexes are identical, which indicates that the heterolytic cleavage of dihydrogen is not problematic for iron catalysts [10b].

Easy accessible deactivation pathways can drastically limit the catalytic productivity and possible catalyst loading, but detailed information on catalyst deactivation steps is often missing. However, the formation of (ligand protected) nanoparticles that remain catalytically active has, in some cases, been reported as well as the detection of well-defined inactive iron(0) complexes. Pincer-type complexes with one ancillary carbonyl ligand (Scheme 1.3) serve as an illustrative example. This type of compounds represents a class of highly active hydrogenation catalysts for both metals, iron and ruthenium. The active species **6** gets presumably deactivated by reduction to the tetra-coordinated complex **7**, which for ruthenium is square planar (S = 0) and can exhibit a distorted tetrahedral environment in case of iron (S = 1). In consequence, de- and reactivation of iron complexes are associated with additional (spin-induced) barriers, which might cause irreversibility of this deactivation pathway in many cases. Based on the standard potential, the formation of ruthenium metal should be facile, but, as pointed out earlier, the formation of tetra-coordinated ruthenium complexes is not favored for ruthenium complexes and the absence of spin-induced barriers allows for fast reactivation. For this type of iron catalysts, the formation of *penta*-coordinated iron(0) complexes **9** together with equimolar amounts of the uncoordinated ligand **8** were observed in several cases [7b, 11].

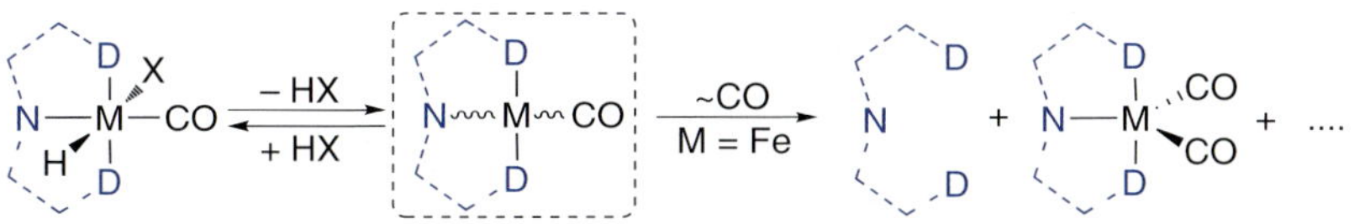

6 **7** **8** **9**

Scheme 1.3 Observed decomposition pathway of iron and ruthenium pincer-type complexes with an ancillary carbonyl ligand via the proposed intermediate **7** (M = Fe, Ru; D = donor group; X = H, alkoxide).

For nonrigid tetradentate ligands, similar observations were made: the catalytically inactive complex $[(\kappa^4\text{-PNNP})Fe(CO)]$ is formed during the catalytic reaction (PNNP = $Ph_2P—C_6H_4—CH{=}N—CH_2CH_2—N{=}CH—C_6H_4—PPh_2$). Interestingly, this complex was demonstrated to serve as a source for catalytically active iron nanoparticles [12].

In view of the standard potentials for the M^{2+}/M^0 couple, the observed deactivation pathways appear counterintuitive, but might be rationalized by the preference for different coordination numbers as well as different spin states of intermediates of iron and ruthenium catalyst systems.

1.3 Mechanistic Scenarios and the Role of Substrates

The apparent dependence of the mechanism of the hydrogen transfer step on the substrate polarity mainly originates from a preference for a certain coordination mode of each substrate: nonpolar substrates such as olefins coordinate in a side-on manner to the central metal atom, which makes a migratory insertion of the side-on coordinated substrate into the M—H bond viable. Polar substrates are often end-on coordinated to the central metal atom, which slows down the migratory insertion step, and a concerted proton–hydride transfer from the catalyst to substrate is usually faster. However, for carbon dioxide and for nitriles, for example, evidence has been provided that both stepwise migratory insertion mechanisms and concerted proton–hydride transfer can occur efficiently depending on the nature of the catalysts. In the following paragraphs, we discuss the available mechanistic information for iron-based catalysts in the context of the most common mechanisms for the hydrogenation of polar and nonpolar substrates. For relevant examples, the mechanistic information is compared to noble metal based catalysts.

1.3.1 Nonpolar Substrates

Iron-catalyzed hydrogenation reactions of olefins and alkynes have been known for several decades and usually proceed via mechanisms in which the hydrogen is transferred in a stepwise manner [13]. For well-established homogeneous noble metal hydrogenation based on rhodium and iridium, two types of textbook mechanisms are known [14]. The difference between these mechanisms is the order of the three basic steps involved in the stepwise hydrogen transfer: oxidative addition of H_2, migratory insertion in the M—H bond, and C—H reductive elimination of the product (Scheme 1.4). Cycle **A** involves a monohydride as active species that allows for the side-on coordination of the unsaturated substrate in *cis* position to the hydrido ligand. Migratory insertion leads to an alkyl complex. The subsequent reaction of this intermediate with H_2 results in product release and regeneration of the active monohydride. As the H_2 activation and product release do not necessarily involve a change in the formal oxidation state, this is a common mechanism for d^6 metal catalysts. A typical example would be ruthenium(II)-based catalysts. The second cycle (**B**) includes oxidative addition of H_2 and coordination of olefin to a metal complex [M]. Notably, the order can vary; in Scheme 1.4, for simplicity, only one case is depicted. The resulting dihydride olefin species allows for migratory insertion of the olefin into the metal

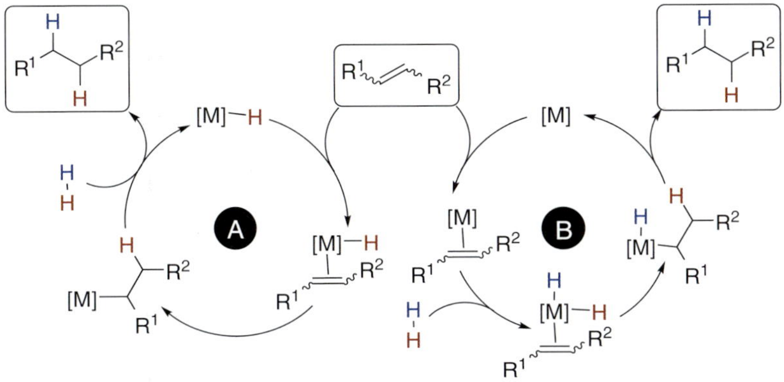

Scheme 1.4 Simplified reaction mechanisms for the hydrogenation of C—C multiple bonds, depicted for olefins.

hydrogen bond. Subsequently, reductive C–H elimination results in the release of the alkane product along with the initial catalyst species [M]. This mechanistic pathway is often assumed for homogeneous catalysts with a d^8 electron count (e.g. Rh^{I}, Ir^{I}). During the catalytic cycle, the change in oxidation states of the catalytic intermediates is by ±2 (e.g. Rh^{I}/Rh^{III} and Ir^{I}/Ir^{III}).

One of the first well-defined iron-based hydrogenation catalysts, whose mode of action has been investigated, is the iron(II) complex $[(PP_3)Fe(H)(H_2)](BPh_4)$ (**10**, PP_3 = $P(CH_2CH_2PPh_2)_3$ Figure 1.3). It was employed as a catalyst for semihydrogenation of alkynes to alkenes [13]. The octahedral arrangement of the four phosphine donor group, one hydrido and one dihydrogen ligand results in a strong ligand field and, as a result, in a diamagnetic complex. Interestingly, the absence of a potentially cooperative ligand site indicates that this is not an essential requirement for this catalytic reaction. Kinetic studies on this system point toward a mechanism via cycle **A**, in which the coordination of the dihydrogen ligand seems to be rate determining.

The employment of the so-called *non-innocent* or *redox-active* ligands [15] in iron catalysts for olefin hydrogenation leads to a significant improvement of catalyst activity. One of the first examples is the iron complex $[(^{iPr}PDI)Fe(N_2)_2]$ (**11**) that contains the 2,6-di-*iso*-propylphenyl-substituted bis(imino)pyridine (^{iPr}PDI)

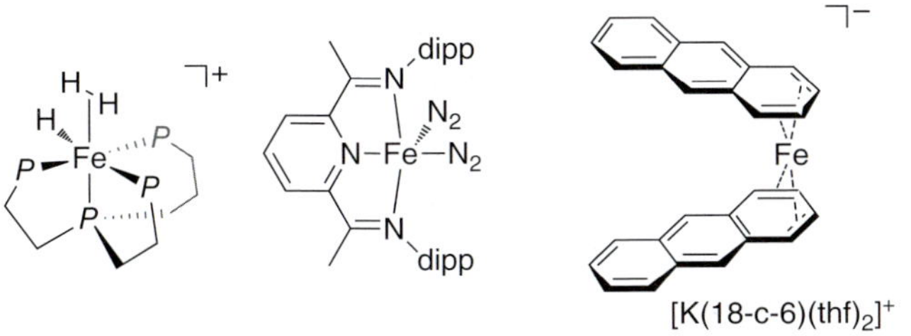

10 **11** **12**

Figure 1.3 Examples of different types of iron catalysts for the hydrogenation C—C multiple bonds.

as non-innocent ligand [16]. Complex **11** efficiently catalyzes the hydrogenation of olefines to alkanes under mild conditions [16a]. Later on, it was shown that this type of hydrogenation can also be conducted for substrates containing functional groups with heteroatoms [17]. Initially, a mechanism proceeding via cycle **B** was suggested after dissociation of N_2 from **11**. Comprehensive mechanistic investigations indicate that for some complexes and catalytic intermediates, reversible reduction of the bisiminopyridine ligand takes place rather than reduction of the central iron atom.

Active iron catalysts for the hydrogenation of olefins can also be formed by the reaction of $FeCl_3$ and $LiAlH_4$ in the presence of suitable arenes [18]. Under reaction conditions, it is assumed that bisligated iron(0) complexes are formed, which allows for olefin coordination and H_2 activation. Notably, the nature of the active catalyst changes during the reaction from a homogeneous system to a heterogeneous, but still active, system. The ferrate complex **12** (Figure 1.3) is an active olefin hydrogenation catalyst as well [19]. Here, the exchange of one arene ligand by an olefin generates the catalytically active species, which subsequently activates H_2. Mechanistic investigations point toward the mechanism shown in cycle **B**.

These examples clearly illustrate that catalytic activity in olefin hydrogenation can be achieved with a variety of iron catalysts, ranging from well-defined non-functionalized catalysts to complexes with non-innocent ligands, nanoparticles, and poorly defined, and also *in situ* generated catalysts. Based on the current mechanistic understanding, it can be concluded that it is not necessary to mimic the conventional reactivity patterns of noble metal catalysts to achieve catalytic activity with iron complexes. Moreover, in some cases, the tendency of 3d medals to undergo one-electron redox steps seems even advantageous.

1.3.2 Polar Substrates

For hydrogenation reactions of polar bonds, completely different reactivity patterns and mechanisms are observed. For many catalysts, a stepwise mechanism via migratory insertion of a polar multiple bond into a metal hydrogen bond was demonstrated to be inefficient. It has been reasoned that polar substrates such as carbonyl compounds preferably coordinate in an *end-on* κO manner to the metal and not a *side-on* manner like apolar substrates such as olefins or alkynes. The migratory insertion of such *end-on* coordinated substrates is often disfavored, due to too high reaction barriers [20]. In the case of simple ruthenium phosphine complexes, it was shown that addition of a chelating primary amine ligand accelerates the rate of ketone hydrogenation by orders of magnitude. The resulting catalytically active species is often called a bifunctional or cooperative catalyst [21], which is based on the finding that the coordinated amine ligand can act as a proton source in a concerted proton–hydride transfer from the catalyst to the polar substrate (cycle **C** in Scheme 1.5). This *in concerto* mode of action is also called metal ligand cooperation. Although, ruthenium complexes with different ligand platforms and functionalities can be active hydrogenation catalysts, which, in principle, can follow different mechanisms, cooperative catalysts turned out to be particularly active catalysts for the hydrogenation of polar multiple bonds.

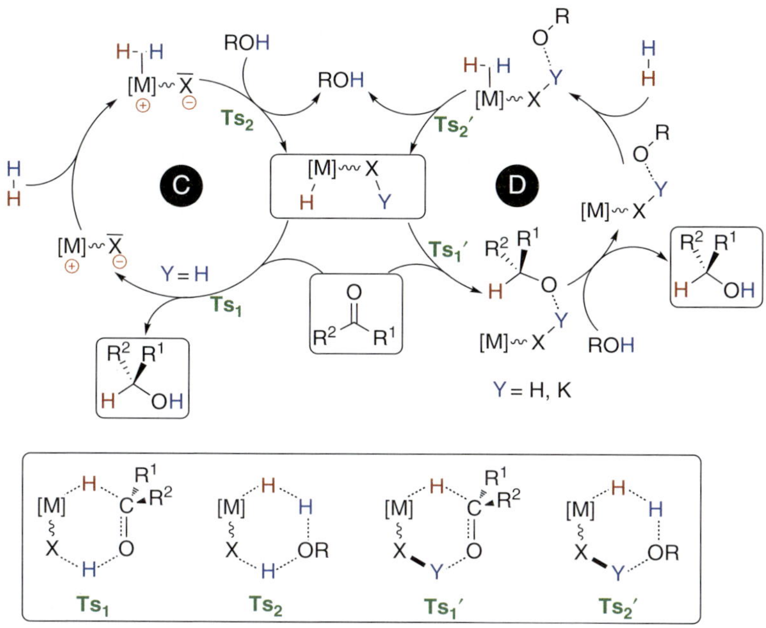

Scheme 1.5 Simplified reaction mechanisms for the hydrogenation of C—O multiple bonds, depicted for olefins.

It is worth noting that the mechanistic view of a concerted proton–hydride transfer (as in **TS_1**) was recently questioned by DFT studies with explicit solvent modeling [22]. It was suggested that for a number of catalysts the cooperative site serves as a binding site for the substrate that allows for pre-coordination of the substrate and the subsequent hydride transfer (**TS_1'**). It has been further indicated that a potassium ion (in many cases, the counterions of the utilized base) can also be involved in this interaction (cycle **D**). The H· · ·O- or K· · ·O-bound alkoxide either serves as the internal base in the heterolytic cleavage of dihydrogen or is protonated and exchanged by the alcohol solvent prior to dihydrogen cleavage.

A closer look at the reported iron catalysts for the hydrogenation of polar multiple bonds reveals that the most active catalysts contain cooperative ligands and that they usually operate via the simplified mechanism shown in Scheme 1.5.

1.3.3 Exceptions

In the previous paragraphs, we have summarized the available mechanistic information on the operating mechanisms of the most active catalysts. The conclusion from this information might be taken as a rule of thumb for the prediction of a viable mechanism and catalyst design, but there are many exceptions and these are briefly discussed in following paragraphs.

Carbon dioxide is readily hydrogenated to formates in the presence of stoichiometric amounts of a base and a suitable catalyst. Despite the polar C—O bond, highly active catalysts were reported with cooperative and noncooperative ligands, operating via different mechanisms [23]. It has even been demonstrated that ruthenium catalysts facilitating an inner-sphere mechanism allow for the direct hydrogenation to methanol [24].

A recent study suggested that an amine-based iron pincer catalyst for the hydrogenation of olefins operates via a stepwise cooperative mechanism [25]. For this catalyst system, this might be the preferred pathway, as the active reducing dihydride species does not have any vacant coordination site, allowing for olefin coordination.

Nitriles are common ligands in coordination chemistry and, like ketones, they usually prefer an *end-on* coordination. Although some cooperative ruthenium catalysts are known to operate via the mechanisms shown in Scheme 1.5 [26], it has also been demonstrated for a pyridine-based ruthenium pincer catalyst that the reaction can proceed via an intermediate with a *side-on* coordinated nitrile and a subsequent migratory insertion [27].

1.4 Iron-Catalyzed Hydrogenation of C—C Multiple Bonds

In the Section 1.3.1, highly active iron catalysts for the hydrogenation of C—C multiple bonds were introduced. The following Sections 1.4.1 and 1.4.2 discusses the most active homogeneous iron-based catalysts for different transformations involving the hydrogenation of C—C multiple bonds, namely, the hydrogenation of olefins and alkynes.

1.4.1 Hydrogenation of Olefins

The hydrogenation of olefins (alkenes) to alkanes is catalyzed by a wide range of well-defined and *in situ* generated iron catalysts. Among the well-defined iron catalysts are complexes with potentially cooperative ligands that either act as internal base, sometimes described as a proton relay (e.g. the amido group in **13** [25] and **14** [28]), or as a hydride acceptor (e.g. the Z-type BR_3-group in **15** [29] (Figure 1.4)), as well as complexes with non-innocent (= redox active) ligands (**11** [16a]), formally an electron relay, or non-functionalized ligands (**12** [19]).

Also, *in situ* generated catalysts, such as the $FeCl_3/LiAlH_4$ system [18], can lead to highly active catalysts in terms of turnover frequency. However, the active species in these systems seem to have a limited lifetime, which is also reflected by the steady transition of a homogeneous to a heterogeneous catalyst system.

A comparison of the different catalyst types reveals that the pincer-type complex **11,** featuring the redox-active bis(imino)pyridine ligand, is by far the most active catalyst for the hydrogenation of mono- and disubstituted olefins (Table 1.2). However, substrates with tri- and tetrasubstituted C-C double bonds are not hydrogenated under the reported conditions [16a]. The *in situ* generated catalyst based on $Fe(hmds)_2/i\text{-}Bu_2AlH$, in contrast, appears to be less active for the hydrogenation of monosubstituted olefins, but is capable of hydrogenating

Figure 1.4 Representative examples of iron catalysts for the hydrogenation of olefins (in addition to the examples in Figure 1.3).

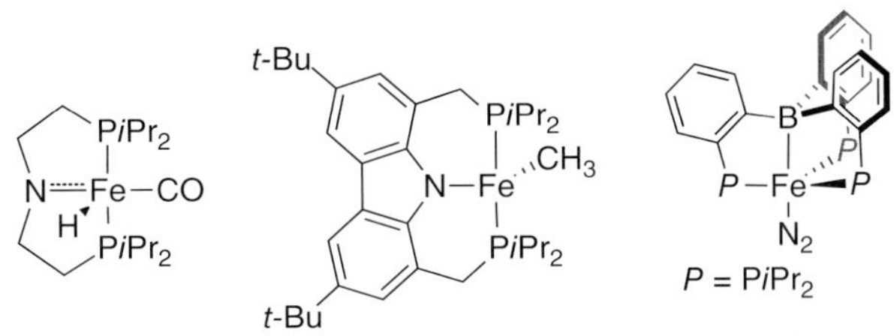

13 14 15

Table 1.2 Comparison of different catalysts for the hydrogenation of styrene.

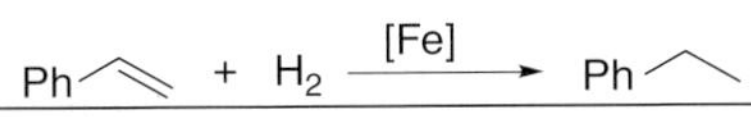

[Fe]	Loading (mol%)	$p(H_2)$ (bar)	Yield (%)	TON	TOF (h^{-1})	References
11	0.3	1	>99	330	1344	[16a]
12	1.0	2	95	95	32	[19]
13	5.0	1	>99	20	1	[25]
14	1.0	8	>99	100	17	[28]
15	3.3	1	95	29	0.27	[29]
$Fe(hmds)_2$/*i*-Bu_2AlH a)	5.0	2	>99	20	6.67	[18]
$FeCl_3/LiAlH_4$	5.0	1	>99	20	3.33	[18]

a) Data is for *para*-butyloxystyrene.

tri- and tetrasubstituted olefines [18]. The diverse types of active catalysts again justify the conclusion that there are no specific catalyst requirements to achieve high activity in olefin hydrogenation.

1.4.2 Hydrogenation of Alkynes

The hydrogenation of alkynes can lead to different reaction products, depending on the type of catalyst and the reaction conditions. In general, the first hydrogenation step yields olefins, which, in the case of internal alkynes as substrates, can result in *E*- and *Z*-isomers (Scheme 1.6). As pointed out in the Section 1.4.1, olefins can get hydrogenated in the presence of a suitable catalyst as well, leading to saturated alkanes. If the main hydrogenation product(s) are olefins, the reaction is commonly called a semihydrogenation, as the main hydrogenation product remains unsaturated and does not get further hydrogenated (Scheme 1.6).

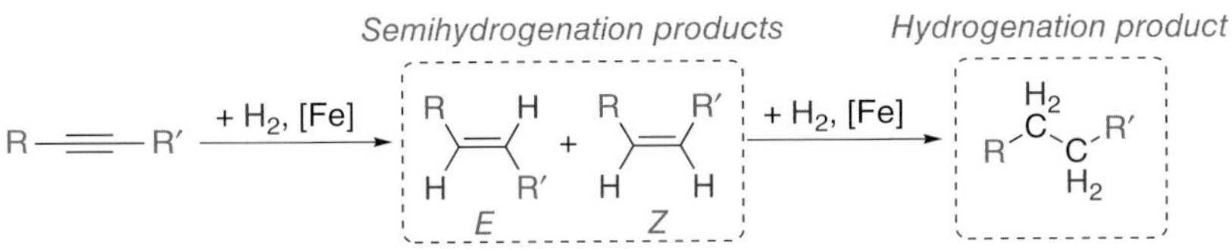

Scheme 1.6 Semihydrogenation and hydrogenation of alkynes.

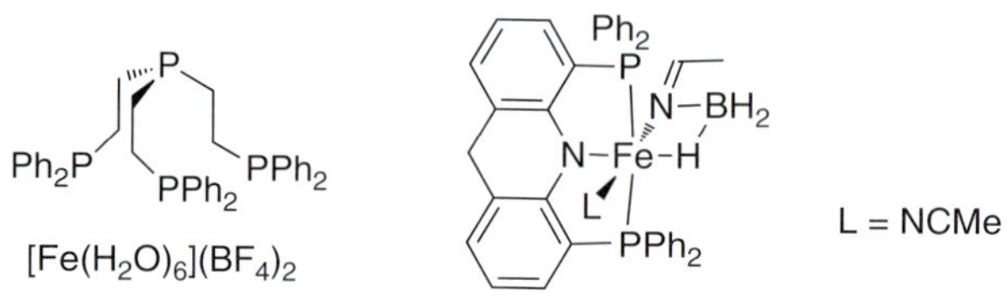

16 **17**

Figure 1.5 Examples of iron catalysts for the hydrogenation of alkynes.

As for olefins, the examples of iron-based catalysts for the (semi)hydrogenation of alkynes are limited and often catalysts for the olefin hydrogenation are active for alkynes as well. The bisiminopyridine complex **11**, for example, is an active catalyst for the hydrogenation diphenylacetylene, too, which leads to the formation of *Z*-stilbene as an intermediate that gets further hydrogenated to dibenzyl over the course of the reaction [16a]. The selective formation of terminal olefins by semi-transfer hydrogenation of alkynes is catalyzed by an *in situ* generated catalyst (**16**), containing the tretraphos ligand and $[Fe(H_2O)_6](BF_4)_2$ (Figure 1.5) [30]. The highly desirable *E*-selective hydrogenation of internal alkynes is facilitated by the iron pincer catalyst **17** [31].

1.5 Iron-Catalyzed Hydrogenation of C—O Multiple Bonds

Organic carbonyl compounds containing a C=O double bond can exhibit distinct differences in their reactivity, as the substrate scope ranges from reactive aldehydes to inert substrates which are ureas, carbamates, and carbonates. The latter ones are sometimes even used as inert solvents for hydrogenation reactions [32]. Figure 1.6 illustrates the decreasing carbonyl reactivity of different selected substrates, with which it becomes more challenging to achieve a catalytic hydrogenation. Recently, significant progress has been made in the development of ruthenium catalysts that are capable of hydrogenating even the most inert substrates among these series, such as carbonates, carbamates, and ureas [32, 33]. With corresponding iron complexes, tremendous achievements in hydrogenation catalysis have been reported as well, but so far the hydrogenation activity has been limited to esters and even amides, as well as more reactive carbonyl compounds (Figure 1.6). These findings might be the result of reduced hydricity observed for iron complexes in comparison to the corresponding ruthenium analogs.

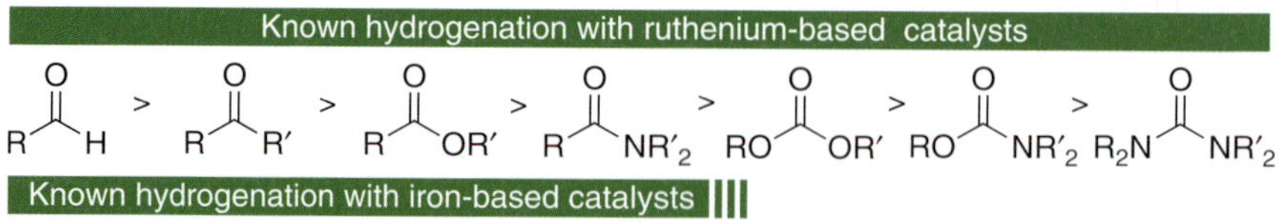

Figure 1.6 General reactivity of selected carbonyl compounds toward nucleophiles and the current state of research for the hydrogenation with iron and ruthenium catalysts.

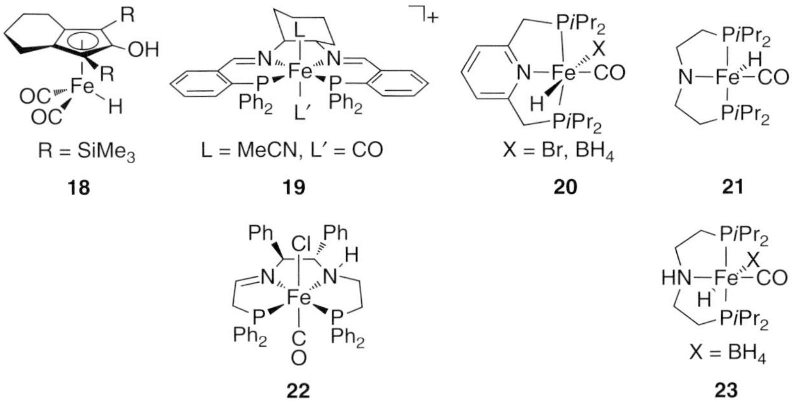

Figure 1.7 Selection of iron catalysts for the hydrogenation of aldehydes and ketones.

Typical, iron-based hydrogenation catalysts for carbonyl and carboxyl compounds are shown in Figure 1.7. For the majority of these catalysts, the underlying construction principles can be summarized as follows: (i) Cooperative ligands, acting as a proton relay, are found in the most active representatives. (ii) At least one polydentate or cyclopentadienyl ligand is present in these complexes. (iii) Donor groups and ligands with a strong field are employed in these ligands. In the following Subsections 1.5.1–1.5.3, we briefly discuss the recent developments and state-of-the-art iron catalysts for each substrate.

1.5.1 Hydrogenation of Aldehydes and Ketones

An important milestone in the development of iron-based hydrogenation catalysts was *Casey*'s report about the catalytic activity of *Knölker*'s complex **18** in the hydrogenation of aldehydes, ketones, and aldimines under mild conditions (3 bar H_2 pressure, rt) [34] (Table 1.3). Although the catalyst loading was quite high (2 mol%) and the yield of 1-phenylethanol was only 83%, it represented the first example of an iron-catalyzed ketone hydrogenation using hydrogen gas. While *in situ* generated transfer hydrogenation catalysts based on iron were reported early, the report of complex **19** as an efficient and enantioselective hydrogenation catalyst for ketones and imines represents an important breakthrough [11b]. The subsequent modification of the employed tetra dentate ligand in the past decade has led to chiral state-of-the-art catalysts for asymmetric transfer hydrogenation reactions, such as **22**, which exhibit in comparable activities and selectivity like noble metal catalysts [35].

The pyridine-based iron pincer complex **20** showed by an order of magnitude higher productivities TON and activities TOF in the hydrogenation of ketones with hydrogen gas (4.1 bar) [36]. Later on, complex **20** and related iron PNP-pincer (pre)catalysts based on pyridine backbones [36], were also employed as hydrogenation catalysts for other polar substrates including aldehydes [37], activated esters [38], and even amides [39]. Also, the efficient dehydrogenation of formic acid in the presence of amine additives was reported using iron

Table 1.3 Comparison of different catalysts for the hydrogenation of acetophenone.

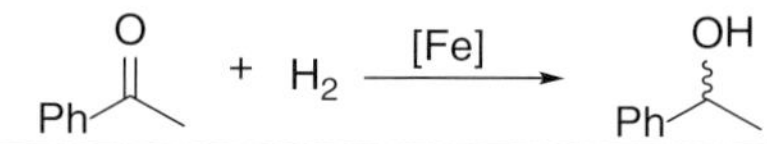

[Fe]	Loading (mol%)	H_2 source	Yield (%)	TON	TOF (h^{-1})	References
18	2.0	H_2 (3 bar)	83	42	2	[34]
19	0.25	*i*-PrOH	91	362	907	[11b]
20	0.05	H_2 (4 bar)	94	1 720	430	[36]
22	0.016	*i*-PrOH	82	5 000	10 000	[35]
23	0.05	H_2 (5 bar)	>99	2 000	500	[4]

pyridine-PNP catalysts [40]. Similar iron pincer complexes (**21** and **23**) were reported later on. The design principle of combining a PNP-pincer ligand with an ancillary carbonyl ligand resulted in highly active and productive catalysts systems for PNP ligands with aliphatic and olefinic backbones. This catalyst family was then applied in hydrogenation and dehydrogenation reactions for a variety of substrates [41]. Applications for this type of (pre)catalysts include hydrogen liberation from formic acid [42] and from aqueous methanol solutions [43], hydrogenation of ketones [11b, 44], esters [42a, 45], and amides [11a, 46], the asymmetric hydrogenation of ketones and imines [11b], the hydrogenation of polarized C-C double bonds of substituted alkenes [25], and the reversible hydrogenation and dehydrogenation of N-heterocycles [47].

Complexes **19** and **22** are chiral and capable of catalyzing the enantioselective hydrogenation of prochiral ketones to chiral alcohols with high selectivity (Figure 1.7). After the initial report on the first iron-based catalyst (**19**) [48] for this reaction, the activity and selectivity have been significantly improved with the modified tetra dentate ligand platform in **22** [42b, 49]. For other non-chiral iron catalysts shown in Figure 1.7, different strategies have been used for the introduction of a center of chirality in the catalyst system. A modified version of *Knölker's* catalyst (**18**) was used in combination with a chiral phosphine ligand (**24**) for the hydrogenation of acetophenone (Figure 1.8), resulting in moderate enantiomeric excess of up to 33% ee [50]. Similarly, the combination of complex (**18**) with chiral Brønsted acids allowed enantioselective hydrogenation reactions of imines to amines [51]. The imine-based iron pincer complex **25** contains two centers of chirality in the pincer backbone and is a catalyst for the enantioselective transfer hydrogenation of ketones and imines. Under the reported conditions, the preparation of *S*-1-phenylethanol with an enantioselectivity of 80% ee from acetophenone was achieved [11b]. The activation of the pre-catalyst presumably involves reduction of the imine function and formation of the same type of active species like with **21** and **23**.

A chiral analog of catalyst **20** has been reported with centers of chirality at the two terminal phosphorus atoms of the pincer ligand in **26**, which allowed for the hydrogenation of acetophenone with moderate enantioselectivity of 48%

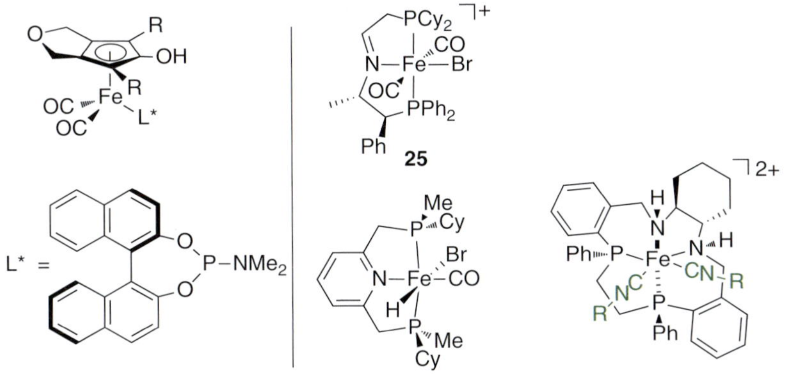

24 **25** **26** **27**

Figure 1.8 Examples of chiral iron catalysts for the enantioselective hydrogenation of ketones. Asterisk indicates a chiral ligand.

ee to *S*-1-phenylethanol [52]. The tetra dentate ligand in complex **27** is similar to the one in complex **19**, but is macrocyclic. Therefore, the macrocyclic ligand provides more rigidity, which in turn increases the stereoselectivity (up to 98% ee for acetophenone hydrogenation) [53].

1.5.2 Hydrogenation of Esters

Some of the iron catalysts for ketone hydrogenation turned out to be active catalysts for the hydrogenation of esters to alcohols (Scheme 1.7). In contrast to the hydrogenation of ketones and aldehydes, two hydrogen transfer steps are required to obtain a mixture of alcohols as products of an ester hydrogenation. After the transfer of the first equivalent of H_2 to the ester, the corresponding hemi-acetal is formed, which is known to be in equilibrium with the corresponding aldehyde and an alcohol. The rapid hydrogenation of the aldehyde intermediate results in the formation of the final products. Notably, for symmetric esters, only one type of alcohol is obtained (e.g. ethanol is obtained for the hydrogenation of ethyl acetate), whereas the hydrogenation of asymmetric esters results in two alcohols (e.g. the hydrogenation of benzyl acetate results in benzyl alcohol and ethanol as products).

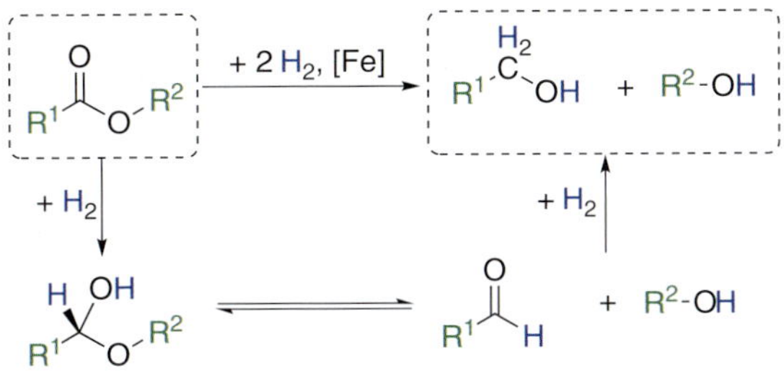

Scheme 1.7 Hydrogenation of esters to alcohols.

The aforementioned amine- and pyridine-based iron pincer complexes were reported to be active catalysts for the hydrogenation of different types of esters. While the pyridine-based iron pincer complex **20** showed significant activity in the hydrogenation of activated trifluoroacetates [38], the amine-based complexes **21** and **23** are capable of hydrogenating nonactivated alkyl- and aryl esters with high productivity and activity [45].

1.5.3 Hydrogenation of Amides

The catalytic hydrogenation of amides can, in principle, lead to different reaction products, depending on the preferred reaction pathway of the hemi-aminal, which is formed as an intermediate of the reaction. The latter is formed after transfer of one equivalent of hydrogen to the amide substrate (Scheme 1.8). Formation of the aldehyde and amine from the hemi-aminal, followed by further reduction of the aldehyde, leads to primary alcohols and amines as major reaction products. Alternatively, the hemi-aminal can eliminate one equivalent of water to give the corresponding imine. The reduction of the latter yields secondary amines (tertiary amides in the case of secondary amides) as major reaction products. The desired formation of alcohols and amines is observed with some of the ruthenium catalysts [33b].

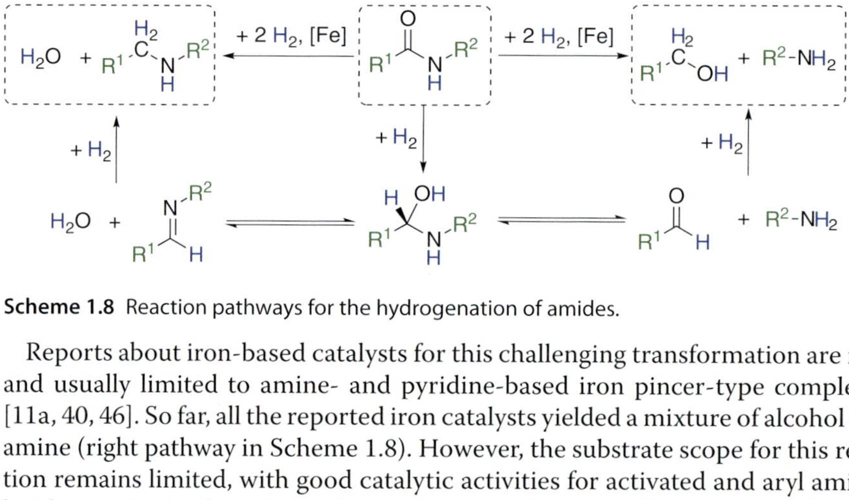

Scheme 1.8 Reaction pathways for the hydrogenation of amides.

Reports about iron-based catalysts for this challenging transformation are rare and usually limited to amine- and pyridine-based iron pincer-type complexes [11a, 40, 46]. So far, all the reported iron catalysts yielded a mixture of alcohol and amine (right pathway in Scheme 1.8). However, the substrate scope for this reaction remains limited, with good catalytic activities for activated and aryl amides but low activities for other substrates. An accelerating effect has been demonstrated for Lewis acid co-catalysts such as formamides [46a], which results in significantly increased activities (TOF) and productivities (TON), as well as a widely applicable protocol.

1.6 Iron-Catalyzed Hydrogenation of C—N Multiple Bonds

Nitriles can be directly hydrogenated to primary amines using iron catalysts. In this two-step process, a primary aldimine (**I**) is generated as an intermediate

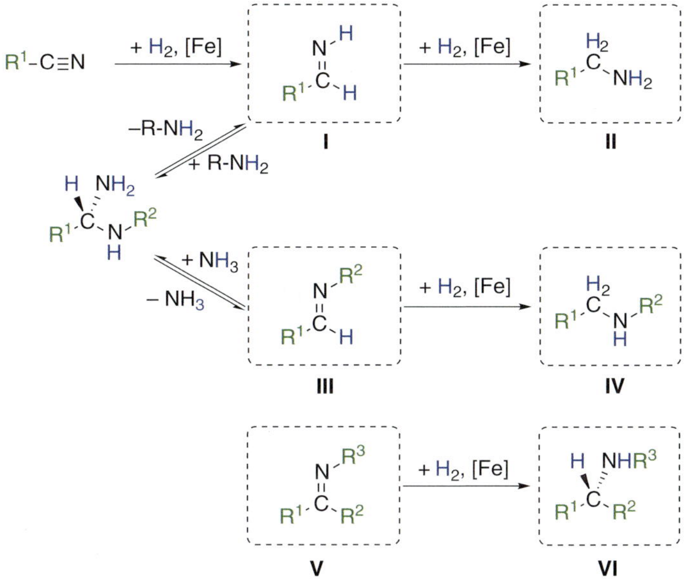

Scheme 1.9 Possible reaction products for the hydrogenation of nitriles and imines.

after the first transfer of dihydrogen, which is further hydrogenated to the corresponding primary amine in the second hydrogenation step (Scheme 1.9). More stable and isolable secondary ketimines (**V**) can be hydrogenated as well using iron catalysts, which in the case of prochiral ketimines yields chiral secondary amines (**VI**), valuable products in the value adding chain of the chemical industry.

The hydrogenation of nitriles with an iron-based catalyst has only recently been reported [54]. Similar to the hydrogenation of C-O double bonds, the bifunctional catalysts, such as **23** [54a], **28** [54b], and **29** [54c, d], have been utilized (Figure 1.8). Mechanistic investigations indicate that a concerted proton–hydride transfer takes place in both hydrogenation steps.

In the presence of a primary amine, the initially formed primary aldimine intermediate **I** can be trapped by formation of the corresponding hemi-aminal and gradual NH_3 elimination. The hydrogenation of the resulting secondary aldimine (**III**) yields secondary amines (**IV**). Using complex **25** as catalyst, the scavenging primary amine can be *in situ* generated by complete hydrogenation of the nitrile or it is added to the reaction mixture (Figure 1.9).

The hydrogenation of ketimines (**V**) is of particular interest, as with different substituents R^1 and R^2 the ketimine is prochiral and the hydrogenation of the latter results in chiral amines. Such an enantioselective hydrogenation of ketimines is, for example, an important step in the synthesis of (S)-Metolachlor, one of the most common herbicides, for which iridium catalysts are preferably used [55].

The reactivity of the ketimine substrate strongly depends on the nitrogen-bound substituent R^3 and most publications on iron-based imine hydrogenation

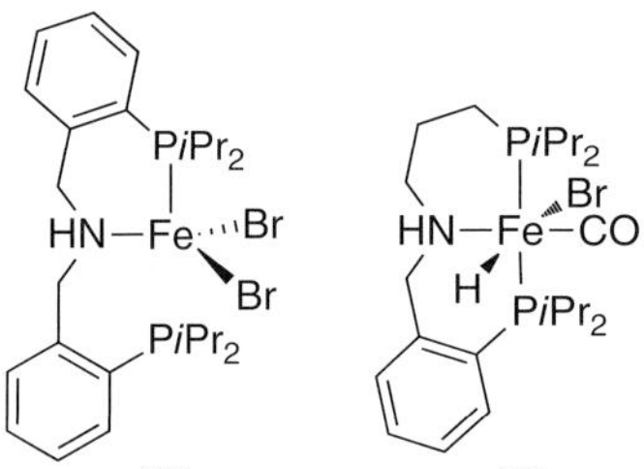

28 29

Figure 1.9 Iron-based catalysts for the hydrogenation of nitriles.

catalysts demonstrate catalytic activity for *N*-diphenylphosphinoyl-imines ($R^3 = P(O)Ph_2$). Among the most active and selective iron catalysts for this reaction is complex **22** [42b, 49, 56], but complex **25** and chiral variants of **18** have also been successfully applied as catalysts in the asymmetric hydrogenation of imines [51].

1.7 Conclusion

A major focus of research in homogeneous catalysis in the past decade was on the replacement of noble metal catalysts by their 3d metal counterparts. Among the catalytic reactions, hydrogenation reactions have received considerable attention due to the importance of these atom economic and industrially important transformations. The rapid development in terms of activity, productivity, and selectivity, as well as the fact that nature uses preferably 3d metals in highly active metalloenzymes, may allow predicting that 3d metal based catalyst systems exhibit the potential to one day be real low cost, sustainable, and environmentally benign alternatives for established noble metal based hydrogenation protocols. In this context, iron is a highly attractive metal due to high abundance, low price, and toxicity. A shared characteristic of the most active iron-based catalysts for the hydrogenation of polar substrates is the presence of a tri- or tetra dentate, cooperative ligand in an iron(II) complex. For the hydrogenation of nonpolar substrates, on the other hand, a variety of different catalyst types were reported to be active hydrogenation catalysts. The majority of well-defined catalysts contains a central iron atom with formal oxidation state +2, while only few iron(0) complexes are among the reported catalyst systems.

Understanding the catalytic mechanisms of these reactions is often challenging. This is in part due to analytic limitations on the one hand, and quantum chemical challenges on the other hand, both affiliated with energetically accessible spin states. However, clear trends in reactivity can be observed for different types of catalysts and comprehensive mechanistic investigations will most probably be the basis for designing novel, highly efficient catalytic protocols. A key to the development of highly efficient Fe-based hydrogenation catalysis lies in the suppression of decomposition pathways of iron-based hydrogenation catalysts. Overcoming this challenge will allow to domesticate iron and use its full catalytic potential.

Abbreviations

t-Bu	*tert*-butyl
Cy	cyclohexyl
DFT	density functional theory
dppe	1,2-bis(diphenylphosphino)ethane
ee	enantiomeric excess
hmds	hexamethyldisilazanide
HOMO	highest occupied molecular orbital
LCAO	linear combination of atomic orbitals
LUMO	lowest unoccupied molecular orbital
Me	methyl
^{iPr}PDI	2,6-di-*iso*-propylphenyl-substituted bis(imino)pyridine
PNP	pincer-type ligand in which the PNP indicates the ligating atoms
Ph	phenyl
PP_3	$P(CH_2CH_2PPh_2)_3$
i-Pr	*iso*-propyl
TOF	turnover frequency
TON	turnover number

References

1 R. Shannon, *Acta Cryst. Sect. A* 1976, 32, 751–767.
2 R.J. Ryther, E. Weitz, *J. Phys. Chem.* 1991, 95, 9841–9852.
3 W. Zuo, A.J. Lough, Y.F. Li, R.H. Morris, *Science* 2013, 342, 1080–1083.
4 S. Chakraborty, P.O. Lagaditis, M. Förster, E.A. Bielinski, N. Hazari, M.C. Holthausen, W.D. Jones, S. Schneider, *ACS Catal.* 2014, 4, 3994–4003.
5 (a) P. Stoppioni, F. Mani, L. Sacconi, *Inorg. Chim. Acta* 1974, 11, 227–230; (b) C. Federsel, A. Boddien, R. Jackstell, R. Jennerjahn, P.J. Dyson, R. Scopelliti, G. Laurenczy, M. Beller, *Angew. Chem. Int. Ed.* 2010, 49, 9777–9780.
6 (a) C.P. Casey, H. Guan, *J. Am. Chem. Soc.* 2009, 131, 2499–2507; (b) C.P. Casey, H. Guan, *J. Am. Chem. Soc.* 2007, 129, 5816–5817.
7 (a) R. Langer, M. A. Iron, L. Konstantinovski, Y. Diskin-Posner, G. Leitus, Y. Ben-David, D. Milstein, *Chem. Eur. J.* 2012, 18, 7196–7209; (b) R. Langer, G. Leitus, Y. Ben-David, D. Milstein, *Angew. Chem. Int. Ed.* 2011, 50, 2120–2124.
8 (a) P.L. Holland, *Acc. Chem. Res.* 2015, 48, 1696–1702; (b) D. Schröder, S. Shaik, H. Schwarz, *Acc. Chem. Res.* 2000, 33, 139–145.
9 E.S. Wiedner, M.B. Chambers, C.L. Pitman, R.M. Bullock, A.J.M. Miller, A.M. Appel, *Chem. Rev.* 2016, 116, 8655–8692.
10 (a) K. Abdur-Rashid, T.P. Fong, B. Greaves, D.G. Gusev, J.G. Hinman, S.E. Landau, A.J. Lough, R.H. Morris, *J. Am. Chem. Soc.* 2000, 122, 9155–9171; (b) R.H. Morris, *J. Am. Chem. Soc.* 2014, 136, 1948–1959.
11 (a) F. Schneck, M. Assmann, M. Balmer, K. Harms, R. Langer, *Organometallics* 2016, 35, 1931–1943; (b) P.O. Lagaditis, P.E. Sues, J.F.

Sonnenberg, K.Y. Wan, A.J. Lough, R.H. Morris, *J. Am. Chem. Soc.* 2014, 136, 1367–1380.

12 J.F. Sonnenberg, N. Coombs, P.A. Dube, R.H. Morris, *J. Am. Chem. Soc.* 2012, 134, 5893–5899.

13 C. Bianchini, A. Meli, M. Peruzzini, P. Frediani, C. Bohanna, M.A. Esteruelas, L.A. Oro, *Organometallics* 1992, 11, 138–145.

14 L.A. Oro, D. Carmona, Rhodium. In: *The Handbook of Homogeneous Hydrogenation* (eds. J.G. de Vries and C.J. Elsevier). Wiley-VCH Verlag GmbH, 2008, 2–30.

15 (a) A.I. Olivos Suarez, V. Lyaskovskyy, J.N.H. Reek, J.I. van der Vlugt, B. de Bruin, *Angew. Chem. Int. Ed.* 2013, 52, 12510–12529; (b) W.I. Dzik, J.I. van der Vlugt, J.N.H. Reek, B. de Bruin, *Angew. Chem. Int. Ed.* 2011, 50, 3356–3358; (c) V. Lyaskovskyy, B. de Bruin, *ACS Catal.* 2012, 2, 270–279.

16 (a) S.C. Bart, E. Lobkovsky, P.J. Chirik, *J. Am. Chem. Soc.* 2004, 126, 13794–13807; (b) S.K. Russell, C. Milsmann, E. Lobkovsky, T. Weyhermüller, P.J. Chirik, *Inorg. Chem.* 2011, 50, 3159–3169; (c) P.J. Chirik, K. Wieghardt, *Science* 2010, 327, 794–795; (d) M.W. Bouwkamp, A.C. Bowman, E. Lobkovsky, P.J. Chirik, *J. Am. Chem. Soc.* 2006, 128, 13340–13341.

17 R.J. Trovitch, E. Lobkovsky, E. Bill, P.J. Chirik, *Organometallics* 2008, 27, 1470–1478.

18 T.N. Gieshoff, M. Villa, A. Welther, M. Plois, U. Chakraborty, R. Wolf, A. Jacobi von Wangelin, *Green Chem.* 2015, 17, 1408–1413.

19 P. Buschelberger, D. Gartner, E. Reyes-Rodriguez, F. Kreyenschmidt, K. Koszinowski, A. Jacobi von Wangelin, R. Wolf, *Chem. Eur. J.* 2017, 23, 3139–3151.

20 (a) R. Noyori, T. Ohkuma, *Angew. Chem. Int. Ed.* 2001, 40, 40–73; (b) S.E. Clapham, A. Hadzovic, R.H. Morris, *Coord. Chem. Rev.* 2004, 248, 2201–2237.

21 (a) C. Gunanathan, D. Milstein, *Science* 2013, 341, 1229712; (b) C. Gunanathan, D. Milstein, *Acc. Chem. Res.* 2011, 44, 588–602; (c) S. Chakraborty, H. Guan, *Dalton Trans.* 2010, 39, 7427–7436; (d) C. Gunanathan, D. Milstein, *Top. Organomet. Chem.* 2011, 37, 55–84; (e) H. Grützmacher, *Angew. Chem. Int. Ed.* 2008, 47, 1814–1818. (f) J.R. Khusnutdinova, M. David, *Angew. Chem. Int. Ed.* 2015, 54, 12236–12273; (g) D. Milstein, *Top. Catal.* 2010, 53, 915–923.

22 (a) P.A. Dub, J.C. Gordon, *ACS Catal.* 2017, 7, 6635–6655; (b) P.A. Dub, N.J. Henson, R.L. Martin, J.C. Gordon, *J. Am. Chem. Soc.* 2014, 136, 3505–3521; (c) P.A. Dub, T. Ikariya, *J. Am. Chem. Soc.* 2013, 135, 2604–2619.

23 W.-H. Wang, Y. Himeda, J.T. Muckerman, G.F. Manbeck, E. Fujita, *Chem. Rev.* 2015, 115, 12936–12973.

24 (a) S. Wesselbaum, V. Moha, M. Meuresch, S. Brosinski, K. M. Thenert, J. Kothe, T. V. Stein, U. Englert, M. Holscher, J. Klankermayer, W. Leitner, *Chem. Sci.* 2015, 6, 693–704; (b) S. Wesselbaum, T. vom Stein, J. Klankermayer, W. Leitner, *Angew. Chem. Int. Ed.* 2012, 51, 7499–7502.

25 R. Xu, S. Chakraborty, S.M. Bellows, H. Yuan, T.R. Cundari, W.D. Jones, *ACS Catalysis* 2016, 6, 2127–2135.

26 J. Neumann, C. Bornschein, H. Jiao, K. Junge, M. Beller, *Eur. J. Org. Chem.* 2015, 2015, 5944–5948.

27 C. Gunanathan, M. Hölscher, W. Leitner, *Eur. J. Inorg. Chem.* 2011, 2011, 3381–3386.

28 J.C. Ott, C.K. Blasius, H. Wadepohl, L.H. Gade, *Inorg. Chem.* 2018, 57, 3183–3191.

29 H. Fong, M.E. Moret, Y. Lee, J.C. Peters, *Organometallics* 2013, 32, 3053–3062.

30 G. Wienhofer, F.A. Westerhaus, R. V. Jagadeesh, K. Junge, H. Junge, M. Beller, *Chem. Commun.* 2012, 48, 4827–4829.

31 D. Srimani, Y. Diskin-Posner, Y. Ben-David, D. Milstein, *Angew. Chem. Int. Ed.* 2013, 52, 14131–14134.

32 (a) E. Balaraman, C. Gunanathan, J. Zhang, L.J.W. Shimon, D. Milstein, *Nat. Chem.* 2011, 3, 609–614; (b) E. Balaraman, Y. Ben-David, D. Milstein, *Angew. Chem. Int. Ed.* 2011, 50, 11702–11705.

33 (a) E. Balaraman, D. Milstein, *Top. Organomet. Chem.* 2014, 48, 19–43; (b) P.A. Dub, T. Ikariya, *ACS Catal.* 2012, 2, 1718–1741.

34 A. Quintard, J. Rodriguez, *Angew. Chem. Int. Ed.* 2014, 53, 4044–4055.

35 R.M. Bullock, *Science* 2013, 342, 1054–1055.

36 T. Zell, D. Milstein, *Acc. Chem. Res.* 2015, 48, 1979–1994.

37 T. Zell, Y. Ben-David, D. Milstein, *Catal. Sci. Technol.* 2015, 5, 822–826.

38 T. Zell, Y. Ben-David, D. Milstein, *Angew. Chem. Int. Ed.* 2014, 53, 4685–4689.

39 J.A. Garg, S. Chakraborty, Y. Ben-David, D. Milstein, *Chem. Commun.* 2016, 52, 5285–5288.

40 T. Zell, B. Butschke, Y. Ben-David, D. Milstein, *Chem. Eur. J.* 2013, 19, 8068–8072.

41 (a) S. Chakraborty, P. Bhattacharya, H. Dai, H. Guan, *Acc. Chem. Res.* 2015, 48, 1995–2003; (b) R.H. Morris, *Acc. Chem. Res.* 2015, 48, 1494–1502.

42 E.A. Bielinski, P.O. Lagaditis, Y. Zhang, B.Q. Mercado, C. Würtele, W.H. Bernskoetter, N. Hazari, S. Schneider, *J. Am. Chem. Soc.* 2014, 136, 10234–10237.

43 E. Alberico, P. Sponholz, C. Cordes, M. Nielsen, H.J. Drexler, W. Baumann, H. Junge, M. Beller, *Angew. Chem.* 2013, 52, 14162–14166.

44 J.F. Sonnenberg, A.J. Lough, R.H. Morris, *Organometallics* 2014, 33, 6452–6465.

45 (a) S. Werkmeister, K. Junge, B. Wendt, E. Alberico, H. Jiao, W. Baumann, H. Junge, F. Gallou, M. Beller, *Angew. Chem. Int. Ed.* 2014, 53, 8722–8726; (b) S. Chakraborty, H. Dai, P. Bhattacharya, N.T. Fairweather, M.S. Gibson, J.A. Krause, H. Guan, *J. Am. Chem. Soc.* 2014, 136, 7869–7872.

46 (a) N.M. Rezayee, D.C. Samblanet, M.S. Sanford, *ACS Catal.* 2016, 6, 6377–6383; (b) U. Jayarathne, Y. Zhang, N. Hazari, W.H. Bernskoetter, *Organometallics* 2017, 36, 409–416.

47 S. Chakraborty, W.W. Brennessel, W.D. Jones, *J. Am. Chem. Soc.* 2014, 136, 8564–8567.

48 (a) C. Sui-Seng, F.N. Haque, A. Hadzovic, A.-M. Pütz, V. Reuss, N. Meyer, A.J. Lough, M. Zimmer-De Iuliis, R.H. Morris, *Inorg. Chem.* 2009, 48, 735–743; (b) C. Sui-Seng, F. Freutel, A.J. Lough, R.H. Morris, *Angew. Chem. Int. Ed.* 2008, 47, 940–943.

49 W. Zuo, S. Tauer, D.E. Prokopchuk, R.H. Morris, *Organometallics* 2014, 33, 5791–5801.
50 A. Berkessel, S. Reichau, A. von der Höh, N. Leconte, J.R.-M. Neudörfl, *Organometallics* 2011, 30, 3880–3887.
51 S. Zhou, S. Fleischer, K. Junge, M. Beller, *Angew. Chem. Int. Ed.* 2011, 50, 5120–5124.
52 R. Huber, A. Passera, A. Mezzetti, *Organometallics* 2018, 37, 396–405.
53 B. Raphael, H. Raffael, M. Antonio, *Angew. Chem. Int. Ed.* 2015, 54, 5171–5174.
54 (a) C. Bornschein, S. Werkmeister, B. Wendt, H. Jiao, E. Alberico, W. Baumann, H. Junge, K. Junge, M. Beller, *Nat. Commun.* 2014, 5, 4111; (b) S. Chakraborty, G. Leitus, D. Milstein, *Chem. Commun.* 2016, 52, 1812–1815; (c) S. Chakraborty, G. Leitus, D. Milstein, *Angew. Chem. Int. Ed.* 2017, 56, 2074–2078; (d) S. Chakraborty, D. Milstein, *ACS Catal.* 2017, 7, 3968–3972; (e) S. Lange, S. Elangovan, C. Cordes, A. Spannenberg, H. Jiao, H. Junge, S. Bachmann, M. Scalone, C. Topf, K. Junge, M. Beller, *Catal. Sci. Technol.* 2016, 6, 4768–4772.
55 H.-U. Blaser, B. Pugin, F. Spindler, M. Thommen, *Acc. Chem. Res.* 2007, 40, 1240–1250.
56 D.E. Prokopchuk, S.A.M. Smith, R.H. Morris, Ligands for iron-based homogeneous catalysts for the asymmetric hydrogenation of ketones and imines. In: *Ligand Design in Metal Chemistry* (eds. M. Stradiotto and R.J. Lundgren) Wiley, 2016, 205–236.

2

Cobalt-Catalyzed Hydrogenations

Felicia Weber and Gerhard Hilt

Carl von Ossietzky Universität, Institut für Chemie, Fakultät V – Mathematik und Naturwissenschaften, Carl-von-Ossietzky-Str. 9-11, 26129 Oldenburg, Germany

2.1 Introduction

The addition of dihydrogen to an unsaturated substrate allows the formation of various saturated functional groups based on the nature of the starting material. The overall addition of dihydrogen can be accomplished in different ways, such as direct hydrogenation utilizing molecular hydrogen by a (transition) metal catalyst or via hydrometalation of the unsaturated substrate with a metal hydride complex followed by a protodemetalation, just to name two possible mechanisms. The first-row transition metals are of considerable interest in the future when substituting the higher homologs in hydrogenation reactions and other transformations, because of the limited access to precious metals. The application of cobalt-based materials as catalysts either in heterogeneous or homogeneous form is therefore a topic of intense research. In this chapter, we present the hydrogenation as well as the transfer hydrogenation of unsaturated functional groups with the primary focus on latest advances in terms of synthetic applications covering the modern literature up to the end of 2017. Recently, several reviews have been published illustrating the topic's general interest for chemists in academia and in industry [1–3].

2.2 Hydrogenation Reactions

2.2.1 Activation of Molecular Hydrogen–Dihydrogen Complexes vs. Dihydride Complexes

One of the most critical steps in a homogeneous hydrogenation reaction is the uptake of molecular hydrogen by the transition metal catalyst and the cleavage of the H—H bond. Generally, a vacant coordination site is needed for the side-on coordination of H_2, and the cleavage of the H—H bond (oxidative addition) leads to a dihydride complex where two coordination sites are occupied by hydride ligands. For the conversion of the substrate, another vacant coordination site is

Homogeneous Hydrogenation with Non-Precious Catalysts, First Edition. Edited by Johannes F. Teichert.

needed. This can be accomplished by abstraction of a proton from the dihydride complex by a base or by dissociation of a ligand. In the case of the first-row transition metal complexes, the stability of hydrogen or dihydride complexes is lower compared to that of the higher transition metal homologs of the second and third row. Therefore, the generation of the corresponding cobalt complexes usually has to be performed at high dihydrogen pressure and the characterization is best performed at low temperatures. Depending on the ligand system surrounding the cobalt atom, either the (di)hydride species, such as **1** [4, 5] or the dihydrogen complexes (**2**) [6–8], can be characterized. The selected complexes shown in Scheme 2.1 are those which have been characterized by various analytical methods and illustrate the broad scope of species that can be generated.

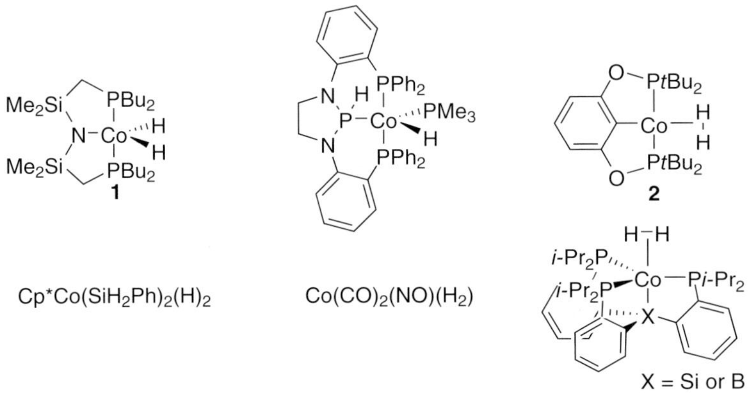

Scheme 2.1 Examples of characterized cobalt-dihydrogen and cobalt-hydride complexes.

The characterization can be performed by regular IR, EPR, or nuclear magnetic resonance (NMR) spectroscopy as well as inelastic neutron scattering analysis [9] or *para*-hydrogen induced polarization of NMR (PHIP) signals [10, 11]. For the analysis of heterogeneous cobalt nanoparticles, near edge X-ray absorption fine structure (NEXAFS) spectroscopy has been applied to identify key intermediates [12]. Also, theoretical investigations have shed light on the bonding situation of the cobalt–hydrogen bond and possible follow-up reactions [13, 14].

Interestingly, complex **1** shows reactivity toward ethylene resulting in the transfer of dihydrogen to this substrate, but this reactivity could not be observed for other alkenes. For a similar cobalt complex, the insertion product of ethylene into a Co—H bond was characterized as a model system for the hydrogenation of alkenes [15]. In any case, for the hydrogenation of substrates with molecular hydrogen, the H—H bond has to be cleaved; and, for the understanding of these processes, it is important to generate and evaluate defined cobalt–hydrogen/hydride complexes.

2.2.2 Hydrogenation of CO_2, Carboxylic Acids, Carboxylic Esters, and Nitriles

Carbon dioxide is an abundant source of carbon in the atmosphere whose negative effects on the world climate is well understood. The reduction of carbon dioxide with hydrogen gas can be catalyzed by cobalt complexes and the reaction conditions can be modified concerning the variance of reduction steps; for example, to generate formate species or methanol by additional reduction. Although second- and third-row transition metals have been used very successfully in all types of hydrogenation reactions [16], chemists are nowadays increasingly interested in finding conditions to apply nonprecious first-row transition metal catalysts, achieving even challenging hydrogenation reactions such as the transformation of carbon dioxide.

In this respect, Linehan analyzed the problems associated with the reduction of CO_2 with H_2 in a seminal work [17]. For the proposed reaction pathway (Scheme 2.2), the free energy profiles were calculated and crucial parameters that have to be considered for a successful hydrogenation were identified. The more reactive the cobalt monohydride species **A** is able to transfer the hydride ion to the substrate, the more basic the applied base has to be in order to remove a proton from the cobalt dihydride species **B**. This implies that the oxidative addition of molecular hydrogen to the low-valent cobalt species and cleavage of the H—H bond to generate the dihydride **B** is not a strongly limiting factor.

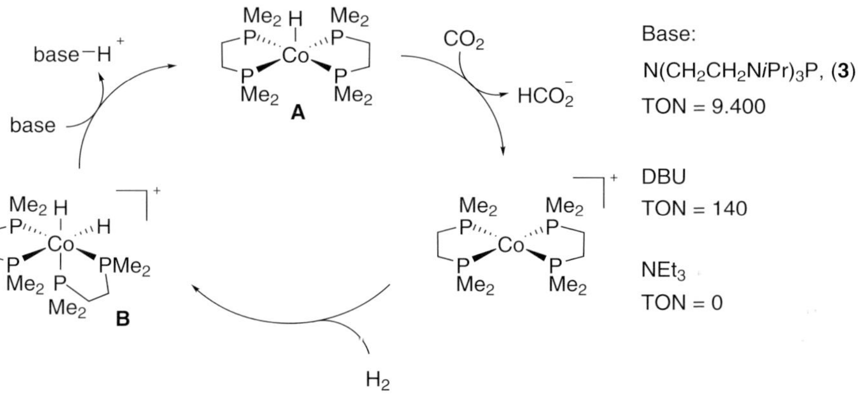

Scheme 2.2 Proposed reaction mechanism for the hydrogenation of CO_2.

Accordingly, the influence of the base to activate the cobalt complex was investigated and the application of *Verkade*'s base (**3**) proved to be highly beneficial for the transformation for the generation of formate as the desired product. The best results were obtained at a moderate pressure of 20 bar with a 0.04 mmolar solution of catalyst **A** and 510 mmolar base (**3**) to generate formate with a turnover frequency (TOF) of 74,000 h^{-1} and a total turnover number (TON) of 9,400 within one hour. As control experiments revealed, weaker bases gave inferior results and proved the concept for this type of cobalt complexes. Based on these results, Linehan optimized the ligand design and applied a tetradentate phosphine ligand, occupying the equatorial coordination sites of the cobalt atom in complex **4**, which allowed the application of weaker bases (Scheme 2.3) [18]. However, this advantage also resulted in a drastically reduced reactivity and only up to 270 turnovers could be achieved before the reaction stopped.

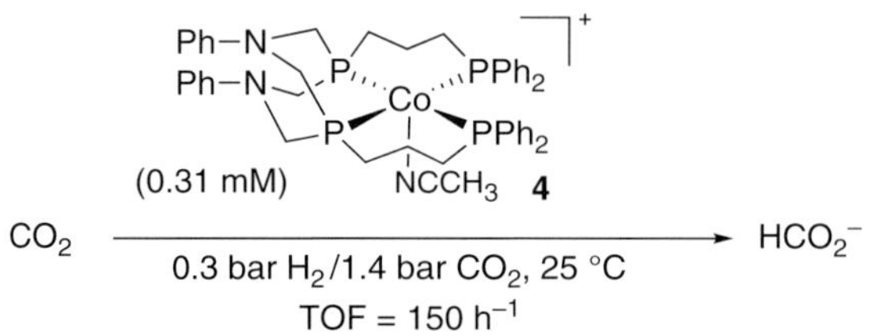

Scheme 2.3 Tetradentate ligand design for the hydrogenation of CO_2.

A pincer-type ligand design was used by Bernskoetter. They reasoned from their experience with iron-catalyzed hydrogenations that the rate-limiting step could be the loss of the formate ion from the metal atom [19]. Therefore, they investigated cobalt complexes, such as **5** (Scheme 2.4), for the desired transformation in different solvents, at various temperatures and applied various counterions. All these factors had an effect on the hydrogenation of CO_2 and led to an optimized catalytic system consisting of cobalt complex **5**, DBU, and LiOTf in THF resulting in a maximal TON of 29,000. Of note, the base was not modified in their investigation and, in comparison with the work of Linehan, a comparably mild base (DBU) and lower dihydrogen pressures were applied to successfully hydrogenate CO_2. The Beller group also investigated this transformation and applied tridentate ligands of the tetraphos type (**6**, Scheme 2.4) on well-defined cobalt complexes [20]. The mechanistic studies supported the proposal that the cobalt dihydride species is the active catalyst to undergo the hydride transfer to CO_2 for these type of complexes. The hydrogenation of CO_2 or sodium bicarbonate led to the desired formate product in up to 94% yield at 60 bar dihydrogen pressure.

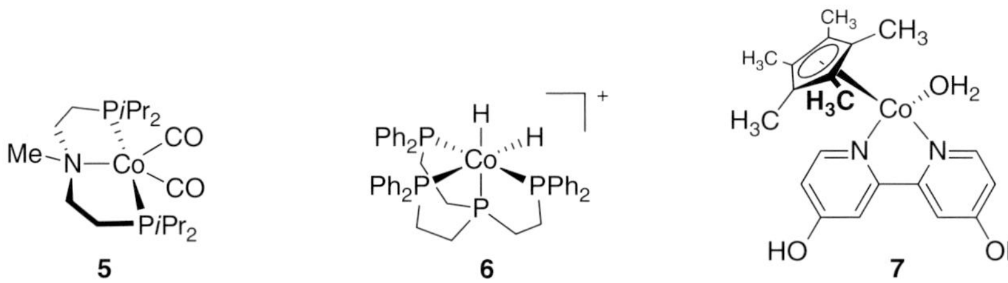

Scheme 2.4 Cobalt complexes for the hydrogenation of CO_2.

Besides the cobalt-phosphine type complexes, cyclopentadienyl cobalt complexes (**7**, Scheme 2.4) were also applied by Fujita for the hydrogenation of CO_2 to generate formate in aqueous media and a maximum TOF of $39\,h^{-1}$ could be achieved [21]. Under basic conditions, the hydroxyl groups on the bipyridine ligand are deprotonated, which increased the stability of these complexes at elevated temperatures required for the hydrogenation of CO_2 of up to 100 °C.

A rather small change of the ligand type from the complex of type **6** was reported by Klankermayer utilizing triphos-type ligand for the hydrogenation of CO_2 in methanol [22]. In their case, the reduction exceeded the formate stage and the methanol led to the formation of dimethoxymethane (DMM) as the final product (Scheme 2.5). Weakly coordinating counterions, such as BF_4^-,

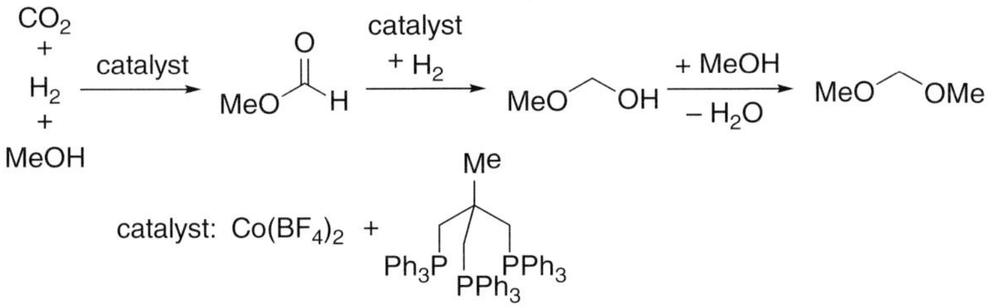

Scheme 2.5 Cobalt-catalyzed reductive formation of dimethoxymethane.

and $HNTf_2$ as additive were beneficial to generate DMM in up to 56% yield at a dihydrogen pressure of 80 bar. Interestingly, other alcohols could also be applied and the formation of the corresponding dialkoxymethane derivatives was observed. More interestingly, when sterically bulkier alcohols such as isopropanol were applied, the formation of methanol as hydrogenation product could be detected with a TON of up to 98.

The hydrogenation of carboxylic acids and esters/lactones to the corresponding alcohol derivatives was realized by de Bruin utilizing the identical catalysts system, as was already investigated by Klankenmayer [22, 23]. The $Co(BF_4)_2$/triphos system (Scheme 2.5) catalyzed the hydrogenation of aliphatic and aromatic carboxylic acids at a dihydrogen pressure of 80 bar at 100 °C with an extraordinary small amount of cobalt catalyst (125 ppm, Scheme 2.6). When unsaturated fatty acids were applied, the C=C was hydrogenated as well. Of note, phthalic acid was hydrogenated to benzofuranone and this lactone was not reactive for further hydrogenation under the reaction conditions.

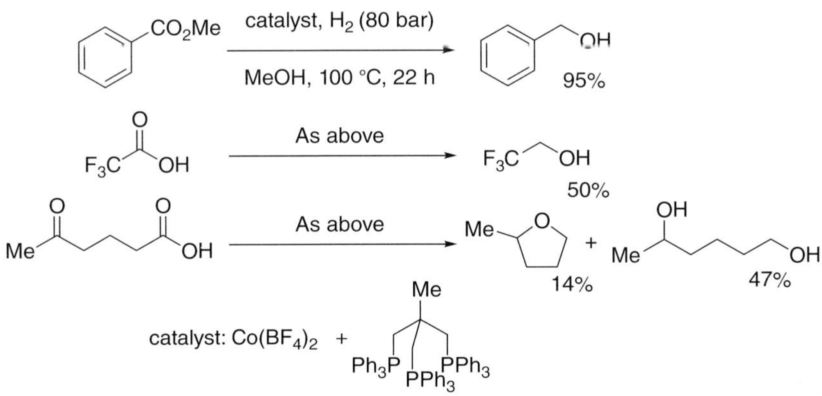

Scheme 2.6 Cobalt-catalyzed hydrogenation of acid derivatives.

When a pincer-type ligand of the PNP type was applied by Jones in the hydrogenation of esters, as shown in Scheme 2.7, the corresponding alcohols were isolated in moderate to excellent yields [24]. In this case, the hydrogenation of additional C=C double bonds of aliphatic as well as aromatic carboxylic esters was encountered where the latter showed higher reactivity. However, the influence of the degree of substitution of the C=C double bond was not investigated in

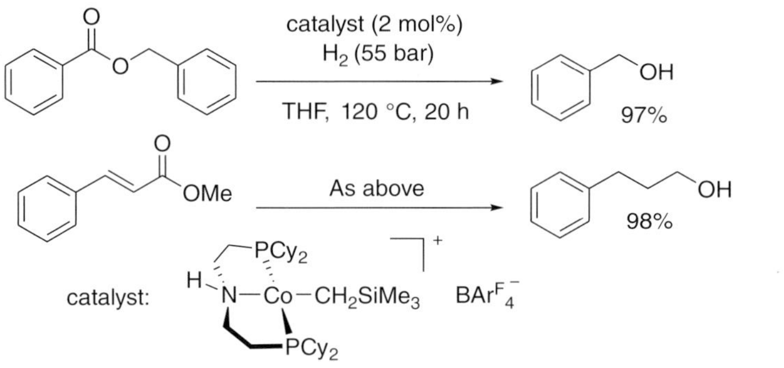

Scheme 2.7 Cobalt-catalyzed hydrogenation of carboxylic esters and acids.

more detail. Interestingly, when benzylic alcohol was added to methyl cyclohexane carboxylate, the transesterification product benzyl cyclohexane carboxylate could also be detected. Also, the decomposition of the catalyst at elevated temperatures and poisoning of the catalyst with CO was observed, indicating that the catalyst system is sensitive to harsh conditions and not entirely stable.

For the hydrogenation of carboxylic esters, the behavior of cobalt complexes with a pincer-type ligand structure motif (**5**, Scheme 2.4) under reductive conditions was further investigated by Beller [25]. When the cobalt dihalide complex (e.g. the dichloride complex **8**, Scheme 2.8) was reacted with a reducing agent, such as $NaBH_4$, the formation of a cobalt(I) species (**9**) was observed and the subsequent addition of CO gas led to an unreactive dicarbonyl complex (**10**), which provided a rationale for the deactivation of cobalt pincer complexes when CO was formed under the reaction conditions, as was reported by Jones [24].

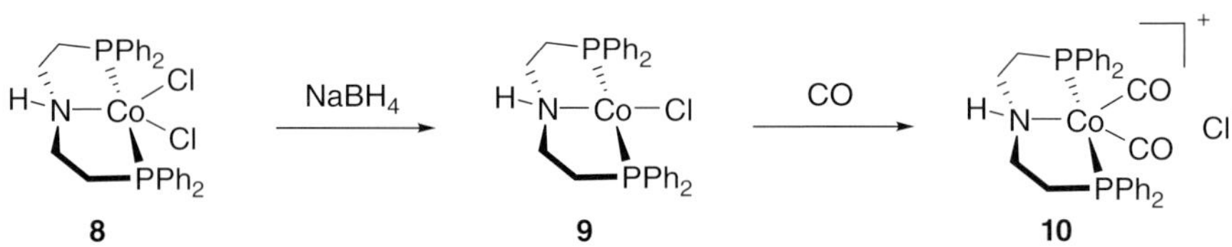

Scheme 2.8 Reductive activation and deactivation by CO of cobalt pincer-type complexes.

For the present cobalt catalyst, the NH group proved to be mandatory, which indicates that an outer-sphere mechanism might be operating. The corresponding NMe complex showed no reactivity. This finding is in contrast to the observations made by Hanson concerning the hydrogenation of ketones with a similar cobalt complex under slightly modified reaction conditions [26]. Therein, an N-methylated cobalt complex hydrogenated acetophenone with the same efficiency as did the cobalt complex with the NH group [27].

Computational studies by Zhao, Ke, [28] as well as by Yang [29] on the topic of a hydrogenation/transfer hydrogenation mechanism of cobalt complexes of type **8** intended to differentiate between an outer-sphere and an inner-sphere mechanism. While the outer-sphere mechanism is believed to operate via a hydrogen transfer where the NH group participates in the reaction, the inner-sphere mechanism proceeds by coordination of the substrate to the cobalt atom and subsequent hydride transfer to the substrate within the coordination sphere of the cobalt complex. The investigation concerning pincer-type ligands of type **8** resulted in the statement that most likely an inner-sphere reaction mechanism is preferred on the basis of the higher deformation energy compared with an outer-sphere mechanism. The authors also realized that the structure of the ligand and the electronic configuration of the metal atom, as a result of the ligand-to-metal donor/acceptor abilities, could have a decisive impact on the reaction mechanism. In a detailed computational study, Hopmann investigated the activation of the pre-catalyst and the inner-sphere reaction mechanism for the hydrogenation of alkenes with chiral pincer-type complexes and presented a model for the chiral induction of those complexes [30].

As a personal note, we would like to add that the different, and sometimes, contradictive results of different groups might easily be a two-sided picture of the same medal. As we know from our own investigations in cobalt-catalyzed reactions, small changes in the ligand geometry and/or electronic properties can cause either a profound change in regioselectivities or a complete change of reaction pathway to different products. As an example, the regioselectivity of cobalt-catalyzed Diels–Alder reactions is strongly influenced by the ligand design [31].

The synthetic application of the complex **8** for the hydrogenation of carboxylic esters afforded the alcohols from the corresponding aliphatic and (hetero)aromatic carboxylic esters under basic conditions and 50 bar dihydrogen pressure at 120 °C after up to 24 hours reaction time in moderate to excellent yields [25]. The catalyst tolerates a number of "uncritical" functional groups, such as methoxy, trifluoromethyl, fluoro, and alkyl substituents at various positions of an aromatic substituent under the reaction conditions, but the tolerance toward more "critical" substituents, such as a nitrile, a bromide, or an iodide substituent on the arene moiety was not reported. Interestingly, internal and isolated CC double bonds were unreactive. Unfortunately, conjugated unsaturated esters were not investigated (Scheme 2.9).

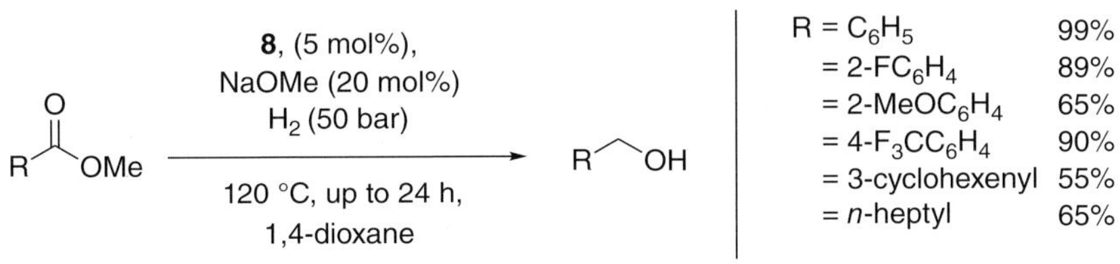

Scheme 2.9 Hydrogenation of esters with the cobalt pincer-type complexes **8**.

The expansion toward the hydrogenation of nitriles with heterogeneous cobalt complexes was realized by Beller [32]. The catalyst was prepared from a mixture of $Co(OAc)_2$, phenanthroline, and α-Al_2O_3 which was pyrolyzed at 800 °C (→ cobalt catalyst **A**). The heterogeneous catalyst was able to reduce a large number of aliphatic and aromatic nitriles to the corresponding amines (Scheme 2.10) and a similarly large number of ketones and aldehydes to the corresponding alcohols.

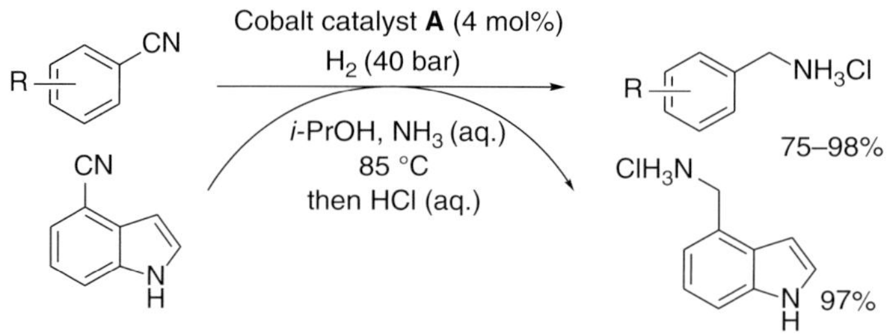

Scheme 2.10 Hydrogenation of nitriles with a heterogeneous cobalt catalyst.

The addition of NH_3 proved to be beneficial in reducing the amount of dibenzylamine by-products formed by addition of the amine product to the imine intermediate, the following elimination of ammonia, and the consequent hydrogenation of the corresponding adduct. In contrast, in the absence of ammonia, the dibenzylamine products were formed predominantly when Raney cobalt was applied as reducing agent at 4.1 bar hydrogen pressure [33].

For the homogeneous hydrogenation of nitriles, Milstein [34] reported the use of a cobalt complex **11** with an unsymmetrical PNN-pincer ligand (Scheme 2.11). The hydrogenations could be performed at 30 bar dihydrogen pressure at elevated temperatures for a prolonged reaction time. However, the desired aryl and alkyl primary amines were isolated in moderate to mostly very good yields. For the activation of the cobalt complex, a mixture of a reducing agent ($NaBHEt_3$) and a base (NaO*t*Bu) were applied, which led to minimized amounts of dibenzyl amine-type side products.

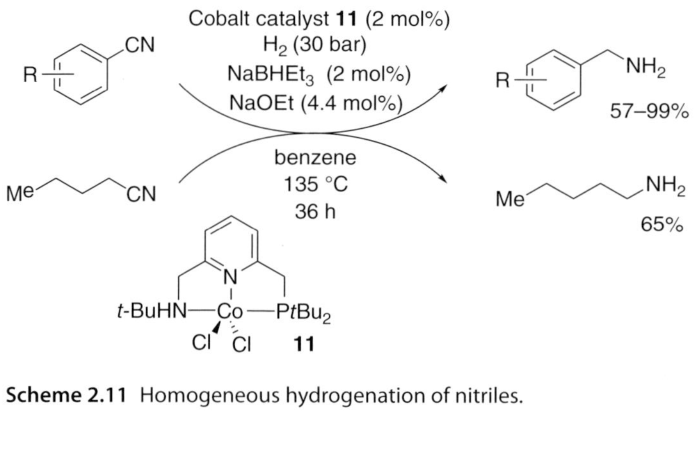

Scheme 2.11 Homogeneous hydrogenation of nitriles.

2.2.3 Hydrogenation of C=O, C=N, C=C, C≡C, and (Hetero)arenes

The hydrogenation of unsaturated functional groups by non-noble metals has been investigated for a long time. Hand in hand with investigations for the cobalt-catalyzed hydroformylation reaction, cobalt complexes with carbonyl ligands and phosphine ligands were applied for the hydrogenation. Among such complexes, species such as the cobalt complex $CoH(N_2)(PPh_3)_3$ reported by Markó [35], as well as a family of cobalt complexes of the general type $CoH(CO)_{4-n}(PnBu_3)_n$, are reported in the literature [36, 37].

Early reports for the hydrogenation of unsaturated groups in an enantioselective manner utilized bis-dimethylglyoximato-cobalt complexes in combination with chiral aminoalcohols as additional ligands and sources of chirality for the enantioselective hydrogenation of α,β-unsaturated esters and ketones, resulting in the formation of the enantiomerically enriched products in moderate to good yields (62–95%) and mostly moderate enantioselectivities (optical yields of up to 78%) [38–41]. Moderately diastereoselective hydrogenations of 2-methylcyclohexanone (*cis*/*trans* up to 27 : 73) utilizing monodentate chiral phosphine ligands, which resulted in enantioselectivities of up to 5% *ee* at yields around 25%, were also reported [42]. These findings represent the first reports in this research area, but the breakthrough was way to go. More recent homogeneous catalysts based on pincer-type ligands (and some others) or heterogeneous catalysts as the abovementioned catalyst system **A** (generated by pyrolysis of $Co(OAc)_2$, phenanthroline and α-Al_2O_3; see Scheme 2.2) were applied [26, 32]. Next to the hydrogenation of nitriles, the latter was further applicable for the hydrogenation of nonconjugated unsaturated ketones which led primarily to the hydrogenation of the C=O group to afford unsaturated alcohols. In contrast to the catalyst systems described, carboxylic esters were unreactive under the reaction conditions. The application of the chemoselective hydrogenation of the steroid **12** to alcohol **13** (Scheme 2.12) illustrates its usefulness for the selective hydrogenation of complex molecules.

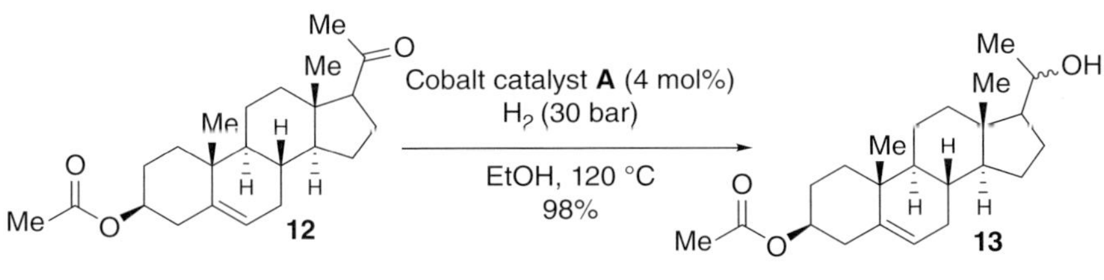

Scheme 2.12 Hydrogenation of steroid **12** with a heterogeneous cobalt catalyst.

However, products such as **13** were formed in moderate diastereoselectivities, only ranging from 1 : 1 to 1 : 2 depending on the steroid substrates.

The Hanson group investigated a pincer-type ligand motif for the hydrogenation of unsaturated functional groups and found high catalytic activity for the hydrogenation of C=C, C=O and C=N double bonds at low dihydrogen pressure when cobalt complex **14** was treated with $H[B(Ar^F{}_4)]$ at relatively low temperatures [26]. Styrenes as well as mono- and disubstituted alkenes could be converted under these mild reaction conditions, whereas trisubstituted double bonds exhibited lower reactivities. The hydrogenation of carbonyl groups, such as in benzaldehyde, proceeded faster than the hydrogenation of a monosubstituted C=C double bond, as in styrene. However, when 2,3-dihydrocarvone was hydrogenated, a chemoselective hydrogenation of the C=C double bond was observed and the saturated carbonyl compound was isolated in 99% yield.

Nevertheless, when cobalt catalyst **14** was applied, a broad variety of ketones, aldehydes, and aromatic imines were converted to the corresponding alcohols or amines in good to excellent yields, while carboxylic esters as well as carboxylic acids were well tolerated (Scheme 2.13), and even small amounts of water (up to 10 mol%) were tolerated by the catalyst system. Besides the high functional group tolerance and the mild reaction conditions, the latter property makes it a prime candidate for larger scale hydrogenations of a robust catalyst system.

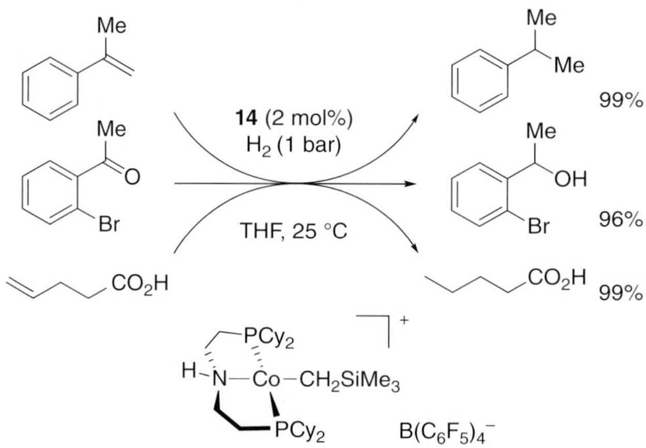

14

Scheme 2.13 Hydrogenation of various unsaturated starting materials under mild conditions utilizing cobalt complex **14**.

Similarly, Kempe introduced a pincer-type ligand with a heterocyclic backbone which could be applied for the hydrogenation of various ketones in a comparable manner (Scheme 2.14) [43]. A number of functional groups were tolerated, such as an acetal group, heterocycles, and a bromoarene substituent to give the alcohols in nearly quantitative yields. The major differences from the Hanson system are the basic medium and the higher dihydrogen pressure needed for the conversion of ketones, but a lower catalyst loading (as low as 0.25 mol%) could be applied with the cobalt catalyst **16**. From a synthetic point of view, an altered chemoselectivity was observed when unsaturated carbonyl compounds were investigated: while the Hanson system resulted in the complete reduction of the starting material to afford the saturated product **17** (Scheme 2.14) [26], the conversion of the unsaturated starting material **18** by the Kempe catalyst **16** afforded the corresponding alcohol **19** in nearly quantitative yield [43]. Conversion of an identical unsaturated starting material for both catalysts to illustrate the different chemoselectivities remains to be investigated.

An astonishingly simple catalyst system consisting of a bidentate phosphine ligand, such as dppe, and a cobalt dialkyl fragment was reported by Chirik. The cobalt complex Co(dppe)(CH_2SiMe_3)$_2$ proved to be an efficient pre-catalyst for the hydrogenation of alkenes in the presence of carbonyl groups (ketones, esters) as well as sulfones [44]. The application of this kind of catalyst design was demonstrated in a high-throughput investigation applying 192 commercially available

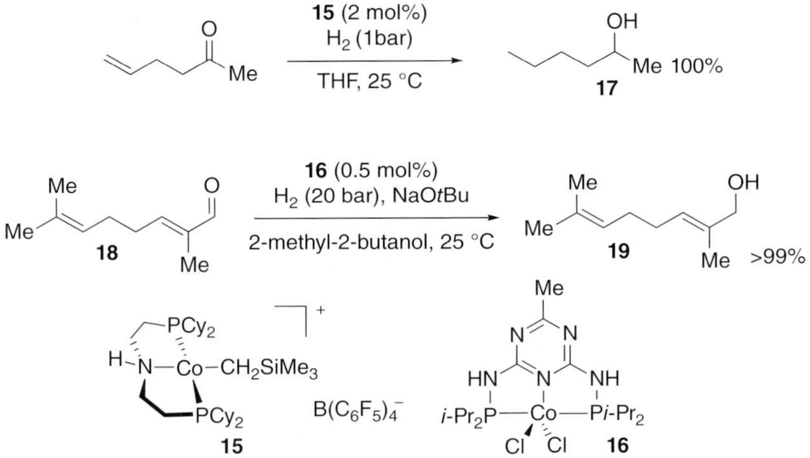

Scheme 2.14 Comparison between neutral and basic conditions in the hydrogenation of carbonyl compounds.

chiral diphosphine ligands, such as (*R*,*R*)-EtDuphos (**20**), for the hydrogenation of enamide derivatives to generate chiral amino acids (Scheme 2.15) [45].

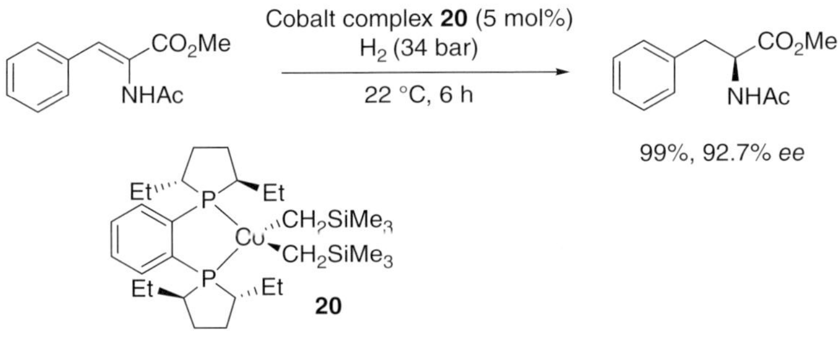

Scheme 2.15 Asymmetric hydrogenation of enamides for the synthesis of chiral amino acids.

For the activation of the catalyst, various organometallic species, such as Grignard reagents, were investigated and besides the well-established $LiCH_2SiMe_3$, the organozinc reagent 2-fluorobenzyl zinc chloride in combination with CoI_2 as the cobalt source gave very good results as well.

For the hydrogenation of ketones, Li reported an enantioselective version utilizing a chiral partially oxidised tetradentate PNNO ligand (Scheme 2.16) [46]. Interestingly, cobalt complex **21** lost its C_2 symmetry upon partial oxidation; however, this did not seem to hamper the complex to catalyze the hydrogenation of prochiral aryl-alkyl ketones with moderate to very good enantioselectivities of up to 95% *ee*. The yields and enantioselectivities for 2-substituted acetophenone derivatives were generally lower, as they were for 3- and 4-substituted aryl derivatives. However, a number of acetophenone derivatives gave good results, with the exception of 2-bromoacetophenone; in this case, the debrominated product, (*S*)-1-phenylethanol, was generated, which indicates that low-valent

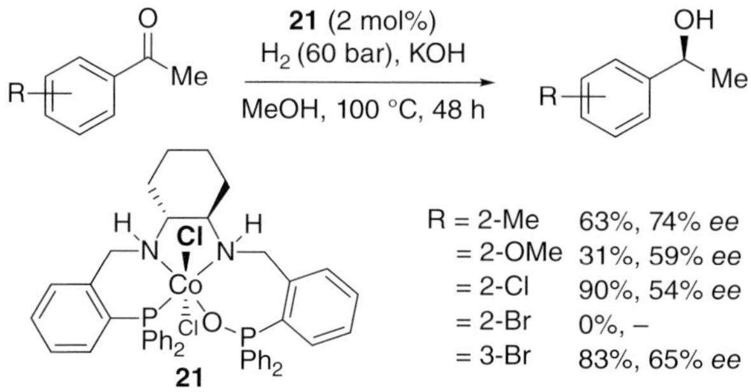

Scheme 2.16 Cobalt-catalyzed hydrogenation of acetophenone derivatives with a chiral PNNO ligand.

cobalt complexes were formed leading to oxidative insertion followed by protolysis of the intermediate. The 3-bromoacetophenone derivative could be hydrogenated in 83% yield and 65% *ee*, which suggested that the proximity of the bromo-substituent to the carbonyl group seems to be decisive for the oxidative addition of the cobalt catalyst into the C—Br bond.

Apart from ketones and aldehydes, the C=N double bond is a suitable functional group for hydrogenation utilizing cobalt-based catalysts. A recent application was reported by Borths, who investigated the hydrogenation of oximes for the synthesis of amines [33]. While Ru/C, Rh/C, Pt/C, Pd/C, and Raney nickel gave only undefined products or the hydroxylamine as the final product, Raney cobalt (Raney Co 2724®) instead produced the desired amines in excellent yield. The investigation was directed toward the production of heterocyclic benzylic amines, which are of general interest for the application as kinase inhibitors. Accordingly, only a number of acylated pyridine and other heterocyclic aromatic derivatives were converted into the corresponding oximes and hydrogenated thereafter utilizing Raney cobalt. The selected examples in Scheme 2.17 illustrate the broad applicability of Raney cobalt in the hydrogenation of heterocyclic aromatic oxime derivatives. It should be noted that the hydrogenated products as well as some starting materials represent

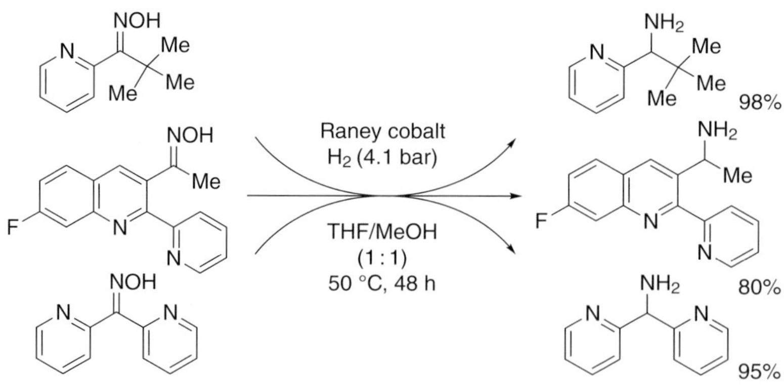

Scheme 2.17 Raney cobalt as reducing agent for the hydrogenation of heterocyclic oximes.

potentially good ligands for transition metal based reducing agents which probably rationalize their deactivation.

While ketoximes gave the desired product in good to excellent yields, the transformation of aldoximes was challenging, leading preferentially to the formation of the corresponding secondary amines. As mentioned earlier, the application of additional ammonia to the reaction mixture might prevent the formation of these undesired side products, but this has to be demonstrated for this particular catalytic system in the future.

Recently, a chiral pincer-type ligand was reported by Lu [47] and Chirik [48, 49] for the enantioselective hydrogenation of prochiral 1,1-diarylethene derivatives (Scheme 2.18) as well as styrene derivatives, indenes and dihydronaphthalenes [50, 51]. For the efficient enantioselectivites an ortho-substituent of one of the aryl moieties had to be applied, such as an alkyl, a methoxy, or a fluoro substituent. The best results were obtained with 2-chloroarene substituents, resulting in enantioselectivities ranging from 86% to 99% *ee*. A wide range of other functional groups were tolerated, but a 2-bromoaryl moiety showed only

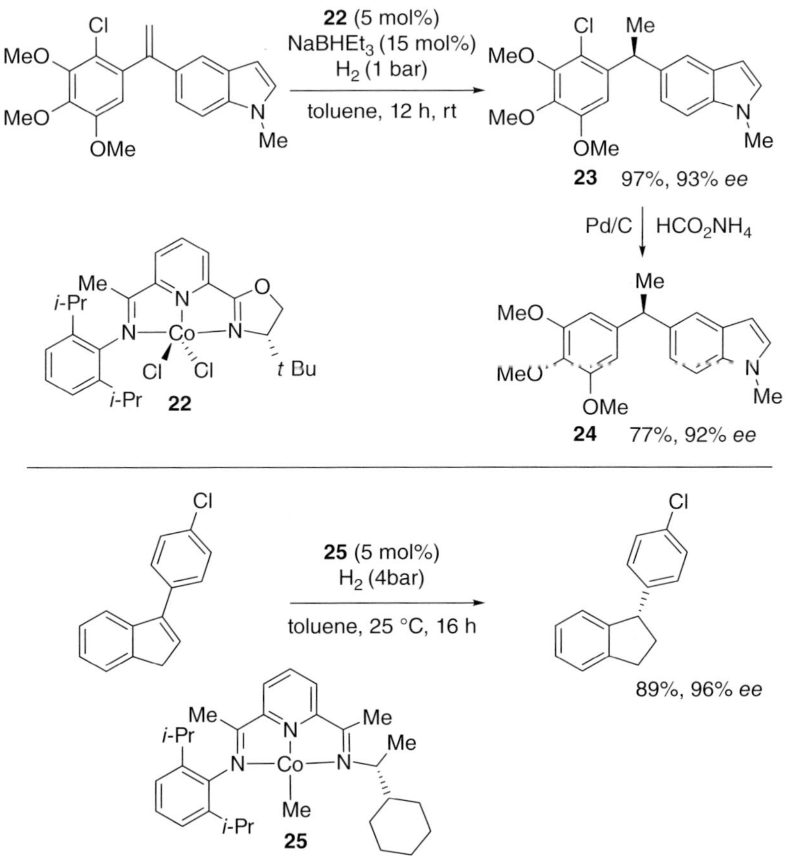

Scheme 2.18 Cobalt-catalyzed hydrogenation as a key step for the synthesis of the anticancer agent **24** and the hydrogenation of indenes.

low conversion. However, the method could be applied for the synthesis of a chiral anticancer agent (**24**), as shown in Scheme 2.17 utilizing cobalt complex **22** which is activated with $NaBHEt_3$ and a low dihydrogen pressure (1 bar) at ambient temperature for the hydrogenation of the prochiral starting material to generate the intermediate **23** in excellent yield and enantioselectivity.

In contrast to Lu's catalyst [47], the chiral cobalt complex **25** could be used without any other activating agent and the enantioselective hydrogenation of prochiral indenes could be performed at low dihydrogen pressures at ambient temperatures.

Based on very detailed investigations on the chemical behavior and the hydrogenation of alkenes to alkanes utilizing cobalt complexes, such as **26** by Chirik and **27** by Fout (Scheme 2.19), similar cobalt catalysts (**28**–**30**) were investigated [52, 53]. Of particular interest is the partial hydrogenation or transfer hydrogenation of alkynes, as was reported by several groups in close succession in 2016/2017 for the stereoselective synthesis of either *Z*-alkenes or *E*-alkenes. While Zhang [54] and Fout [55] investigated cobalt catalysts for the formation of *Z*-alkenes, Luo and Liu reported a stereodivergent catalyst system which gave either the *Z*- or the *E*-alkene with high selectivities depending on the substituents of the phosphine donor [56]. While the sterically demanding P*t*Bu_2 groups in **28** favored the *Z*-conformer, the less bulky P*i*Pr_2 groups in **29** preferably formed the *E*-alkene. A further increase of the *E*-selectivity could be achieved when an unsymmetrical ligand was applied and the cobalt complex **30** led to outstanding results for an *E*-selective transfer hydrogenation, as shown in Scheme 2.19. Among these catalysts, the Zhang system is by far the most simple and cost-efficient system because the materials are readily available and rather inexpensive [54]. This catalyst system consists of a simple cobalt salt – $Co(OAc)_2(H_2O)_4$ – $NaBH_4$ as activating agent – and ethylenediamine as ligand to afford *Z*-alkenes in excellent yields and stereoselectivities. As the only limitation thus far, the authors reported mostly diaryl- or aryl-alkyl alkynes (with a broad functional group tolerance) as substrates. Only a single result concerning dialkylalkynes and a thiophene group as a model for a heterocyclic arene substituent were reported.

An unusual approach for a ligand system for the stabilization of low-valent cobalt metal atom was reported by Peters. Cobalt-hydride complexes, such as

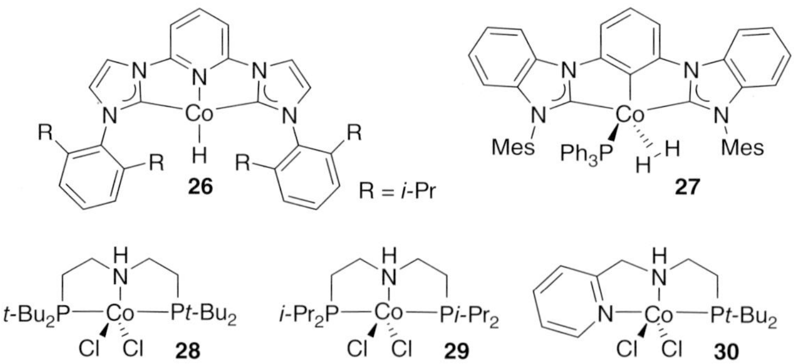

Scheme 2.19 Cobalt catalyst for the hydrogenation of alkenes and partial hydrogenation of alkynes.

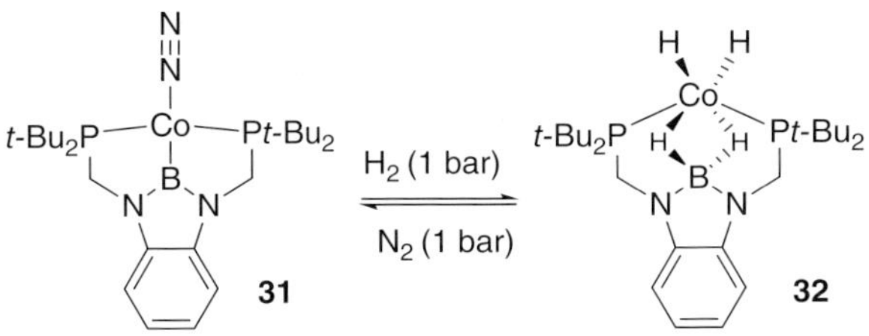

Scheme 2.20 Reversible formation of a cobalt hydride complex with a boron moiety.

32, with a boryl-moiety as a ligand could be formed reversibly (Scheme 2.20) [57]. Further investigations on this type of ligand system revealed the formation of a dimeric cobalt complex which proved to be reactive in hydrogenation reactions [58].

The cobalt complex **32** and its dimeric counterpart were applied in the hydrogenation of terminal alkenes, such as in styrene, 1-octene, and several internal alkenes. However, the hydrogenation of internal alkenes with the dinitrogen complex **31** gave no conversion.

An even more drastic reduction of ligand complexity was reported by Wolf, von Wangelin, [59, 60]. The rationale for this ligand design is that a low-valent cobalt atom in a homogeneous solution attracts unsaturated substrates as the ligand to be stabilized. When the cobalt complex **33** (Scheme 2.21) was applied in the hydrogenation of a large number of mono- and disubstituted alkenes (trisubstituted alkenes were unreactive), the reaction mixture remained homogeneous over the course of the transformation until the starting material was mostly consumed. Only thereafter was the formation of heterogeneous species observed. Also, aryl-alkyl and dialkyl ketones as well as diarylaldimines could be hydrogenated in good to excellent yields.

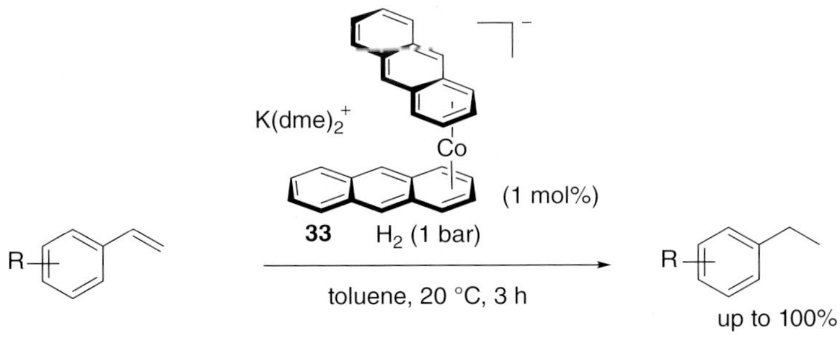

Scheme 2.21 Arene cobalt complexes as pre-catalyst for the hydrogenation of alkenes.

The overall hydrogenation of alkenes was performed under mild and neutral conditions, because neither acidic nor basic additives were needed and an activation of the cobalt complex by a reducing agent such as a borohydride was obsolete.

An alternative source of hydrogen as reducing agent was reported by Zhang for the transfer hydrogenation of alkenes with isopropanol as the hydrogen source [61]. In this challenging transfer hydrogenation reaction, the well-known pincer-type ligand motif in **34** was applied and more drastic conditions had to be applied to achieve complete conversions. The transfer hydrogenation could be used for the transformation of terminal aliphatic and aromatic alkenes and sterically less hindered (a)cyclic internal alkenes, such as cyclooctene, to generate the alkanes in up to 99% yield, but the functional group tolerance was only moderate (Scheme 2.22).

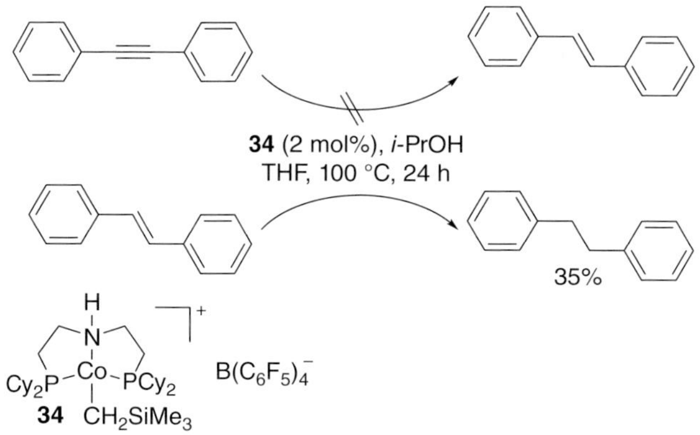

Scheme 2.22 Transfer hydrogenation of alkenes with cobalt complex **34**.

Interestingly, the authors were able to demonstrate that diphenylacetylene could be reduced under hydrogenation reaction conditions to *E*/*Z*-stilbene, while the transfer hydrogenation conditions gave no reaction (Scheme 2.22). On the other hand, in the hydrogenation of an *E*/*Z*-stilbene mixture, the cobalt complex failed, whereas the transfer hydrogenation produced 1,2-diphenylethane, although in relatively low yield of 24% (from *Z*-stilbene) and 35% (from *E*-stilbene). Accordingly, we have to conclude that the source of hydrogen led to different reactivities and therefore the reactions were unlikely to proceed via identical cobalt-hydride species.

An interesting cascade reaction was recently published by Lu, applying cobalt complexes, such as **35**, with tridentate donor ligands for the hydrosilylation/hydrogenation as well as the hydroboration/hydrogenation sequence. The hydrosilylation of mostly terminal alkynes with aromatic substituents and the subsequent hydrogenation were performed in a one-pot sequence to generate chiral benzyl-substituted silanes in good to excellent yields and enantioselectivities (Scheme 2.23) [62]. The first hydrosilylation proceeds with high regioselectivity to the branched vinyl-silane, which is then hydrogenated via catalyst **35** to the final product **36**. Several functional groups were tolerated, including heterocycles, such as indole, as well as esters, chloro- and bromo-substituents.

In a similar manner, Lu reported the hydroboration/hydrogenation sequence (Scheme 2.24) for the synthesis of chiral alkylboranes [63]. Unlike the hydrosilylation, the hydroboration proceeded best with internal alkynes and the regioselectivity was explained by the preferred insertion of the alkyne into the cobalt–hydrogen bond to yield the alkenyl-cobalt species with the aryl-substituent next to the cobalt atom. This was followed by a reductive

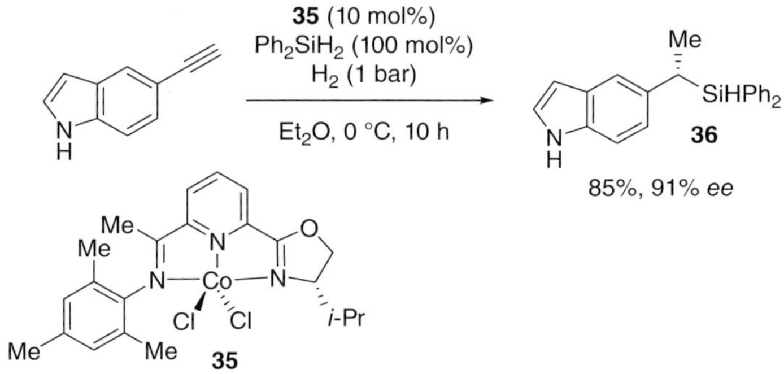

Scheme 2.23 Hydrosilylation/hydrogenation sequence to generate chiral benzylic silanes.

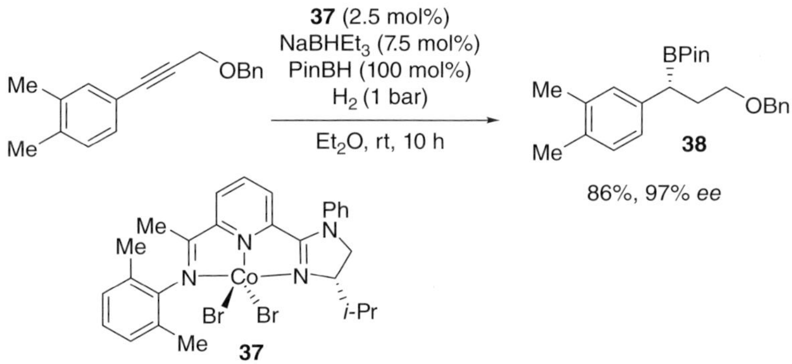

Scheme 2.24 Hydroboration/hydrogenation sequence to generate chiral benzylic boronic esters.

elimination to generate the vinyl-boron intermediate which was then hydrogenated by the catalyst **37** to form the final product **38**. Terminal alkynes, such as phenylacetylene, gave the linear product and eventually 2-phenylethylboronic ester as the final product.

The incorporation of hydroxyl groups or ether functionalities in many examples seemed to be of advantage in order to direct the hydroboration toward the products with the boron functionality next to the aryl group, but this regioselectivity was also encountered for unfunctionalized simple aryl-alkyl alkynes, such as 1-phenyl-but-1-yne. Nevertheless, the method represents a simple way to generate chiral silanes and boranes for follow-up reactions, such as the oxidation to chiral secondary alcohols.

A few early reports were published about the partial hydrogenation of arenes, such as anthracene and pyrene [64, 65]. Recent reports describe the partial hydrogenation of selected bicyclic heterocycles to generate products such as **39**, **40**, and **41**, utilizing a heterogeneous catalyst system for the transfer hydrogenation at elevated temperatures (Scheme 2.25). The transfer hydrogenations were performed utilizing a cobalt catalyst **B** derived from $Co(OAc)_2 \cdot 4H_2O$ and melamine after pyrolysis at 500–700 °C and formic acid/trimethylamine as hydrogen source [66].

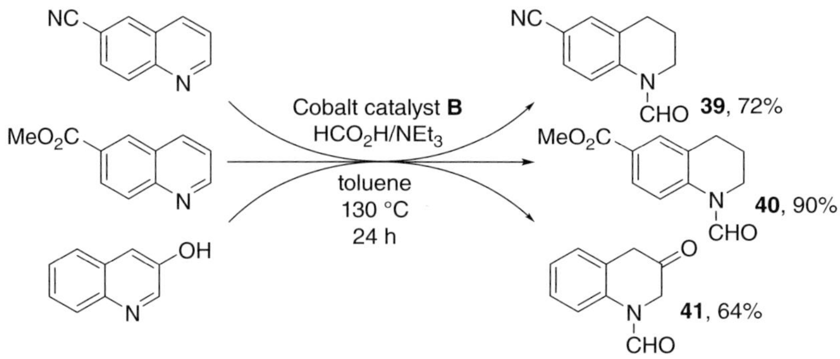

Scheme 2.25 Partial transfer hydrogenation of quinoline derivatives.

Interestingly, a large number of N-containing heterocycles could be partially hydrogenated, and in each case, the N-heterocycle was reduced, even if an electron-deficient arene was present in the starting material, such as in **40**. Also, the quinoline nitrogen atom was transformed into a formamide derivative, which obviously proved to be unreactive under the reaction conditions. In a similar way, the nitrile functional group was unreactive and product **39** was formed in a good yield of 72%. The hydroxy-substituted quinoline was converted to the product **41** without a significant further hydrogenation of the carbonyl group. Accordingly, this very interesting version of a partial hydrogenation reaction of N-heterocycles could be a versatile tool to address interesting partially hydrogenated structures for follow-up reactions in the future.

In contrast, the complete hydrogenation of alkyl-substituted aromatic compounds reported by Pan and Lin (Scheme 2.26) [67], such as mesitylene to trimethylcyclohexane (**42**), or heterocycles, to afford products such as **43** or **44** were performed utilizing a heterogeneous titanium(III)-oxo cluster to support a cobalt-hydride catalyst (cobalt catalyst **C**).

The catalyst reached a TON of up to 4,200. Unfortunately, the main focus of the report was directed toward the characterization of the metal-organic framework rather than on a systematic investigation toward the application in organic transformation, thereby ignoring interesting issues such as stereochemistry and

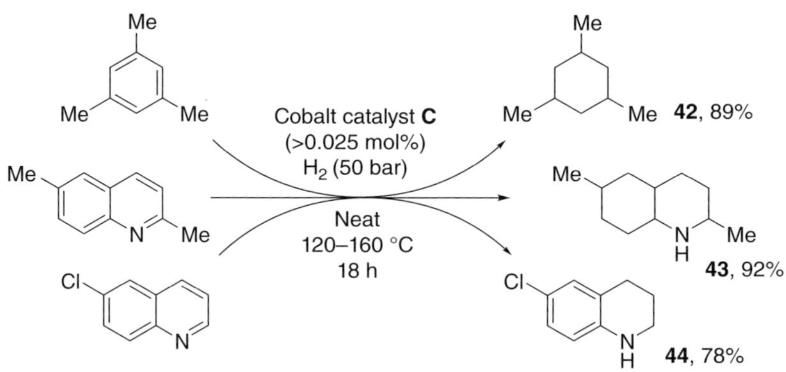

Scheme 2.26 Complete/partial hydrogenation of benzene and heteroarene derivatives.

control of partial vs. complete hydrogenation of the starting materials. Nevertheless, these few examples illustrate that a heterogeneous cobalt catalyst can play in increasingly important role in the hydrogenation of aromatic compounds in the future.

2.3 Conclusion

Over the past decades, an increasing number of cobalt complexes with cobalt–hydrogen bonds were synthesized and characterized. Although at first a curiosity, these Co–H species are nowadays generated on a regular basis and their applications in organic synthesis are explored in detail. The majority of the homogeneous cobalt catalyst systems utilized for the hydrogenation of unsaturated functional groups consists of a cobalt complex with a pincer-type ligand or a triphos-type ligand and activating agents, such as a reducing agent to introduce a Co—H bond *in situ*. From there these complexes are able to activate dihydrogen and recent examples showed that the dihydrogen pressure can be as low as 1 bar at ambient temperatures. In recent years, simple starting materials, such as CO_2 and CO as well as an increasing number of carboxylic acid derivatives could be hydrogenated successfully. Ketones, imines, oximes, alkenes, alkynes, and (hetero)arenes were hydrogenated and the various cobalt complexes exhibited different chemoselectivities which make them interesting for the hydrogenation of more complex starting materials. Eventually, cobalt complexes with chiral ligands were employed to generate products in high enantiomeric excess to substitute the precious transition metal catalysts which established their role in such reactions decades ago. However, more work has to be devoted in the future to further increase the reactivity of cobalt catalysts with simple and readily available (chiral) ligand systems to establish cobalt as the metal of choice for hydrogenation reactions.

References

1 P.J. Chirik, Iron- and cobalt-catalyzed alkene hydrogenation: catalysis with both redox-active and strong field ligands. *Acc. Chem. Res.* 2015, 48, 1687–1695.

2 R. Kempe, F. Kallmeier, Manganese complexes for (De)hydrogenation catalysis: a comparison to cobalt and iron catalysts. *Angew. Chem. Int. Ed.* 2018, 57, 46–60.

3 Y.-Y. Li, S.-L. Yu, W.-Y. Shen, J.-X. Gao, Iron-, cobalt-, and nickel-catalyzed asymmetric transfer hydrogenation and asymmetric hydrogenation of ketones. *Acc. Chem. Res.* 2015, 48, 2587–2598.

4 A.M. Poitras, S.E. Knight, M.W. Bezpalko, B.M. Foxman, C.M. Thomas, Addition of H2 across a cobalt-phosphorus bond. *Angew. Chem. Int. Ed.* 2018, 57, 1497–1500.

5 M. Ingleson, H. Fan, M. Pink, J. Tomaszewski, K.G. Caulton, Three-coordinate Co(I) provides access to unsaturated dihydrido-Co(III) and seven-coordinate Co(V). *J. Am. Chem. Soc.* 2006, 128, 1804–1805.

6 R.M. Bullock, A mercurial route to a cobalt dihydrogen complex. *Angew. Chem. Int. Ed.* 2011, 50, 4050–4052.
7 G.E. Gadd, R.K. Upmacis, M. Poliakoff, J.J. Turner, Complexes of iron and cobalt containing coordinated molecular dihydrogen: infrared evidence for Fe(CO)(NO)2(H2) and Co(CO)2(NO)(H2) in liquefied xenon solution. *J. Am. Chem. Soc.* 1986, 108, 2547–2552.
8 D.L.M. Suess, C. Tsay, J.C. Peters, Dihydrogen binding to isostructural S = 1/2 and S = 0 cobalt complexes. *J. Am. Chem. Soc.* 2012, 134, 14158–14164.
9 A.J. Ramirez-Cuesta, P.C.H. Mitchell, S.F. Parker, An inelastic neutron scattering study of the interaction of dihydrogen with the cobalt site of a cobalt aluminophosphate catalyst: two-dimensional quantum rotation of adsorbed dihydrogen. *J. Mol. Catal. A Chem.* 2001, 167, 217–224.
10 C. Godard, S.B. Duckett, S. Polas, R. Tooze, A.C. Whitwood, An NMR study of cobalt-catalyzed hydroformylation using para-hydrogen induced polarisation. *Dalton Trans.* 2009, 2496–2509.
11 C. Godard, S.B. Duckett, S. Polas, R. Tooze, A.C. Whitwood, Detection of intermediates in cobalt-catalyzed hydroformylation using para-hydrogen-induced polarization. *J. Am. Chem. Soc.* 2005, 127, 4994–4995.
12 S.K. Beaumont, S. Alayoglu, C. Specht, W.D. Michalak, V.V. Pushkarev, J. Guo, N. Kruse, G.A. Somorjai, Combining in situ NEXAFS spectroscopy and CO2 methanation kinetics to study Pt and Co nanoparticle catalysts reveals key insights into the role of platinum in promoted cobalt catalysis. *J. Am. Chem. Soc.* 2014, 136, 9898–9901.
13 L. Versluis, T. Ziegler, Theoretical study on HP-induced acetaldehyde elimination from CH3(0)CCo(CO)3. *Organometallics* 1990, 9, 2985–2992.
14 M. Solà, T. Ziegler, Theoretical study on acetaldehyde and ethanol elimination from the hydrogenation of $CH_3(O)CCo(CO)_3$. *Organometallics* 1996, 15, 2611–2618.
15 V.C. Gibson, K.P. Tellmann, M.J. Humphries, D.F. Wass, Bis (imino) pyridine cobalt alkyl complexes and their reactivity towards ethylene: a model system for beta-hydrogen chain transfer. *Chem. Commun.* 2002, 2316–2317.
16 J.G. de Vries, C.J. Elsevier, *The Handbook of Homogeneous Hydrogenation.* Wiley, 2007.
17 M.S. Jeletic, M.T. Mock, A.M. Appel, J.C. Linehan, A cobalt-based catalyst for the hydrogenation of CO2 under ambient conditions. *J. Am. Chem. Soc.* 2013, 135, 11533–11536.
18 S.A. Burgess, K. Grubel, A.M. Appel, E.S. Wiedner, J.C. Linehan, Hydrogenation of CO2 at room temperature and low pressure with a cobalt tetraphosphine catalyst. *Inorg. Chem.* 2017, 56, 8580–8589.
19 A.Z. Spentzos, C.L. Barnes, W.H. Bernskoetter, Effective pincer cobalt precatalysts for Lewis acid assisted CO_2 hydrogenation. *Inorg. Chem.* 2016, 55, 8225–8233.
20 C. Federsel, C. Ziebart, R. Jackstell, W. Baumann, M. Beller, Catalytic hydrogenation of carbon dioxide and bicarbonates with a well-defined cobalt dihydrogen complex. *Chem. Eur. J.* 2012, 18, 72–75.

21 Y.M. Badiei, W.H. Wang, J.F. Hull, D.J. Szalda, J.T. Muckerman, Y. Himeda, E. Fujita, Cp*Co(III) catalysts with proton-responsive ligands for carbon dioxide hydrogenation in aqueous media. *Inorg. Chem.* 2013, 52, 12576–12586.

22 B.G. Schieweck, J. Klankermayer, Tailor-made molecular cobalt catalyst system for the selective transformation of carbon dioxide to dialkoxymethane ethers. *Angew. Chem. Int. Ed.* 2017, 56, 10854–10857.

23 T.J. Korstanje, J.I. van der Vlugt, C.J. Elsevier, B. de Bruin, Hydrogenation of carboxylic acids with a homogeneous cobalt catalyst. *Science* 2015, 350, 298–302.

24 J. Yuwen, S. Chakraborty, W.W. Brennessel, W.D. Jones, Additive-free cobalt-catalyzed hydrogenation of esters to alcohols. *ACS Catal.* 2017, 7, 3735–3740.

25 K. Junge, B. Wendt, A. Cingolani, A. Spannenberg, Z. Wei, H. Jiao, M. Beller, Cobalt pincer complexes for catalytic reduction of carboxylic acid esters. *Chem. Eur. J.* 2018, 24, 1046–1052.

26 G. Zhang, B.L. Scott, S.K. Hanson, Mild and homogeneous cobalt-catalyzed hydrogenation of C=C, C=O, and C=N bonds. *Angew. Chem. Int. Ed.* 2012, 51, 12102–12106.

27 G. Zhang, K.V. Vasudevan, B.L. Scott, S.K. Hanson, Understanding the mechanisms of cobalt-catalyzed hydrogenation and dehydrogenation reactions. *J. Am. Chem. Soc.* 2013, 135, 8668–8681.

28 C. Hou, J. Jiang, Y. Li, Z. Zhang, C. Zhao, Z. Ke, Unusual non-bifunctional mechanism for Co-PNP complex catalyzed transfer hydrogenation governed by the electronic configuration of metal center. *Dalton Trans.* 2015, 44, 16573–16585.

29 Y. Jing, X. Chen, X. Yang, Computational mechanistic study of the hydrogenation and dehydrogenation reactions catalyzed by cobalt pincer complexes. *Organometallics* 2015, 34, 5716–5722.

30 K.H. Hopmann, Cobalt-bis(imino)pyridine-catalyzed asymmetric hydrogenation: electronic structure, mechanism, and stereoselectivity. *Organometallics* 2013, 32, 6388–6399.

31 G. Hilt, 1,4-Cyclohexadienes – easy access to a versitile building block via transition-metal-catalysed Diels–Alder reactions. *Chem. Rec.* 2014, 14, 386–396.

32 F. Chen, C. Topf, J. Radnik, C. Kreyenschulte, H. Lund, M. Schneider, A.E. Surkus, L. He, K. Junge, M. Beller, Stable and inert cobalt catalysts for highly selective and practical hydrogenation of C≡N and C=O bonds. *J. Am. Chem. Soc.* 2016, 138, 8781–8788.

33 K.D. Baucom, A.S. Guram, C.J. Borths, Effective conversion of heteroaromatic ketones into primary amines via hydrogenation of intermediate ketoximes. *Synlett* 2015, 26, 201–204.

34 A. Mukherjee, D. Srimani, S. Chakraborty, Y. Ben David, D. Milstein, Selective hydrogenation of nitriles to primary amines catalyzed by a cobalt pincer complex. *J. Am. Chem. Soc.* 2015, 137, 8888–8891.

35 É. Balogh-Hergovich, G. Speier, L. Markó, Kinetics of the homogeneous hydrogenation of cyclohexene catalyzed by hydridocobalt complexes formed in situ from CoH(N2)(PPh3)3. *J. Organomet. Chem.* 1974, 66, 303–310.

36 G.F. Pregaglia, A. Andreetta, G.F. Ferrari, R. Ugo, Catalysis by phosphine cobalt carbonyl complexes I. Synthesis and catalytic properties of (tributylphosphine)cobalt (I) hydride carbonyl complexes. *J. Organomet. Chem.* 1971, 30, 387–405.
37 G.F. Ferrari, A. Andreetta, G.F. Pregaglia, R. Ugo, Catalysis by phosphine cobalt carbonyl complexes V. Kinetics of homogeneous cyclohexene hydrogenation catalyzed by CoH(CO)2[P(n-C4H9)3]2. *J. Organomet. Chem.* 1972, 43, 213–223.
38 Y. Ohgo, S. Takeuchi, Y. Natori, J. Yoshimura, Asymmetric reactions. X. Asymmetric hydrogenation catalyzed by bis(dimethylglyoximato)cobalt(II)-chiral cocatalyst (amino alcohol) system. *Bull. Chem. Soc. Jpn.* 1981, 54, 2124–2135.
39 Y. Ohgo, Y. Natori, S. Takeuchi, J. Yoshimura, Asymmetric hydrogenations catalyzed by bis(dimethylglyoximato)cobalt(II) - achiral base complex and chiral aminoalcohol conjugated systems. An oxido-reductase model with enantioselectivity. *Chem. Lett.* 1974, 1327–1330.
40 K. Kobayashi, T. Okamoto, T. Oida, S. Tanimoto, Cobaltmediated reduction of CN bond. Synthesis of methyl N-p-toluenesulfonyl-1-phenylglycinate catalyzed by bis (dioximato)cobalt-quinine complexes. *Chem. Lett.* 1986, 15, 2031–2034.
41 K. Yoshinaga, T. Kito, H. Oka, S. Sakaki, K. Ohkubo, Catalytic ability of cobalt(II) and nickel(II) chiral diphosphine complexes for asymmetric hydrogenation of prochiral unsaturated esters. *J. Catal.* 1984, 87, 517–519.
42 V. Massonneau, P. Le Maux, G. Simonneaux, Asymmetric synthesis by chiral cobalt catalysis: homogeneous reduction of ketones to optically active alcohols. *J. Organomet. Chem.* 1985, 288, C59–C60.
43 S. Rösler, J. Obenauf, R. Kempe, A highly active and easily accessible cobalt catalyst for selective hydrogenation of CO bonds. *J. Am. Chem. Soc.* 2015, 137, 7998–8001.
44 M.R. Friedfeld, G.W. Margulieux, B.A. Schaefer, P.J. Chirik, Bis(phosphine)cobalt dialkyl complexes for directed catalytic alkene hydrogenation. *J. Am. Chem. Soc.* 2014, 136, 13178–13181.
45 M.R. Friedfeld, M. Shevlin, J.M. Hoyt, S.W. Krska, M.T. Tudge, P.J. Chirik, Cobalt precursors for high-throughput alkene hydrogenation catalysts. *Science* 2013, 342, 1076–1080.
46 D. Zhang, E.-Z. Zhu, Z.-W. Lin, Z.-B. Wei, Y.-Y. Li, J.-X. Gao, Enantioselective hydrogenation of ketones catalyzed by chiral cobalt complexes containing PNNP ligand. *Asian J. Org. Chem.* 2016, 5, 1323–1326.
47 J. Chen, C. Chen, C. Ji, Z. Lu, Cobalt-catalyzed asymmetric hydrogenation of 1,1-diarylethenes. *Org. Lett.* 2016, 18, 1594–1597.
48 S. Monfette, Z.R. Turner, S.P. Semproni, P.J. Chirik, Enantiopure C1-symmetric bis(imino)pyridine cobalt complexes for asymmetric alkene hydrogenation. *J. Am. Chem. Soc.* 2012, 134, 4561–4564.
49 M.R. Friedfeld, M. Shevlin, G.W. Margulieux, L.C. Campeau, P.J. Chirik, Cobalt-catalyzed enantioselective hydrogenation of minimally functionalized alkenes: isotopic labeling provides insight into the origin of stereoselectivity and alkene insertion preferences. *J. Am. Chem. Soc.* 2016, 138, 3314–3324.

50 Y. Matsui, M. Orchin, Rhodium and cobalt catalysis of the hydroformylation and hydrogenation of 1,1-diphenylethylene. *J. Organomet. Chem. Organomet. Chem.* 1983, 246, 57–60.

51 Q. Knijnenburg, A.D. Horton, H. van der Heijden, T.M. Kooistra, D.G.H. Hetterscheid, J.M.M. Smits, B. de Bruin, P.H.M. Budzelaar, A.W. Gal, Olefin hydrogenation using diimine pyridine complexes of Co and Rh. *J. Mol. Catal. A Chem.* 2005, 232, 151–159.

52 R.P. Yu, J.M. Darmon, C. Milsmann, G.W. Margulieux, S.C.E. Stieber, S. DeBeer, P.J. Chirik, Catalytic hydrogenation activity and electronic structure determination of bis(arylimidazol-2-ylidene)pyridine cobalt alkyl and hydride complexes. *J. Am. Chem. Soc.* 2013, 135, 13168–13184.

53 K. Tokmic, C.R. Markus, L. Zhu, A.R. Fout, Well-defined cobalt(I) dihydrogen catalyst: experimental evidence for a Co(I)/Co(III) redox process in olefin hydrogenation. *J. Am. Chem. Soc.* 2016, 138, 11907–11913.

54 C. Chen, Y. Huang, Z. Zhang, X.-Q. Dong, X. Zhang, Cobalt-catalyzed (Z)-selective semihydrogenation of alkynes with molecular hydrogen. *Chem. Commun.* 2017, 53, 4612–4615.

55 K. Tokmic, A.R. Fout, Alkyne semihydrogenation with a well-defined non-classical Co-H2 catalyst: a H2 spin on isomerization and E-selectivity. *J. Am. Chem. Soc.* 2016, 138, 13700–13705.

56 S. Fu, N.Y. Chen, X. Liu, Z. Shao, S.P. Luo, Q. Liu, Ligand-controlled cobalt-catalyzed transfer hydrogenation of alkynes: stereodivergent synthesis of Z- and E-alkenes. *J. Am. Chem. Soc.* 2016, 138, 8588–8594.

57 T.P. Lin, J.C. Peters, Boryl-mediated reversible H2 activation at cobalt: Catalytic hydrogenation, dehydrogenation, and transfer hydrogenation. *J. Am. Chem. Soc.* 2013, 135, 15310–15313.

58 T.P. Lin, J.C. Peters, Boryl-metal bonds facilitate cobalt/nickel-catalyzed olefin hydrogenation. *J. Am. Chem. Soc.* 2014, 136, 13672–13683.

59 D. Gärtner, A. Welther, B.R. Rad, R. Wolf, J.A. von Wangelin, Heteroatom-free arene-cobalt and arene-iron catalysts for hydrogenations. *Angew. Chem. Int. Ed.* 2014, 53, 3722–3726.

60 P. Büschelberger, D. Gärtner, E. Reyes-Rodriguez, F. Kreyenschmidt, K. Koszinowski, J.A. von Wangelin, and R. Wolf, Alkene metalates as hydrogenation catalysts. *Chem. Eur. J.* 2017, 23, 3139–3151.

61 G. Zhang, Z. Yin, J. Tan, Cobalt(II)-catalysed transfer hydrogenation of olefins. *RSC Adv.* 2016, 6, 22419–22423.

62 J. Guo, X. Shen, Z. Lu, Regio- and enantioselective cobalt-catalyzed sequential hydrosilylation/hydrogenation of terminal alkynes. *Angew. Chem. Int. Ed.* 2017, 56, 615–618.

63 J. Guo, B. Cheng, X. Shen, Z. Lu, Cobalt-catalyzed asymmetric sequential hydroboration/hydrogenation of internal alkynes. *J. Am. Chem. Soc.* 2017, 139, 15316–15319.

64 H.M. Feder, J. Halpern, Mechanism of the cobalt carbonyl-catalyzed homogeneous hydrogenation of aromatic hydrocarbons. *J. Am. Chem. Soc.* 1975, 97, 7186–7188.

65 T.A. Weil, S. Friedman, I. Wender, Reactions catalyzed by Co2(CO)8. Selective deuterium incorporation into some polycyclic hydrocarbons. *J. Org. Chem.* 1974, 39, 48–50.
66 F. Chen, B. Sahoo, C. Kreyenschulte, H. Lund, M. Zeng, L. He, K. Junge, M. Beller, Selective cobalt nanoparticles for catalytic transfer hydrogenation of N-heteroarenes. *Chem. Sci.* 2017, 8, 6239–6246.
67 P. Ji, Y. Song, T. Drake, S.S. Veroneau, Z. Lin, X. Pan, W. Lin, Titanium(III)-oxo clusters in a metal-organic framework support single-site Co(II)-hydride catalysts for arene hydrogenation. *J. Am. Chem. Soc.* 2018, 140, 433–440.

3

Homogeneous Nickel-Catalyzed Hydrogenations

Marlene Böldl and Ivana Fleischer

University of Tübingen, Institute of Organic Chemistry, Chemistry Department, Auf der Morgenstelle 18, 72076 Tübingen, Germany

3.1 Introduction

When looking at non-precious metals in catalysis, the application of nickel has a long-standing tradition, especially in heterogeneous catalysis. The first appearance of Ni as catalyst marks a historic discovery in the field of catalysis in general, which was acknowledged by the Nobel Prize in Chemistry in 1912. It was awarded to Paul Sabatier for the development of hydrogenation of ethylene using a heterogeneous Ni-based catalyst at 300 °C [1]. He also disclosed the Ni-catalyzed hydrogenation of carbon dioxide to methane, known as the Sabatier process [2]. Later, Murray Raney created a hydrogenation catalyst by treating nickel–aluminum alloy with sodium hydroxide [3]. Referred to as Raney nickel, it is broadly used in organic synthesis and in industrial processes, mainly as the hydrogenation catalyst [4]. After initial investigations of $Ni(CO)_4$-promoted cyclooligomerizations by Reppe and Vetter [5], significant contributions in the field of homogeneous nickel catalysis were made by Wilke [6]. Among other achievements, he synthesized $Ni(COD)_2$ and investigated nickel-catalyzed olefin oligomerizations [7].

Since these key achievements, organonickel chemistry emerged as a powerful tool for catalysis and various remarkable synthetic applications have been developed [8]. Especially in the past years, nickel was mostly seen as a low-cost alternative to palladium and platinum in coupling reactions [9], since the reserves of precious metals in the earth's crust are depleting. Nickel is about 750 times more abundant than is palladium, and hence roughly 2000 times less expensive in its elemental form on a mole to mole basis (Figure 3.1). This makes nickel an attractive catalyst for a more sustainable chemistry [5]. A disadvantage in this regard is its toxicity [10].

However, the recent developments of a wide range of innovative nickel-catalyzed reactions were not only attributed to the lower costs but mainly to the essential properties of nickel. As a first-row metal, Ni-catalysts are often more reactive than Pd-catalysts in certain steps of the catalytic cycle and hence can be applied under milder reaction conditions or can react with unreactive

Homogeneous Hydrogenation with Non-Precious Catalysts, First Edition. Edited by Johannes F. Teichert.

Atom radius 135pm
Electronegativity 1.91
Oxidation states : –1, 0, 1, 2, 3, 4

10.53 €/1 kg (01/2018)
80 ppm of earth crust mass
$Ni(OAc)_2$ rat (oral) LD_{50} = 350 mg/kg

Facile oxidative addition
Facile migratory insertion
Accessible radical pathways

Figure 3.1 Selected aspects and properties of nickel.

organic molecules. On the other hand, high reactivity makes it difficult to predict mechanisms and to control the reactions of interest. Nickel is also a relatively electropositive late transition metal, which facilitates two-electron oxidative additions. This effect is predominantly applied in cross-coupling of electrophiles that are considerably less reactive in palladium catalysis [11]. In addition, compared to palladium catalysis, in which most catalytic cycles are based on two-electron chemistry including Pd(0)/Pd(II) cycles, in nickel catalysis Ni(0)/Ni(I)/Ni(II)/Ni(III) oxidation states are feasible. Besides the various oxidation states, methods based on not only two-electron transfer processes but also one-electron transfer steps are common [12].

Other features of nickel in comparison to other group 10 metals are the strong olefin bonding ability, the facile β-migratory insertion, and a conversely slow β-hydride elimination, as well as a small atomic radius that often leads to relative short Ni-ligand bonds. To sum up, nickel demonstrates valuable and unique reactivity and behavior, which are advantageous for the evolution of an entire set of chemical transformations or for improvement of already established reactions, such as homogeneous catalyzed hydrogenations.

The use of nickel in homogeneous catalyzed hydrogenation reactions seems reasonable, since nature uses nickel as catalyst for the activation of molecular dihydrogen. Prokaryotic [NiFe]-hydrogenases reversibly convert dihydrogen at atmospheric pressure and ambient temperatures into protons and electrons at high rates (Scheme 3.1) [13]. Several investigations on the structure and reactivity of model complexes for these enzymes were conducted [14]. Together with the studies on formation, structure, and reactivity of nickel hydride complexes [15], they might serve as a source of knowledge and inspiration to develop new reduction catalysts.

In general, transition metal complexes can activate dihydrogen by various mechanisms (Scheme 3.1) [16]. Platinum group metals often undergo oxidative addition of dihydrogen, giving dihydride complexes. Heterolytic splitting of dihydrogen, on the other hand, occurs either in the presence of an external base or it can be promoted by the ligand. This type of activation is often found in first-row metal complexes, such as nickel complexes. A polarity inverted heterolysis of dihydrogen is feasible for metals bound to Lewis acidic ligands [17].

Herein, we discuss the most significant reports on the application of nickel complexes in hydrogenation reactions. The overview is structured according to

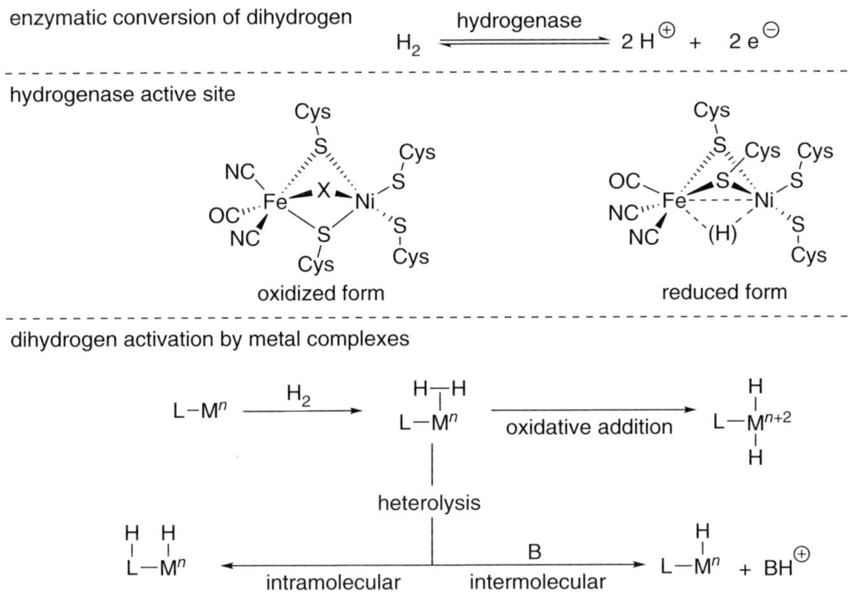

Scheme 3.1 Reversible activation of dihydrogen by enzymes and mode for dihydrogen cleavage by homogeneous metal complexes.

the employed substrate, whereby the most attention is paid to the reduction of alkenes.

3.2 Hydrogenation of Alkenes

3.2.1 Hydrogenation of Alkyl- and Aryl-Substituted Alkenes

In early work on hydrogenation reactions of alkenes, dienes, and other compounds, Ziegler-type nickel and other group 8–10 transition metal based catalysts were applied [18]. In general, they are prepared by treatment of the appropriate metal salt with a reducing agent such as trialkylaluminium or complex hydrides. Although these catalysts have been reported as homogeneous systems, Ziegler proposed the formation of colloidal particles. Several more recent studies suggest that the nature of the catalyst (homogeneous or heterogeneous) highly depends on the reaction conditions and components. Furthermore, both types of catalyst can be copresent and simultaneously active [19]. In addition, isolated studies applying nickel salts modified with monodentate phosphine or dinitrogen ligands appeared; however, mostly low activity and limited scope were reported [20].

Pioneering work in the field of homogeneous nickel hydrogenation catalysis was conducted by Bouwman and coworkers [21]. A combination of bidentate phosphine ligands and nickel(II) acetate showed good activity in the hydrogenation of 1-octene (**1a**) at room or elevated temperature applying 50 bar of dihydrogen (Scheme 3.2). Turnover numbers up to 460 (one hour reaction time)

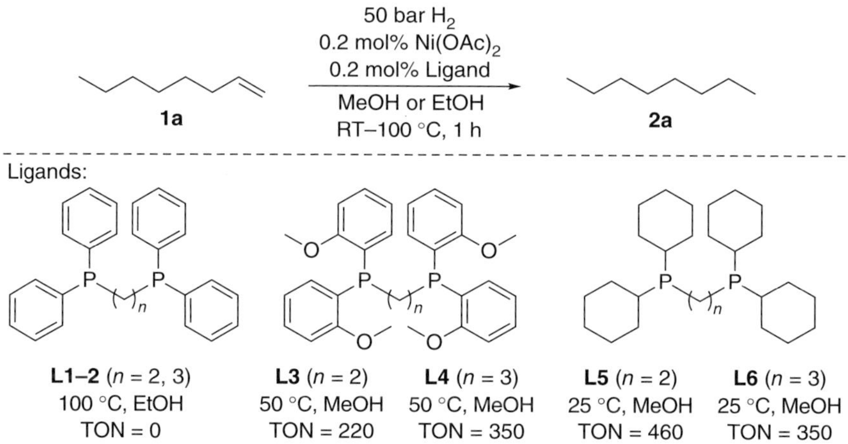

Scheme 3.2 Seminal results on nickel-catalyzed hydrogenation of 1-octene using bidentate phosphines as ligands.

were obtained with electron-donating bisphosphine ligands **L3–6**, while their simple diphenylphosphine-analoga **L1–2** were not active. However, since no hydrogenation activity and negligible alkene isomerization was observed with the *p*-methoxy-substituted ligand even at 100 °C, the influence of the substituent was ascribed rather to steric than to electronic effects. Further optimization of the reaction conditions (solvent, concentration) with **L4** led to an increase of the turnover frequency to 1480 h^{-1} (Scheme 3.2). Applying these conditions to other substrates, an efficient hydrogenation of ethylene (TOF = 9000 h^{-1}) and low conversion of cyclooctene (TOF = 30 h^{-1}) were observed [22]. Styrene and 1-octyne did not react.

The following studies on the structure of the nickel complexes and their respective activity led to the assumption that the steric hindrance protects the ligand from being oxidized by Ni(II) (Scheme 3.3) [23]. Moreover, it was found that ethylene-bridged ligand complex ***i*** derived from **L3** undergoes ligand exchange reactions and is thus in equilibrium with bis-chelated complex ***ii***, which in turn is unable to catalyze the hydrogenation due to the lack of available coordination sites. This was not observed for propylene-bridged ligand **L4** and cyclohexyl-substituted ligands **L5** and **L6**, presumably because of the larger bite angle and the inability of the relatively small Ni(II) ion to bind two such ligands in a square planar complex. In the proposed catalytic cycle, the catalytically active complex ***i*** is converted to nickel hydride ***iii*** by heterolytic cleavage of dihydrogen. The ligand S might act as a base in this process and bind the proton. After coordination of the alkene, migratory insertion leads to alkyl complex ***v***, which undergoes hydrogenolysis in the rate-determining step to release the product and hydride ***iii***.

In addition, the same group systematically investigated the effect of anion, solvent, and ligand structure also by means of competing isomerization of 1-octene (**1a**) [24]. They presynthesized a series of defined complexes [P_2NiX_2] and found that highest conversion and selectivity for the hydrogenation product were

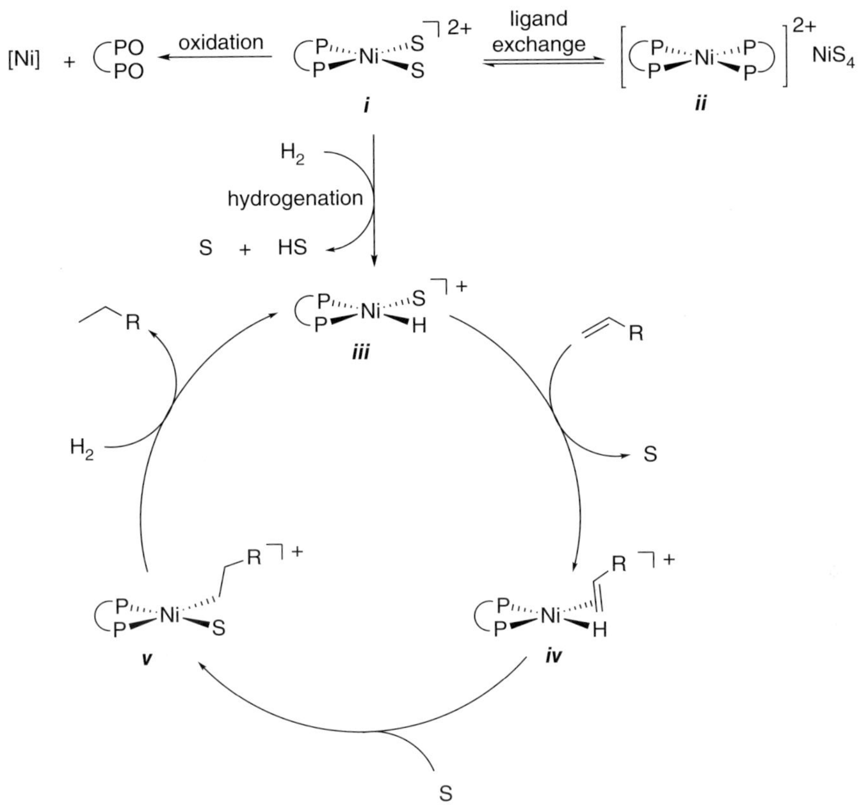

Scheme 3.3 Proposed mechanism of the Ni-catalyzed hydrogenation. (S = polar solvent).

obtained with increased steric bulk of the ligand aryl groups and backbone in the presence of weakly coordinating anions and a base cocatalyst. This isomerization side reaction also occurs in the absence of dihydrogen and leads to a mixture of internal octenes. During the course of these studies, nickel ammonia complex **C1** was unexpectedly formed, when an anion exchange for hexafluorophosphate was attempted (Scheme 3.4). This complex exhibited unprecedented activity in

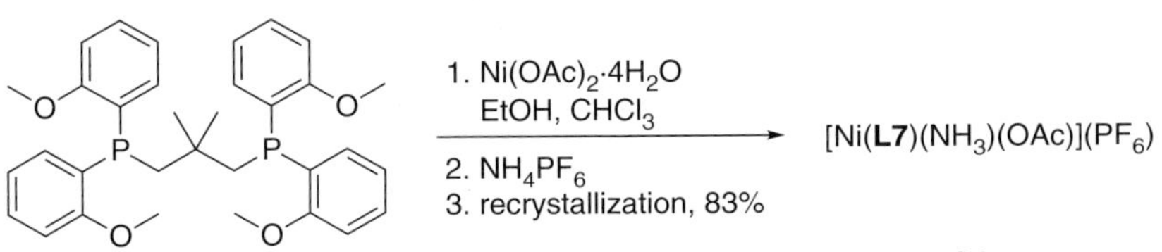

L7 **C1**

hydrogenation:
1-octene (**1a**, 54 mmol), **C1** (5.6 μmol), MeOH/CH_2Cl_2 (1 : 1, 0.2 mM for **C1**)
H_2 (50 bar), 50 °C, 1 h
46% conversion, TOF = 4450 h^{-1}, 96% selectivity

Scheme 3.4 Synthesis of complex **C1** and conditions for the hydrogenation of 1-octene (**1a**).

the selective hydrogenation of 1-octene with TOF of 4500 h^{-1} (50 °C, 50 bar H_2) with only a low extent of isomerization (4%). However, no other substrates were tested.

It should be noted that tetradentate *N,N,O,O*- [25], tridentate *N,N,O*-ligand [26], and N-heterocyclic carbene [27] modified Ni-complexes were also applied in the homogeneous hydrogenation of olefins. Hanson and coworkers introduced aliphatic phosphorus nitrogen phosphorus (PNP)-pincer ligand-based cationic and neutral nickel(II) complexes as hydrogenation catalysts [28]. Employing relatively high amounts of catalyst (10 mol%) at 80 °C and 4 bar of H_2, the highest activity was observed with the cationic complex **C2** (Scheme 3.5). Styrene (**1b**) and 3,3-dimethyl-1-butene (**1d**) were converted in quantitative yields, while lower conversion was observed for 1-octene (**1a**) and the more challenging 1,1-disubstituted α-methylstyrene (**1c**). Moreover, the catalyst displayed low activity in the reduction of carbonyl groups. Mechanistic investigations led to a proposed pathway including insertion of the alkene into the Ni—H bond and subsequent reaction of the resulting metal alkyl complex with dihydrogen. However, the initially assumed involvement of the pincer N—H bond in the catalysis was ruled out, since the N–Me analog showed similar reactivity. Thus, the function of the pincer ligand was suggested to be the stabilization of Ni(II)–H complex.

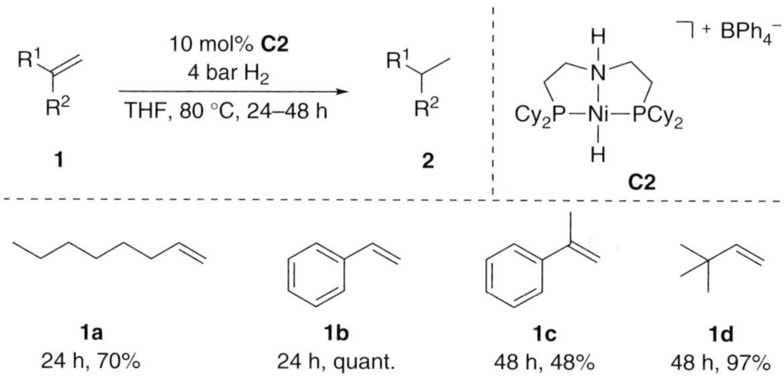

Scheme 3.5 Hydrogenation of alkenes using a Ni(II) PNP-pincer ligand hydride complex.

A unique catalyst design was reported by Stryker and coworkers [29]. They synthesized a square-shaped tetrametallic Ni(I) cluster **C3** with bridging phosphoranimide ligands in a two-step procedure from nickel(II) bromide (Scheme 3.6). The metal centers are linear, coplanar, and bind to two electron-rich nitrogen atoms each. Such complexes are used as soluble surrogates of heterogeneous catalysts, such as Raney-Ni. Complex **C3** was successfully employed as catalyst in the hydrogenation of allylbenzene (**1e**) under mild reaction conditions (room temperature, 1 bar H_2). Full conversion was observed after 12 hours, corresponding to a TOF of 17 h^{-1}. A similar result and complete hydrogenation was obtained with an alkyne. The result of the poisoning test performed using mercury was negative, suggesting the presence of a homogeneous catalyst.

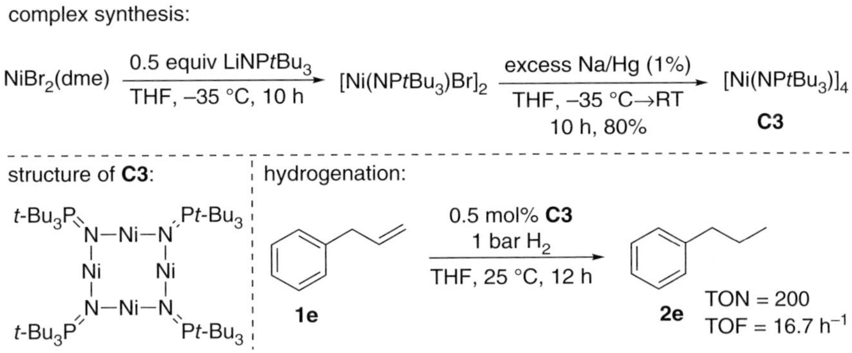

Scheme 3.6 Synthesis of a nickel(I) cluster and its use as catalyst for the hydrogenation of allylbenzene (**1e**).

The concept of ligand–metal cooperativity [30] was applied in the activation of dihydrogen and in the catalytic hydrogenation by Peters and coworkers [31]. Their approach is based on the use of a chelating borane ligand coordinated to a Ni(0) center, affording complexes able to undergo reversible two-electron oxidative addition of dihydrogen, a process related to frustrated Lewis pairs (Scheme 3.7). First, complex **C4** was prepared by a two-step process based on comproportionation and reduction.

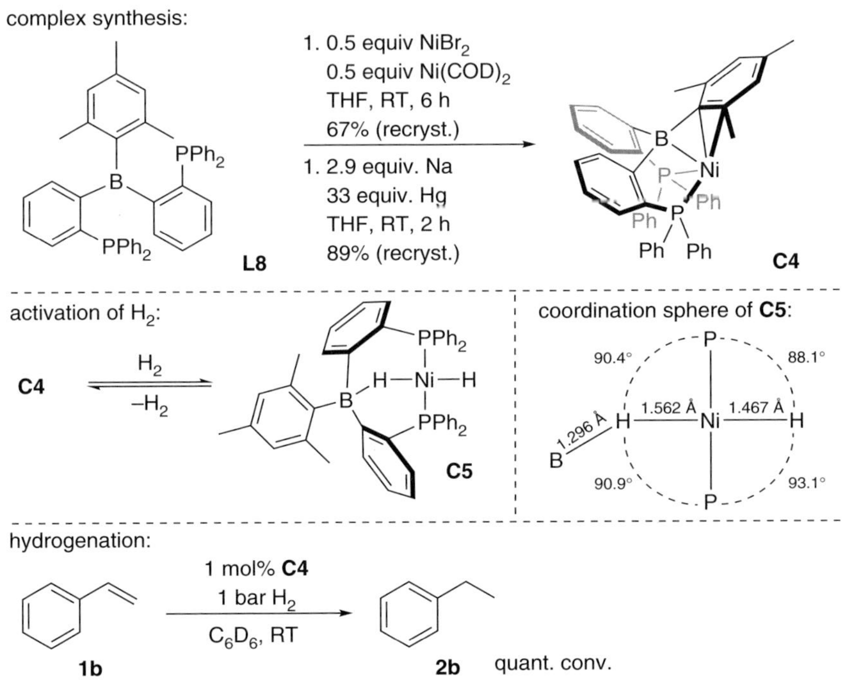

Scheme 3.7 Diphosphane–borane ligand supported nickel complex: synthesis, activation of dihydrogen and hydrogenation of styrene (**1b**).

Initial experiments with a **C4** analogon bearing a phenyl substituent on the boron atom showed no reactivity with dihydrogen. However, decoration with the more sterically demanding mesityl group in **C4** led to the reactive diamagnetic complex **C5** upon treatment with dihydrogen at room temperature. This reaction is reversible and the initial complex **C4** can be reobtained from **C5** by degassing. Noteworthy is the polarity inversion during the heterolysis of H_2 by **C4**: the Lewis basic Ni atom binds the proton, while the Lewis acidic boron center accepts the hydride. The structure of **C5** was determined using nuclear magnetic resonance (NMR) and density functional theory (DFT) calculations. The authors suggest a square planar geometry with phosphorus atoms *trans* to each other and a bridging hydride ligand. Based on these experiments, hydrogenation of styrene (**1b**) was conducted, and full conversion was observed using 1 mol% of **C4**.

In order to gain deeper insight into the cooperativity and to increase the catalytic efficacy, the same group investigated more complexes with boron-based ligands of type **L9** (Scheme 3.8) [32]. First, treatment of **L9** with $Ni(cod)_2$ led to the formation of a dimeric Ni(I)–Ni(I) complex **C6**, which readily and quantitatively reacted with dihydrogen. Notably, this represents the first case of reversible oxidative addition of dihydrogen to a Ni(I) center, which was also proposed to occur in hydrogenase active sites [13b, 33]. However, **C6** did not exhibit any

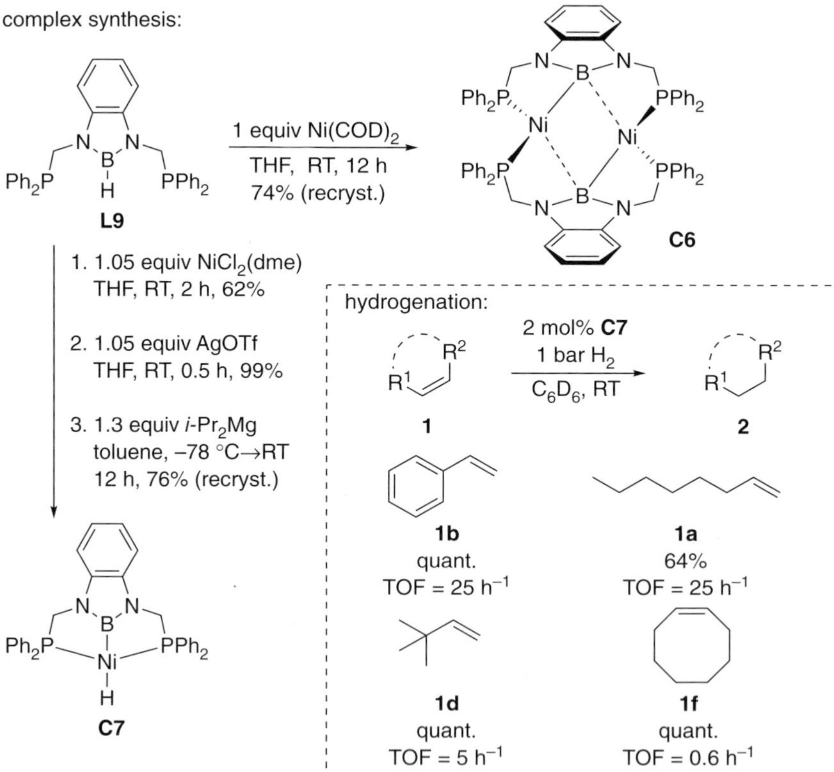

Scheme 3.8 Boryl-nickel cooperativity demonstrated in hydrogenation of alkenes using **C7**.

hydrogenation activity in the presence of 4 bar dihydrogen and 1 bar ethylene. Therefore, a monomeric complex **C7** was synthesized by complexation of **L9** with $NiCl_2$(dme) and subsequent reduction. This complex showed good hydrogenation activity with styrene (**1b**) and 1-octene (**1a**), the latter was partially isomerized to internal olefins. However, these could be hydrogenated at higher temperature (60 °C) within two hours. Reduced activity was observed in the hydrogenation of 3,3-dimethyl-1-butene (**1d**) and cyclooctene (**1f**).

The homogeneous nature of the catalyst was corroborated by a negative qualitative poisoning test with mercury [34]. Two mechanistic scenarios were considered: (i) migratory insertion of the alkene into the Ni—H bond followed by a σ-bond metathesis with H_2 or (ii) radical stepwise hydrogen atom transfer to the unsaturated substrate. The radical pathway was ruled out on the basis of kinetic experiments and the observation that a *syn*-addition of D_2 to norbornene was observed. Notably, the performance of the boryl-based catalyst was superior to that of its phosphorus carbon phosphorus (PCP)- and PNP-pincer catalyst analogs [28]. This is probably caused by the strong *trans*-influence of the boryl ligand, which weakens the Ni—H and Ni-alkyl bonds.

Following this approach, on cooperative activation of dihydrogen, Lu and coworkers investigated the influence of larger ions in group 13 on the reactivity of corresponding Ni(0)-complexes (Scheme 3.9) [35]. Specifically, Al- [36], Ga-, and In-containing complexes **C8** were synthesized in a two-step procedure from ligands **L10**. Structural investigations revealed that larger ions push the nickel center above the plane of the phosphorus atoms. In addition, electrochemical studies led to the conclusion that the Lewis acidity of the M(III) center increases with its size due to a better orbital overlap with Ni(0). These two effects lead to a stronger affinity to bind small molecules such as N_2 and H_2. This reactivity, however, does not translate into improved catalytic activity. This was first tested in the hydrogenation of styrene using 5 mol% of metal complex **C8** under 1 bar of H_2 at room temperature. The best performance was noted for the Ga-based catalyst **C8b**, while lower conversion was obtained using **C8c** with the larger In(III)-ion. In addition, it was shown that the H_2-adduct of this complex was stable toward reduced pressure and it was furthermore observed as the resting state of the catalytic cycle. This and the observed primary kinetic isotope effect of 1.7 indicate that the cleavage of the H—H bond is part of the rate-determining step. Notably, the Al-based complex **C8a** and the Lewis acid free analog **C8d** were not active.

Furthermore, the substrate scope was studied using **C8b** as catalyst. Good to high activity was observed for unhindered terminal olefins **1a, g, h** and *cis*-cyclooctene (**1f**). The hydrogenation of *trans*-2-octene (**1i**) was much slower and only a low yield of the hydrogenated product was obtained with allylbenzene (**1e**). No desired product was formed in case of sterically demanding substrates **1k** and **1l**. The carbonyl groups and an alkyne were not hydrogenated under the reaction conditions. Interestingly, complex **C8c** exhibited higher isomerization activity with substrates such as **1a** and **1e**.

Compared to other ligand structures, N-heterocyclic silylenes are much less frequently used in homogeneous catalysis [37]. However, they are highly attractive due to their properties such as strong σ-donor ability. Based on

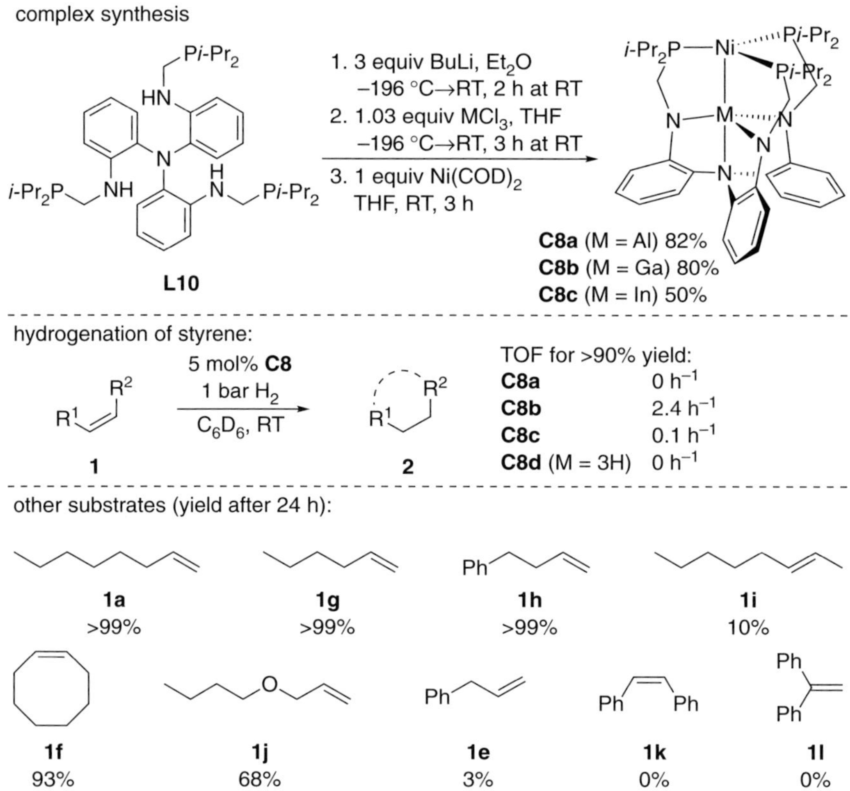

Scheme 3.9 Synthesis of bimetallic complexes of Ni and Al, Ga, or In. Hydrogenation of alkenes using the most active Ga-based complex **C8b**.

the long-standing interest and experience with this type of ligands, Driess and coworkers developed a bidentate bis(silylene)xanthene ligand **L11** and investigated the properties of Ni complexes thereof (Scheme 3.10) [38]. The complexation with $Ni(cod)_2$ resulted in the formation of 16 valence electron Ni(0) complex **C9** with isomerized 1,3-cyclooctadiene ligand. This complex was treated with trimethylphosphine to provide **C10** in 62% yield. Both complexes **C9** and **C10** were subjected to dihydrogen atmosphere (1 bar) at room temperature to furnish **C11** and **C12**, respectively. For the first time, a dinuclear complex **C11** bearing a Ni_2Si_2 moiety was formed by a reaction with H_2. The formation of the dihydrido complex **C12** was the result of reversible dihydrogen activation. A fast exchange between the hydrido and Si(II) ligands on the nickel atom atoms was proposed on the basis of NMR data.

The catalytic hydrogenation activity was tested with complexes **C9–11** employing norbornene (**1m**) as substrate (Scheme 3.11). Full conversion was observed with **C9** after 24 hours, while **C10** displayed lower activity, probably due to strongly binding phosphine ligands. The dinuclear complex **C11** provided a 65% yield of norbornane (**2m**). Therefore, **C11** can be excluded as a hydrogenation intermediate using **C9**. Notably, a mixture of Ni(0) precursor and the free

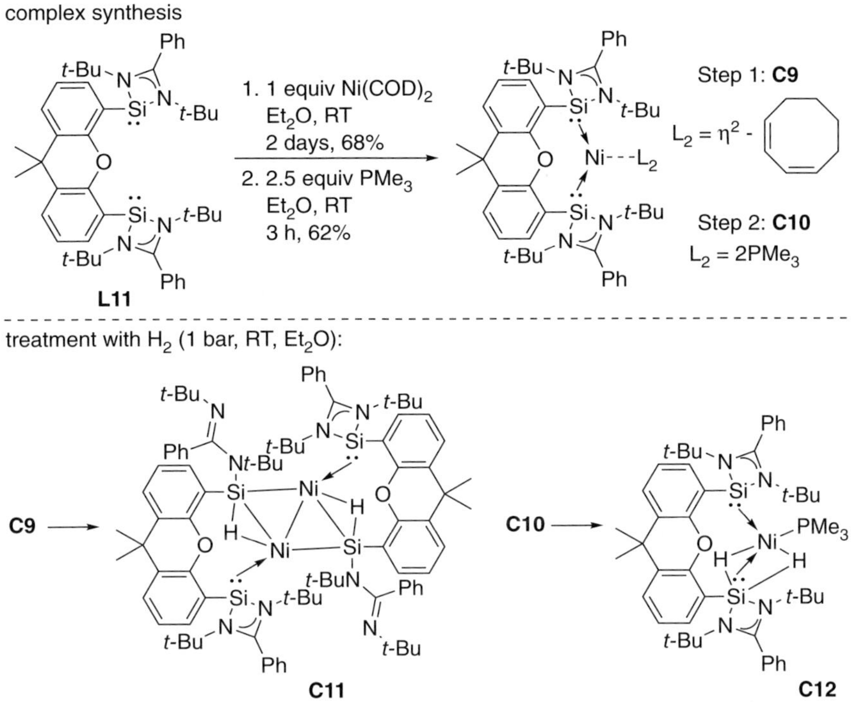

Scheme 3.10 Nickel(0) complexes of bis(N-heterocyclic silylene) ligand **L11** and their reaction with dihydrogen.

ligand **L11** also showed excellent activity. In contrast, the use of diphosphine Xantphos together with $Ni(COD)_2$ only furnished a 6% yield. Next, a broad substrate screening was conducted using **C9**. The aliphatic terminal, internal, and cyclic alkenes **1a, g, n–r** were quantitatively hydrogenated in 12–24 hours, with the exception of cyclooctene (**1f**), which required 48 hours. Notably, the highly challenging tetrasubstituted tetramethylethylene (**1s**) was reduced in 11.5% yield in 12 hours. Various styrene derivatives **1b, t, u** were fully converted to the corresponding ethylarenes; only 4-cyano substitution in **1v** led to a slower reaction. In addition, various dienes and unsaturated carbonyl compounds **1x–z** were successfully employed. Using only 0.1 mol% of the catalyst **C9**, TON of 880 and 1000 and TOF of 160 and 250 h^{-1} were determined for the hydrogenation of 1-dodecene (**1n**) and styrene (**1b**).

A mechanistic scenario was proposed on the basis of DFT calculations. Accordingly, the decisive step (not rate determining) is the activation of dihydrogen in complex ***i*** with coordinated alkene and dihydrogen. The calculations suggest that Si(II) atoms are involved in the activation process, which leads to intermediate ***ii***. This hitherto unknown cooperative effect by silylene-based ligands is supported by the structure of complex **C12**, which features a strong Si–H interaction. Then, one hydride is transferred to the olefin and the second one to silicon, leading to complex ***iii*** with an opened N-heterocyclic ring. This was also observed in

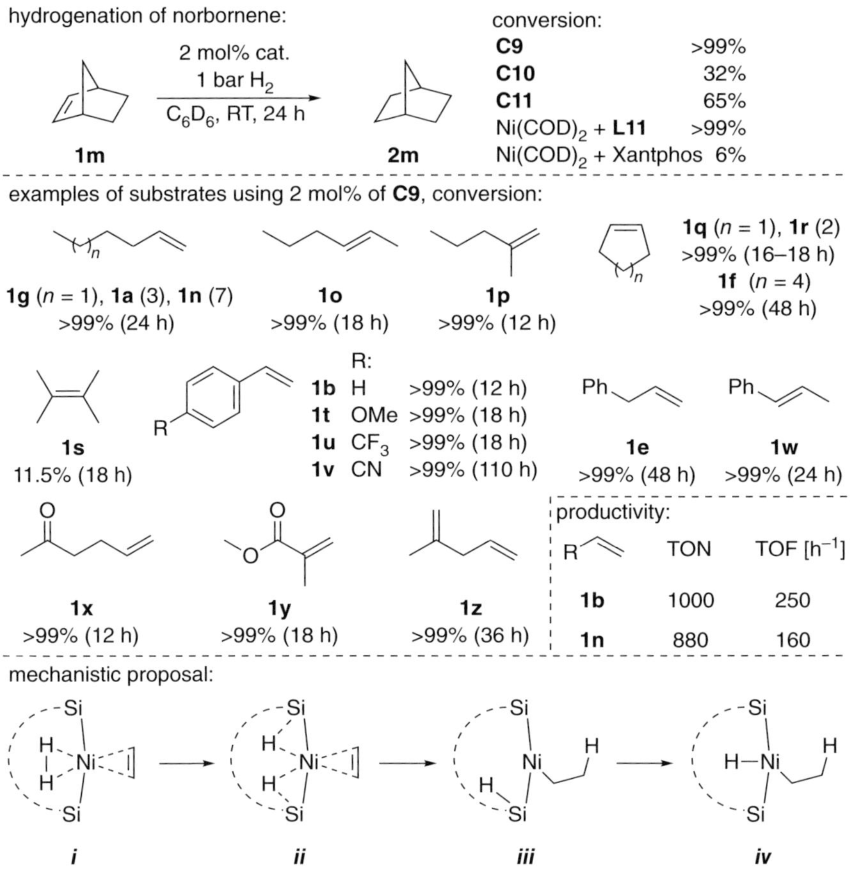

Scheme 3.11 Homogeneous hydrogenation of alkenes using nickel-bis-silylene complex **C9**.

complex **C11**. The hydride then migrates to nickel to give ***iv***, which undergoes isomerization and reductive elimination.

Another Ni-based catalyst system able to hydrogenate hindered alkenes was reported by Chirik and coworkers [39]. They developed a robust and easy-to-handle catalyst based on Ni(II) carboxylate salt and an α-diimine ligand, which was demonstrated to operate under mild reaction conditions without the need for strictly inert atmosphere or specialized high-pressure equipment (Scheme 3.12). A catalyst comprising Nickel(II) octanoate, ligands **L12** and **L13**, and pinacol borane with 4 bar of dihydrogen at 50 °C for 12 hours were found to be optimal conditions for the hydrogenation of challenging substrates such as trisubstituted methylcyclohexene (**1aa**). Similar conditions were employed to conduct the substrate screening, which also featured several demanding tri- and tetrasubstituted alkenes. In most cases, ligand **L12**, which also showed better performance under milder conditions, was used. All monosubstituted alkenes were fully converted in 1.5–5 hours. The same was observed for a series of trisubstituted alkenes with endo- or exocyclic double bonds, with the exception of a few substituted indene derivatives, such as **1ad–af**, which

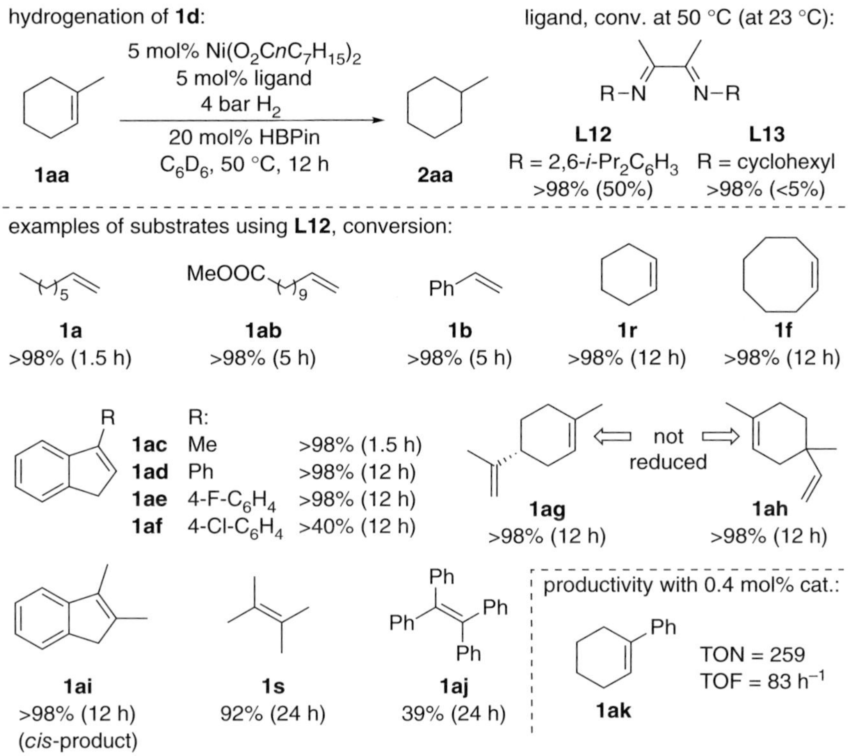

Scheme 3.12 Air-stable catalyst system consisting of Ni(II) precursor and diimine ligands **L12–13** for the hydrogenation of alkenes, including tetrasubstituted double bonds.

required longer reaction times. Interesting results were obtained using cyclic dienes: In 1,4-disubstituted cyclohexenes **1ag** and **1ah** containing both, one trisubstituted endocyclic and one less hindered exocyclic alkene, only the latter was hydrogenated. Notably, a TON of 250 was determined for the reduction of 1-phenyl-1-cyclohexene (**1ak**). With this substrate, a benchtop hydrogenation on large scale with the catalyst previously exposed to air demonstrated the practicability of this protocol. Moreover, 1,2-disubstituted indenes such as **1ai** were successfully subjected to hydrogenation and a full conversion to the *cis*-substituted products was obtained after 12 hours. Also tetrasubstituted acyclic alkenes **1s** and **1aj** were amenable to the standard reaction conditions; however, longer reaction times were needed.

In order to understand the interactions between the catalyst and substrate, a series of deuteration experiments were conducted. Using tetrasubstituted indene **1ai** and D_2, an exclusive formation of the *cis*-product as the only isotopolog was observed (Scheme 3.13). This substrate does not undergo β-hydride elimination from the intermediate Ni-alkyl complex. On the other hand, a chain-walking mechanism for the isomerization of the double bond was confirmed for less hindered substrates, such as **1aa**, as deuterium incorporation was found on all sp^3-carbon atoms and also in the recovered unreacted starting material.

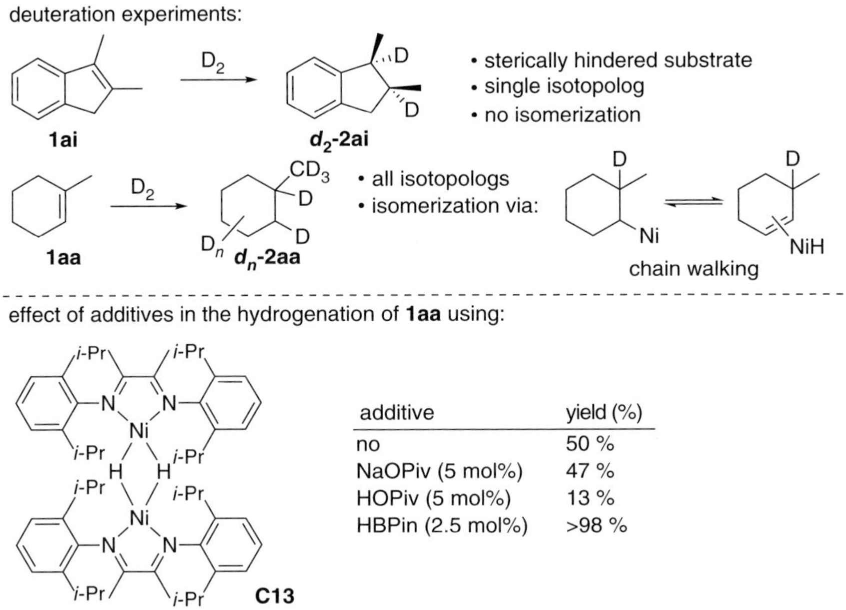

additive	yield (%)
no	50 %
NaOPiv (5 mol%)	47 %
HOPiv (5 mol%)	13 %
HBPin (2.5 mol%)	>98 %

Scheme 3.13 Deuteration experiments and effect of additives for mechanistic understanding: chain-walking mechanism and the importance of borane.

In a stoichiometric experiment using ligand **L12**, borane, and Ni(II) salt, the formation of dimeric complex **C13** was confirmed by NMR. This complex was then prepared by an alternative synthetic approach and used as the catalyst in the hydrogenation of **1aa** in order to study the effect of additives. It was found that while carboxylates did not improve the activity, the addition of borane had a significant positive effect on the catalyst. Further experiments using D_2 showed that borane plays a role in the activation of dihydrogen, since formation of hydrogen deuteride ($^1H^2H$) (HD) and deuterated borane was observed.

3.2.2 Hydrogenation of Electron-Deficient Alkenes

The hydrogenation of α-mono- or β-disubstituted α,β-unsaturated carbonyl compounds offers a great opportunity to develop enantioselective approaches toward α- and/or β-substituted aldehydes, ketones, or esters. Initially, Zhou and coworkers reported a series of methods for enantioselective transfer hydrogenation using Ni(II) salts together with a chiral bidentate ligand [40]. They were able to generate various α- or β-substituted hydrocinnamates and related compounds in good yields and with high ee values employing formic acid or DMF as hydride sources.

Pioneering work used dihydrogen as the terminal reductant and a thorough mechanistic study was conducted by Chirik and coworkers [41]. They carried out a high-throughput screening of 192 chiral bidentate ligands in combination with $Ni(OAc)_2$ for the hydrogenation of ethyl β-methylcinnamate (**3a**) and identified Me-DuPhos (**L14**) to be the most promising ligand as judged by conversion and stereoselectivity. Further optimization revealed a remarkable improvement in reactivity and selectivity in the presence of Bu_4NI, which was superior to other halide additives. The same effect was observed when a combination of NiI_2 and Bu_4NAc was employed. The substrate evaluation showed tolerance for both electron-donating and electron-withdrawing substituents on the aryl group, as well as potentially coordinating groups such as thioether **3c** (Scheme 3.14). Furthermore, only a small extent of protodehalogenation was obtained with chloro- and bromo-substituted compounds **3g, h**.

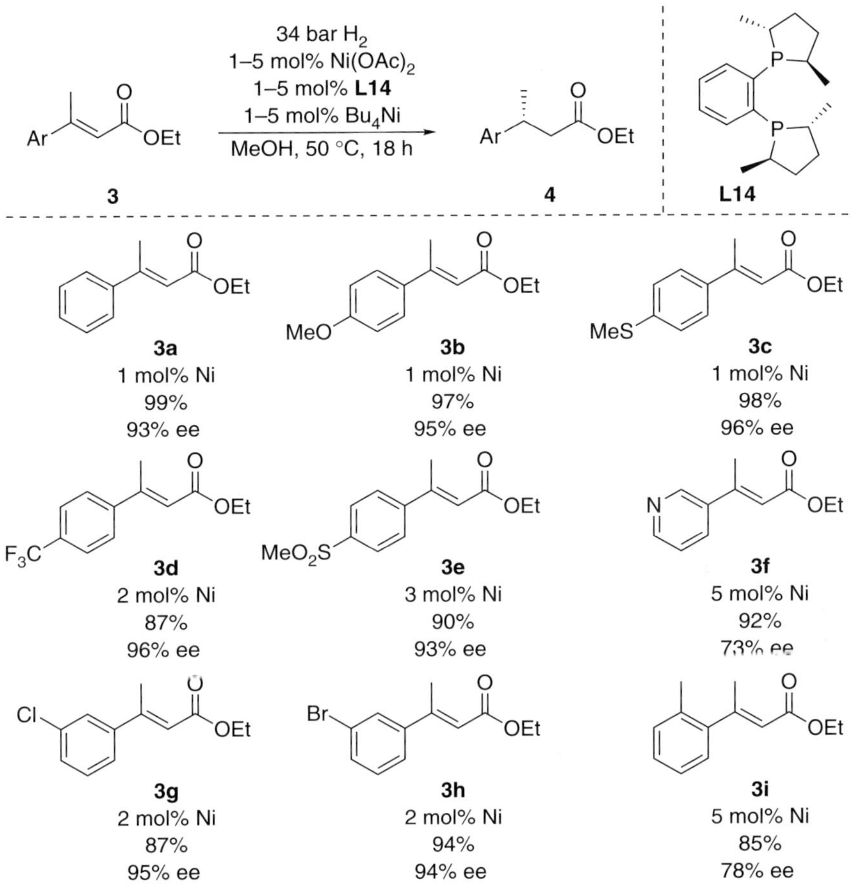

Scheme 3.14 Enantioselective hydrogenation of α,β-unsaturated esters **3** using a bidentate phosphine ligand **L14**.

Mechanistic studies including kinetic experiments, determination of the active catalyst composition, isotope labeling studies, and observation of a negative non-linear effect led to the proposal of a catalytic cycle based on the heterolytic cleavage of dihydrogen (Scheme 3.15). Trimeric complex $(\mathbf{L14})_3Ni_3(OAc)_5I$ (***i***) was suggested to be the resting state of the catalyst in equilibrium with other mono- and dimeric species. The H_2 cleavage is assisted by the acetate ligand and represents the rate-determining step of the cycle. After the coordination of substrate

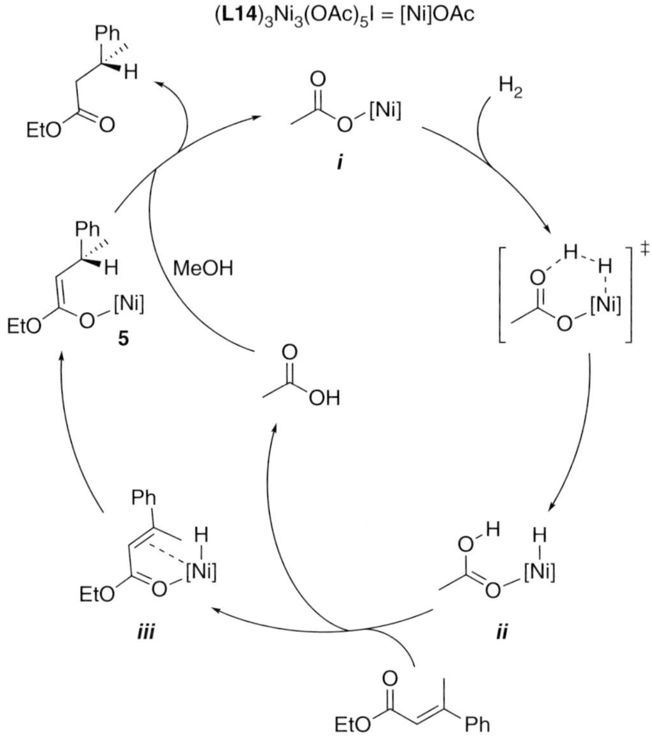

Scheme 3.15 Proposed catalytic cycle for the asymmetric nickel-catalyzed hydrogenation of α,β-unsaturated esters based on heterolytic dihydrogen cleavage.

and release of acetic acid, the enantioselective hydrometalation occurs providing the enolate **5** that is protonated by the solvent or acetic acid.

In addition, enantioselective hydrogenations of β-acylamino-substituted nitroolefins **6** [42] and acrylates **8** [43] were developed (Scheme 3.16). In both cases, the catalyst was based on $Ni(OAc)_2$ and ligand **L15** ((*S*)-Binapine) and the reactions were carried out in trifluoroethanol. Slightly more forcing reaction conditions (50 °C, 50 bar) were applied for the conversion of acrylates **8**. High yields and enantioselectivities were obtained with various substrates and both reactions were demonstrated on gram scale. A mechanism similar to the above-described hydrogenation of α,β-unsaturated esters was suggested and corroborated by DFT calculations.

3.3 Hydrogenation of Alkynes

The most synthetically useful reductive transformation of alkynes is the chemo- and stereoselective semihydrogenation of alkynes to *E*- or *Z*-configured alkenes [44]. Since the introduction of Lindlar catalyst for *Z*-selective hydrogenation [45], numerous catalysts were developed. On the other hand, only a few (mainly homogeneous) catalysts are able to provide *E*-alkenes with high selectivity.

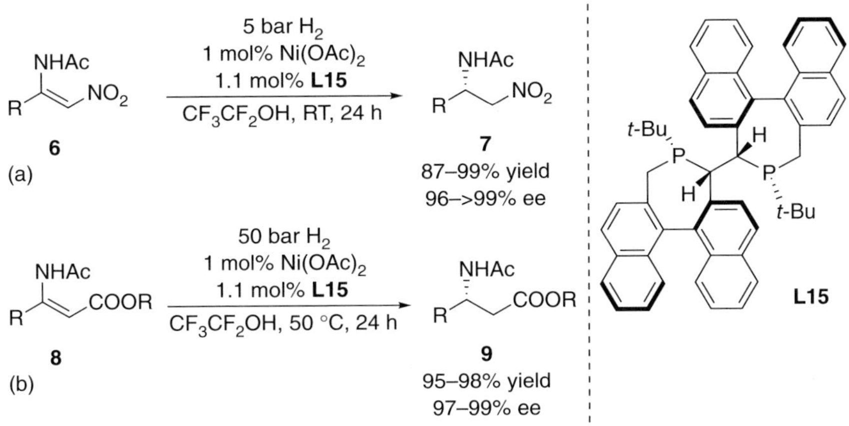

Scheme 3.16 Asymmetric hydrogenations of: (a) β-acylamino-substituted nitroolefins **6** and (b) acrylates **8**.

Among nickel-catalyzed transformations, transfer hydrogenation protocols with hypophosphorous [46] or formic acid [47] as dihydrogen surrogates were developed at first. Teichert and coworkers recently disclosed an *E*-selective semihydrogenation of alkynes **10** with a catalyst consisting of nickel(II) iodide and the bidentate phosphine ligand **L16** (Scheme 3.17) [48].

The reaction was performed at 100 °C in 1,4-dioxane with 15 bar dihydrogen pressure. Outstanding *E*-selectivity of >99% was observed with diaryl alkynes **10a–h** as substrates. Most of them were converted in good yields and only in a few cases were small amounts of the alkane detected. Electron-withdrawing and electron-donating substituents in *para* position did not affect the reaction outcome. Of note, reducible functionalities such as carbonyl or cyano group remained untouched and no dehalogenation was observed with chloride present. As a limitation, 2- and 4-pyridyl-substituted alkynes **10h** did not undergo hydrogenation, probably due to coordination to the metal. Several aryl, alkyl-substituted alkynes, such as **10i** and **10j**, were subjected to the hydrogenation with success, however with lower yields, as well as lower chemo- and stereoselectivities. Product mixtures were obtained when the double bond was prone to the isomerization. Notably, the macrocyclic alkyne **12** was selectively reduced in 45% yield. Preliminary mechanistic investigations revealed that the reaction probably occurs via initial *cis*-hydrogenation to give the *Z*-alkene, which then undergoes insertion into the Ni—H bond, followed by β-hydride elimination to furnish the thermodynamically more stable stereoisomer.

3.4 Hydrogenation of Carbonyl Groups

3.4.1 Hydrogenation of Ketones

The reduction of carbonyl groups represents a valuable transformation widely applied in synthetic laboratories and industry [49]. The discovery of Ru-catalyzed asymmetric hydrogenation of ketones, which provides secondary alcohols in a

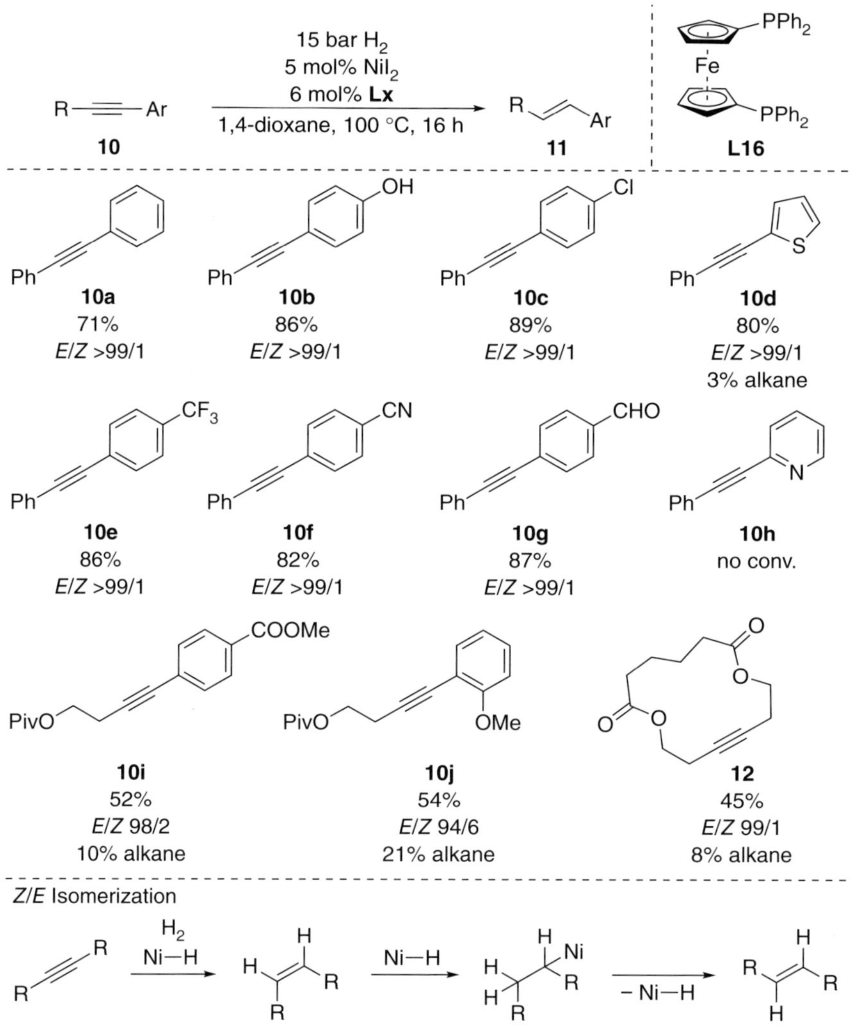

Scheme 3.17 E-selective semihydrogenation of alkynes **10** using *in situ* formed catalyst from NiI_2 and bidentate ligand **L16**.

highly atom-economic manner, was acknowledged by the Nobel Price to Ryoji Noyori [50]. Compared to other homogeneous metal catalysts, there are only a few examples of Ni-based complexes that are able to hydrogenate carbonyl compounds. One of them was reported in 2008 by Hamada and coworkers (Scheme 3.18) [51]. They developed a method for dynamic kinetic resolution of α-amino-β-ketoester hydrochlorides **13** providing the *anti*-aminoalcohols **14** with good diastereo- and enantioselectivities. The most successful catalyst consisted of a Ni(II) precursor and a Josiphos-type ligand **L17** in the presence of a base. A series of acetophenone derivatives **13** was converted in good yields; however, electron-rich or sterically demanding substrates required longer reaction times. Halide substituents in **13d**, **e**, and ester functionalities were

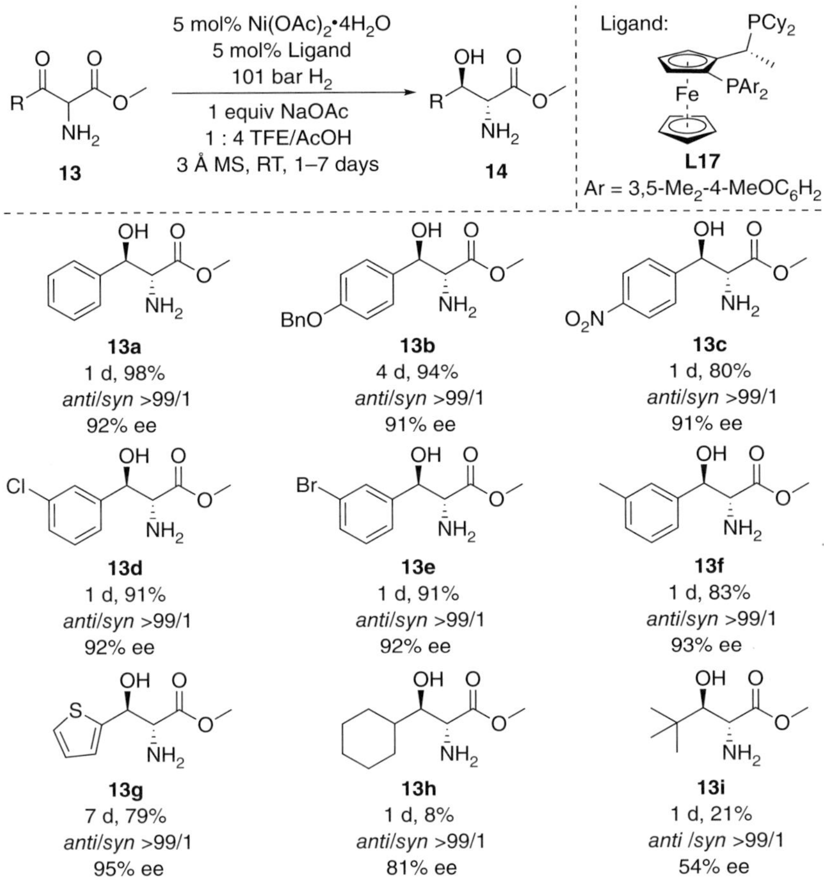

Scheme 3.18 Dynamic kinetic resolution of α-amino-β-ketoesters **13** by Ni-catalyzed hydrogenation.

tolerated by the catalyst. Inferior results in terms of yield and ee values were obtained with alkyl-substituted ketones **13h, i.** The strategy was extended to the hydrogenation of α-aminoketone hydrochlorides that furnished similarly excellent stereoselectivities [52].

Other studies focused on the non-asymmetric hydrogenation of aceto- and benzophenones with emphasis on the investigation of structure and reactivity of new nickel complexes [53]. In addition, transfer hydrogenation methodologies were developed for the reduction of carbonyl groups using alcohols [54], hydrosilanes [55], or boranes [56] as hydride sources. An asymmetric transfer hydrogenation of hydrazones and ketimines was achieved using formic acid [57].

3.4.2 Hydrogenation of Carbon Dioxide

The hydrogenation of the greenhouse gas and waste product CO_2 to formic acid or methanol could evolve into a future technology for a sustainable energy storage and the production of value-added chemicals [58]. Already in 1976, [Ni(dppe)$_2$]

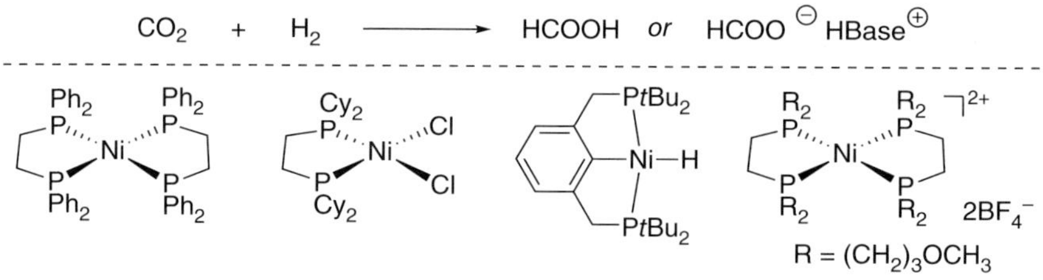

C14 *Ref. 52*	**C15** *Ref. 53*	**C16** *Ref. 54*	**C17** *Ref. 55*
25 bar CO_2	200 bar CO_2/H_2	$NaHCO_3$ (no CO_2)	17 bar CO_2
25 bar H_2	40 bar H_2	55 bar H_2	17 bar H_2
Et_3N, H_2O, C_6H_6, RT	DBU, 50 °C	MeOH, 150 °C	$NaHCO_3$, H_2O, 100 °C
TON = 7	TON = 4400	TON = 3038	TON = -
TOF = 0.35 h^{-1}	TOF = 20.4 h^{-1}	TOF = 152 h^{-1}	TOF = 0.6 h^{-1}

Scheme 3.19 Catalysts for the hydrogenation of carbon dioxide.

(**C14**) was used to produce formic acid from carbon dioxide (Scheme 3.19). A TON of 7 was achieved by performing the reaction at room temperature with 25 bar of dihydrogen in the presence of a base [59]. Despite this early report and the success of other precious and non-precious metals, not much advancement has been made in this field until recently. As a part of a combinatorial screening of various metal complexes in the hydrogenation of CO_2 to formic acid, **C15** was identified to provide high turnover numbers, however with low productivity over time [60]. Higher turnover frequencies were obtained with pincer hydride complex **C16**, although in this case, sodium bicarbonate was used as substrate and sodium formate was produced [61]. Unfortunately, the catalyst system was not active in the direct conversion of carbon dioxide. In addition, the water-soluble Ni(II) complex **C17** was employed to hydrogenate CO_2 to formate with average TOF of 0.6 h^{-1} at 100 °C [62].

There are several reasons for the lack of highly active Ni-based catalysts for CO_2 hydrogenation. Nickel hydride complexes are, in general, weak hydride donors [15]. Those complexes which are stronger H-donors and could potentially react with CO_2 often form strong Ni—O bonds in formate complexes. This renders the release of formate and catalytic turnover difficult. A solution was shown using reductants such as hydrosilanes and boranes; however, such processes produce more waste than the hydrogenation itself [63]. Recently, a highly active catalyst **C8b** for the hydrogenation of CO_2 at ambient reaction conditions was reported by Lu and coworkers (Scheme 3.20) [64]. Their catalyst design features the placement of a σ-donor Ga(III) ligand *trans* to the binding site at Ni. As already demonstrated in the study on hydrogenation of alkenes [35], this complex is able to bind H_2 at 1 bar pressure and the Lewis acidic Ga(III) ion is assumed to stabilize the nickel hydride species ***ii*** whose thermodynamic hydricity was estimated to be ~31 kcal/mol in MeCN. ***ii*** was thus able to transfer the hydride to CO_2 and release the formate ***iii*** efficiently with for a nickel catalyst unprecedented values of TON (3150) and TOF (9700 h^{-1}). The hydrogenation was performed at 20 °C under 34 bar pressure of H_2/CO_2 (1 : 1) in the presence of proazaphosphatrane base **15**.

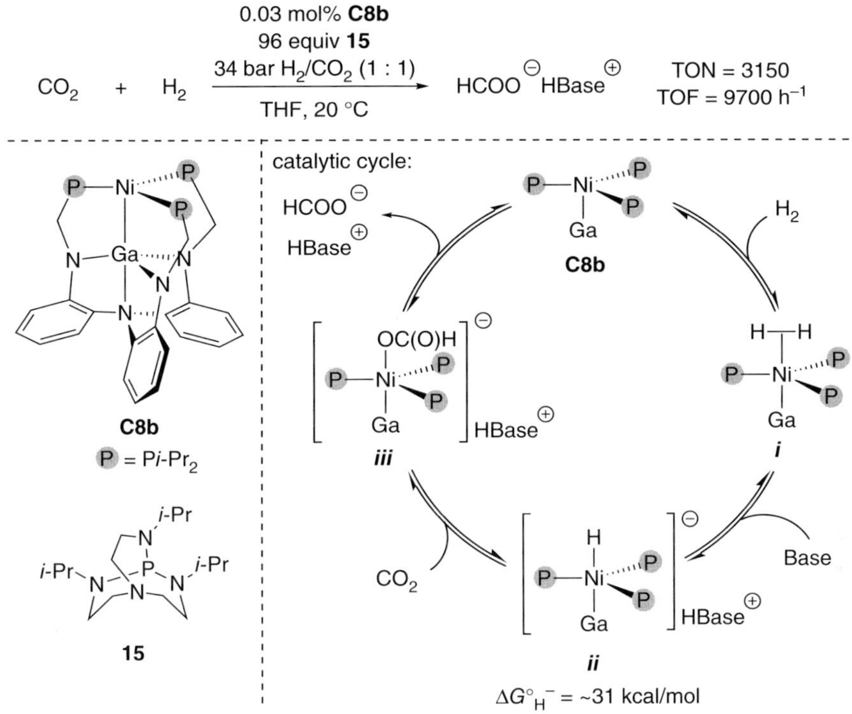

Scheme 3.20 Hydrogenation of carbon dioxide catalyzed by a nickel–gallium complex **C8b**.

3.5 Conclusions

Much progress has been achieved in the field of homogeneous nickel-catalyzed hydrogenation in the past few years. Selective and efficient catalysts have been developed for the reduction of various functional groups. This includes easy-to-handle, air-stable modular catalyst systems, which are of particular interest for synthetic applications. Future developments will most likely include more efforts toward understanding the mechanisms of these catalysts, also with the aim of improving their performance. More detailed investigations of the truly homogeneous nature of the catalysts are also necessary. A great challenge is the development of chiral catalysts for the enantioselective hydrogenation of substrates such as simple tri- or tetrasubstituted alkenes or ketones. It is possible that nickel hydrogenation catalysts will become competitive to the state-of-art noble metal catalysts that are also used in industry.

References

1 P. Sabatier, J.-B. Senderens, *C.R. Acad. Sci.* 1897, 124, 616–618.
2 P. Sabatier, J.-B. Senderens, *C.R. Acad. Sci.* 1902, 134, 689–691.
3 M. Raney, *Ind. Eng. Chem.* 1940, 32, 1199–1203.

4 C.R. Sarko, M. DiMare, M. Yus, F. Alonso, in *Encyclopedia of Reagents for Organic Synthesis*, 2014, 1–8. Wiley https://doi.org/10.1002/047084289X.rn011.pub2.

5 W. Reppe, H. Vetter, *Justus Liebigs Ann. Chem.* 1953, 582, 133–161.

6 G. Wilke, *Angew. Chem. Int. Ed. Engl.* 1988, 27, 185–206.

7 (a) P.W. Jolly, I. Tkatchenko, G. Wilke, *Angew. Chem. Int. Ed.* 1971, 10, 329–330; (b) *Angew. Chem. Int. Ed.* 1971, 10, 328–329; (c) B. Bogdanovic, M. Kroner, G. Wilke, *Justus Liebigs Ann. Chem.* 1966, 699, 1–23.

8 (a) V.P. Ananikov, *ACS Catal.* 2015, 5, 1964–1971. (b) S.Z. Tasker, E.A. Standley, T.F. Jamison, *Nature* 2014, 509, 299–309; (c) Y. Tamaru, *Modern Organonickel Chemistry*. Wiley-VCH Verlag, 2005.

9 (a) B.M. Rosen, K.W. Quasdorf, D.A. Wilson, N. Zhang, A.-M. Resmerita, N.K. Garg, V. Percec, *Chem. Rev.* 2011, 111, 1346–1416; (b) I.P. Beletskaya, V.P. Ananikov, *Chem. Rev.* 2011, 111, 1596–1636; (c) E. Negishi, *Acc. Chem. Res.* 1982, 15, 340–348.

10 K.S. Egorova, V.P. Ananikov, *Organometallics* 2017, 36, 4071–4090.

11 (a) S. Bajo, G. Laidlaw, A.R. Kennedy, S. Sproules, D.J. Nelson, *Organometallics* 2017, 36, 1662–1672; (b) S.Z. Tasker, E.A. Standley, T.F. Jamison, *Nature* 2014, 509, 299.

12 U. Jahn, Radicals in transition metal catalyzed reactions? Transition metal catalyzed radical reactions?: A fruitful interplay anyway. In: *Radicals in Synthesis III* (eds. M. Heinrich and A. Gansäuer) Berlin, Heidelberg: Springer-Verlag, 2012, 323–451.

13 (a) P.M. Vignais, B. Billoud, *Chem. Rev.* 2007, 107, 4206–4272; (b) D. Schilter, J.M. Camara, M.T. Huynh, S. Hammes-Schiffer, T.B. Rauchfuss, *Chem. Rev.* 2016, 116, 8693–8749.

14 (a) Y. Ohki, K. Yasumura, K. Kuge, S. Tanino, M. Ando, Z. Li, K. Tatsumi, *Proc. Natl. Acad. Sci.* 2008, 105, 7652–7657; (b) G. Schulte, X.-L. Luo, R.H. Crabtree, M. Zimmer, *Angew. Chem., Int. Ed. Engl.* 1991, 30, 193–194; (c) M. Rakowski DuBois, D.L. DuBois, *Chem. Soc. Rev.* 2009, 38, 62–72; (d) M. Rakowski Dubois, D.L. Dubois, *Acc. Chem. Res.* 2009, 42, 1974–1982.

15 N.A. Eberhardt, H. Guan, *Chem. Rev.* 2016, 116, 8373–8426.

16 S.J.C. Robinson, D.M. Heinekey, *Chem. Commun.* 2017, 53, 669–676.

17 Y. Li, C. Hou, J. Jiang, Z. Zhang, C. Zhao, A.J. Page, Z. Ke, *ACS Catal.* 2016, 6, 1655–1662.

18 M.F. Sloan, A.S. Matlack, D.S. Breslow, *J. Am. Chem. Soc.* 1963, 85, 4014–4018.

19 W.M. Alley, I.K. Hamdemir, K.A. Johnson, R.G. Finke, *J. Mol. Catal. A: Chem.* 2010, 315, 1–27.

20 (a) D. Chatterjee, H.C. Bajaj, S.B. Halligudi, K.N. Bhatt, *J. Mol. Catal.* 1993, 84, L1–L5; (b) H. Itatani, J.C. Bailar, *J. Am. Chem. Soc.* 1967, 89, 1600–1602.

21 (a) I.M. Angulo, A.M. Kluwer, E. Bouwman, *Chem. Commun.* 1998, 2689–2690; (b) E. Bouwman, Nickel. In: *The Handbook of Homogeneous Hydrogenation* (eds. J.G. de Vries and C.J. Elsevier) Wiley-VCH, 2007, 93–109.

22 I.M. Angulo, E. Bouwman, *J. Mol. Catal. A: Chem.* 2001, 175, 65–72.

23 (a) I.M. Angulo, E. Bouwman, M. Lutz, W.P. Mul, A.L. Spek, *Inorg. Chem.* 2001, 40, 2073–2082; (b) I.M. Angulo, E. Bouwman, R. van Gorkum, S.M. Lok, M. Lutz, A.L. Spek, *J. Mol. Catal. A: Chem.* 2003, 202, 97–106.

24 (a) I.M. Angulo, S.M. Lok, V.F. Quiroga Norambuena, M. Lutz, A.L. Spek, E. Bouwman, *J. Mol. Catal. A: Chem.* 2002, 187, 55–67; (b) T.J. Mooibroek, E.C.M. Wenker, W. Smit, I. Mutikainen, M. Lutz, E. Bouwman, *Inorg. Chem.* 2013, 52, 8190–8201.

25 A.E.-M.M. Ramadan, W. Sawodny, H.Y.F. El-Baradie, M. Gaber, *Transition Met. Chem.* 1997, 22, 211–215.

26 C. González-Arellano, E. Gutiérrez-Puebla, M. Iglesias, F. Sánchez, *Eur. J. Inorg. Chem.* 2004, 2004, 1955–1962.

27 J. Wu, J.W. Faller, N. Hazari, T.J. Schmeier, *Organometallics* 2012, 31, 806–809.

28 K.V. Vasudevan, B.L. Scott, S.K. Hanson, *Eur. J. Inorg. Chem.* 2012, 2012, 4898–4906, S4898/4891-S4898/4899.

29 J. Camacho-Bunquin, M.J. Ferguson, J.M. Stryker, *J. Am. Chem. Soc.* 2013, 135, 5537–5540.

30 J.R. Khusnutdinova, D. Milstein, *Angew. Chem. Int. Ed.* 2015, 54, 12236–12273.

31 W.H. Harman, J.C. Peters, *J. Am. Chem. Soc.* 2012, 134, 5080–5082.

32 T.P. Lin, J.C. Peters, *J. Am. Chem. Soc.* 2014, 136, 13672–13683.

33 S.O.N. Lill, P.E.M. Siegbahn, *Biochemistry* 2009, 48, 1056–1066.

34 R.H. Crabtree, *Chem. Rev.* 2012, 112, 1536–1554.

35 R.C. Cammarota, C.C. Lu, *J. Am. Chem. Soc.* 2015, 137, 12486–12489.

36 P.A. Rudd, S. Liu, L. Gagliardi, V.G. Young, C.C. Lu, *J. Am. Chem. Soc.* 2011, 133, 20724–20727.

37 S. Raoufmoghaddam, Y.-P. Zhou, Y. Wang, M. Driess, *J. Organomet. Chem.* 2017, 829, 2–10.

38 Y.W. Wang, A. Kostenko, S.L. Yao, M. Driess, *J. Am. Chem. Soc.* 2017, 139, 13499–13506.

39 N.G. Léonard, P.J. Chirik, *ACS Catal.* 2017, 342–348.

40 (a) S.Y. Guo, J.R. Zhou, *Org. Lett.* 2016, 18, 5344–5347; (b) S.Y. Guo, P. Yang, J.R. Zhou, *Chem. Commun.* 2015, 51, 12115–12117; (c) P. Yang, H.Y. Xu, J.R. Zhou, *Angew. Chem. Int. Ed.* 2014, 53, 12210–12213.

41 M. Shevlin, M.R. Friedfeld, H.M. Sheng, N.A. Pierson, J.M. Hoyt, L.-C. Campeau, P.J. Chirik, *J. Am. Chem. Soc.* 2016, 138, 3562–3569.

42 W.C. Gao, H. Lv, T.H. Zhang, Y. Yang, L.W. Chung, Y.-D. Wu, X. Zhang, *Chem. Sci.* 2017, 8, 6419–6422.

43 X.X. Li, C. You, S.L. Li, H. Lv, X. Zhang, *Org. Lett.* 2017, 19, 5130–5133.

44 (a) C. Oger, L. Balas, T. Durand, J.-M. Galano, *Chem. Rev.* 2013, 113, 1313–1350; (b) M. Crespo-Quesada, F. Cárdenas-Lizana, A.-L. Dessimoz, L. Kiwi-Minsker, *ACS Catal.* 2012, 2, 1773–1786.

45 H. Lindlar, *Helv. Chim. Acta* 1952, 35, 446–450.

46 T. Chen, J. Xiao, Y. Zhou, S. Yin, L.-B. Han, *J. Organomet. Chem.* 2014, 749, 51–54.

47 E. Richmond, J. Moran, *J. Org. Chem.* 2015, 80, 6922–6929.

48 N.O. Thiel, B. Kaewmee, J.F. Teichert, ChemRXiv 2019, DOI:10.26434/chemrxiv.9715955.v1.

49 J. Magano, J.R. Dunetz, *Org. Process Res. Dev.* 2012, 16, 1156–1184.

50 (a) R. Noyori, *Angew. Chem. Int. Ed.* 2002, 41, 2008–2022; (b) R. Noyori, T. Ohkuma, M. Kitamura, H. Takaya, N. Sayo, H. Kumobayashi, S. Akutagawa, *J. Am. Chem. Soc.* 1987, 109, 5856–5858.

51 Y. Hamada, Y. Koseki, T. Fujii, T. Maeda, T. Hibino, K. Makino, *Chem. Commun.* 2008, 6206–6208.

52 T. Hibino, K. Makino, T. Sugiyama, Y. Hamada, *ChemCatChem* 2009, 1, 237–240.

53 (a) S. Chakraborty, P.E. Piszel, W.W. Brennessel, W.D. Jones, *Organometallics* 2015, 34, 5203–5206; (b) Z.L. Chen, M.L. Zeng, Y.Z. Zhang, Z. Zhang, F. Liang, *Appl. Organomet. Chem.* 2010, 24, 625–630; (c) A. Flores-Gaspar, P. Pinedo-González, M.G. Crestani, M. Muñoz-Hernández, D. Morales-Morales, B.A. Warsop, W.D. Jones, J.J. García, *J. Mol. Catal. A: Chem.* 2009, 309, 1–11.

54 (a) S. Sabater, M.J. Page, M.F. Mahon, M.K. Whittlesey, *Organometallics* 2017, 36, 1776–1783; (b) N. Castellanos-Blanco, A. Arevalo, J.J. Garcia, *Dalton Trans.* 2016, 45, 13604–13614; (c) S. Iyer, J.P. Varghese, *J. Chem. Soc. Chem. Commun.* 1995, 465–466.

55 (a) N.C. Mamillapalli, G. Sekar, *Chem. Commun.* 2014, 50, 7881–7884. (b) L. Postigo, B. Royo, *Adv. Syn. Catal.* 2012, 354, 2613–2618; (c) S. Chakraborty, J.A. Krause, H. Guan, *Organometallics* 2009, 28, 582–586; (d) S.N. MacMillan, W. Hill Harman, J.C. Peters, *Chem. Sci.* 2014, 5, 590–597; (e) F.-F. Wu, J.-N. Zhou, Q. Fang, Y.-H. Hu, S. Li, X.-C. Zhang, A.S.C. Chan, J. Wu, *Chem. Asian J.* 2012, 7, 2527–2530; (f) B.L. Tran, M. Pink, D.J. Mindiola, *Organometallics* 2009, 28, 2234–2243; (g) L.P. Bheeter, M. Henrion, L. Brelot, C. Darcel, M.J. Chetcuti, J.-B. Sortais, V. Ritleng, *Adv. Syn. Catal.* 2012, 354, 2619–2624.

56 F. Chen, Y. Zhang, L. Yu, S. Zhu, *Angew. Chem. Int. Ed.* 2017, 56, 2022–2025.

57 H.Y. Xu, P. Yang, P. Chuanprasit, H. Hirao, J.S. Zhou, *Angew. Chem. Int. Ed.* 2015, 54, 5112–5116.

58 J. Artz, T.E. Müller, K. Thenert, J. Kleinekorte, R. Meys, A. Sternberg, A. Bardow, W. Leitner, *Chem. Rev.* 2018, 118, 434–504.

59 Y. Inoue, H. Izumida, Y. Sasaki, H. Hashimoto, *Chem. Lett.* 1976, 863–864.

60 C.-C. Tai, T. Chang, B. Roller, P.G. Jessop, *Inorg. Chem.* 2003, 42, 7340–7341.

61 S. Enthaler, A. Bruck, A. Kammer, H. Junge, E. Irran, S. Gülak, *ChemCatChem* 2015, 7, 65–69.

62 S.A. Burgess, A.J. Kendall, D.R. Tyler, J.C. Linehan, A.M. Appel, *ACS Catal.* 2017, 7, 3089–3096.

63 (a) Z. Lu, T.J. Williams, *ACS Catal.* 2016, 6, 6670–6673; (b) S. Chakraborty, J. Zhang, J.A. Krause, H. Guan, *J. Am. Chem. Soc.* 2010, 132, 8872–8873.

64 R.C. Cammarota, M.V. Vollmer, J. Xie, J. Ye, J.C. Linehan, S.A. Burgess, A.M. Appel, L. Gagliardi, C.C. Lu, *J. Am. Chem. Soc.* 2017, 139, 14244–14250.

4

Homogeneous Hydrogenation with Copper Catalysts

Niklas O. Thiel, Felix Pape, and Johannes F. Teichert

Technische Universität Berlin, Institut für Chemie, Straße des 17. Juni 115, 10623 Berlin, Germany

4.1 Introduction

The chemistry of copper hydride complexes finds widespread application in synthetic chemistry. Emanating from the stoichiometric use of reasonably stable and easy-to-use copper hydride phosphine complexes, such as $[Ph_3PCuH]_6$, which is also known as *Stryker's reagent*, nowadays many catalytic protocols are available for the mild and (enantio)selective transfer of nucleophilic hydrides from copper to a variety of acceptors. Arguably, the 1,4-reduction of α,β-unsaturated carbonyl or carboxyl compounds (termed conjugate reduction) is the most prominent among these [1–5]. As stoichiometric hydride sources, mainly hydrosilanes (and also, to a much lesser extent, hydroboranes [6] and -stannanes [7]) are being utilized for this so-called copper hydride catalysis [8]. Bearing in mind the utility of copper hydride catalysis for synthesis, it comes as a surprise that similar processes employing dihydrogen (H_2), ultimately leading to much more atom-economic [9] and sustainable processes [10] and circumventing the production of e.g. silicon-based by-products, have rarely been investigated.

4.1.1 Early Studies on Copper-Catalyzed Hydrogenations

Copper hydride was the first metal hydride to be characterized in 1844 [11]. Also, in the early days of homogeneous catalysis with metal complexes, it is noteworthy that copper complexes have been studied as hydrogenation catalysts as early as 1938 [12, 13].[1] In the first study, copper(II) salts [e.g. $Cu(OAc)_2$] were reduced to copper(I) compounds *in situ* with H_2 as reductant in basic solvents such as quinoline at low H_2 pressure (<1 bar) and high temperatures (>100 °C). Based on kinetic measurements, it was suggested that the actual catalyst was a (dimeric) copper(I) complex and the reaction was autocatalytic [12]. By employing copper(I) acetate (CuOAc) (instead of the previously employed copper(II) acetate) in quinoline, no autocatalytic reaction was observed. Moreover, 1,4-benzoquinone could be hydrogenated to hydroquinone [13]. Further details

1 As a side note, this publication coins the term homogeneous catalysis for the first time.

Homogeneous Hydrogenation with Non-Precious Catalysts, First Edition. Edited by Johannes F. Teichert.

about the mechanism of dihydrogen activation by copper(I) salts in amine solutions were extracted from kinetic data [14]. In these cases, a *homolytic* dihydrogen activation to essentially give two hydrides from H_2 was suggested. In parallel, homogeneous hydrogenation with copper(II) salts in aqueous solutions were investigated [15]. Kinetic measurements led to the conclusion that – in contrast to organic solvents (e.g. quinoline) – a copper(II) catalyst is present in aqueous solutions. Finally, *heterolytic* dihydrogen activation to give a copper hydride and a proton was proposed [16–19], which is exemplified by transition state **1** (Scheme 4.1) [17, 18]. The heterolytic dihydrogen activation by a copper(I) alcoholate ("hydrogenolysis") was also suggested for the synthesis of the isolable copper hydride hexamer $[Ph_3PCuH]_6$ (*Stryker's reagent* [1–3, 20–22] or, less common, *Osborn complex*) [23] from H_2 [19, 24]. While at that time no further data was available for the actual activation mode of H_2 with homogeneous copper complexes, the analogy with the activation of hydrosilanes by complexes bearing a copper–oxygen bond should be drawn here. The generally accepted mechanism for the formation of copper hydride compounds bears similarity to the H_2 activation proposed much earlier: Copper complexes bearing a Cu—O bond (**2**) can undergo a σ-bond metathesis [25] with a hydride donor (**3**) to give the copper hydride (**4**) and a heteroatom-based ether (**5**) as by-product (Scheme 4.1). To translate the catalytic processes that are known using hydrosilanes to H_2, the considerably higher H—H bond strength (104 kcal/mol as compared to 92 kcal/mol for the Si—H bond) [26] has to be overcome by the catalyst used.

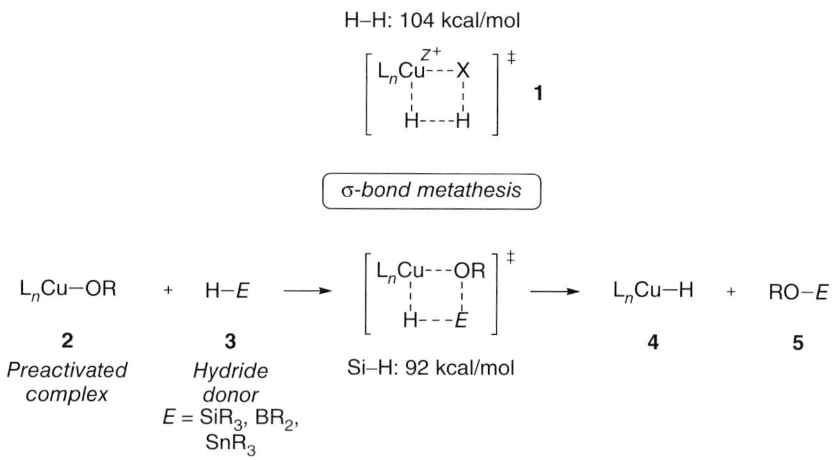

Scheme 4.1 Synthetic strategy toward copper hydride compounds. X = anion or ligand.

4.2 Hydrogenation of (α,β-Unsaturated) Carbonyl and Carboxyl Compounds

4.2.1 Conjugate Reduction

Based on the success of $[Ph_3PCuH]_6$ as a selective stoichiometric reducing agent for enones and enals [20, 22], it was also used as catalyst for the related

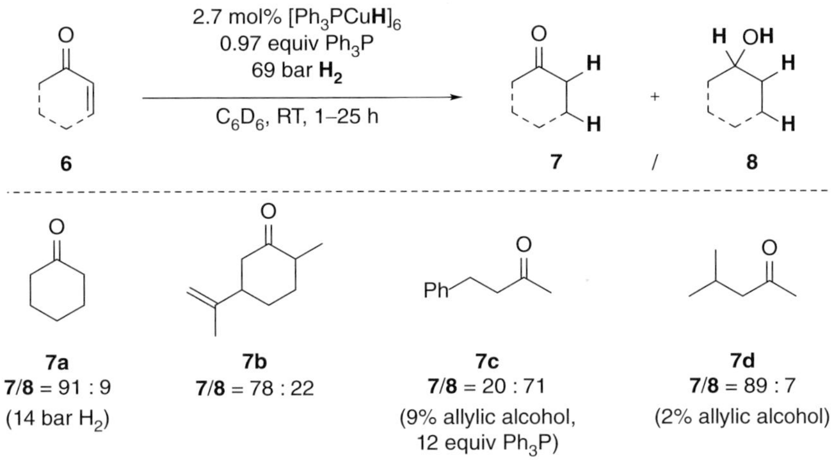

Scheme 4.2 Copper-catalyzed hydrogenation of (a)cyclic enones.

copper-catalyzed homogeneous hydrogenation of α,β-unsaturated carbonyl compounds such as enones **6** at room temperature and elevated H_2 pressure (69 bar) (Scheme 4.2) [27]. The addition of a stoichiometric amount of Ph_3P ensured homogeneity and reactivity. As an indication of a truly homogeneous process, turnover was also observed in the presence of mercury [28]. Although cyclic and acyclic ketones **7** could be obtained selectively, the reaction conditions (H_2 pressure, reaction time, solvent) had to be adjusted for every substrate in order to suppress the formation of allylic alcohols (not shown) or the corresponding saturated alcohols **8**, respectively. On the other hand, no reduction of isolated double bonds was observed under homogeneous conditions (**7b**, from carvone, Scheme 4.2), underscoring the high chemoselectivity for electron-poor alkenes of the presumed copper hydride complexes involved in the catalysis.

The suggested catalytic cycle for the hydride-mediated conjugate reduction with H_2 as terminal reductant commences with the conjugate addition of copper hydride **A** and enone **B** in order to form the intermediate copper-*O*-enolate[2] **C** (Scheme 4.3) [1–3, 27]. Heterolytic dihydrogen activation by the intermediate copper–oxygen bond through a proposed σ-bond metathesis transition state **D** gives rise to enol **E** and regenerates the copper hydride **A**. Consequently, enol **E** is transformed into ketone **F** through tautomerization.

Improving on these seminal results, an asymmetric copper-catalyzed conjugate reduction of enones with H_2 was developed in 2009 [30]. A catalyst mixture consisting of $[Cu(NO_3)(PPh_3)_2]$ and commercially available chiral diphosphine ligand (*R*)-DTBM-Segphos [(*R*)-**13**] was used at 50 bar H_2 and only slightly elevated temperatures. Activation of the pre-catalyst was achieved by addition of an excess of NaO*t*Bu (to generate the Cu—O bond required for heterolytic H_2 activation). Although high enantioselectivities (*e.r.* ≥99 : 1) were obtained in the conjugate reduction of isophorone (**9**), the catalytic activity (<80% conv.) and the regioselectivity (**11** up to 4%, **12** up to 16%) was rather limited at this stage (Scheme 4.4).

2 The nature of copper enolates (*C*-enolate vs. *O*-enolate) has been the subject of debate. For a recent contribution, see [29].

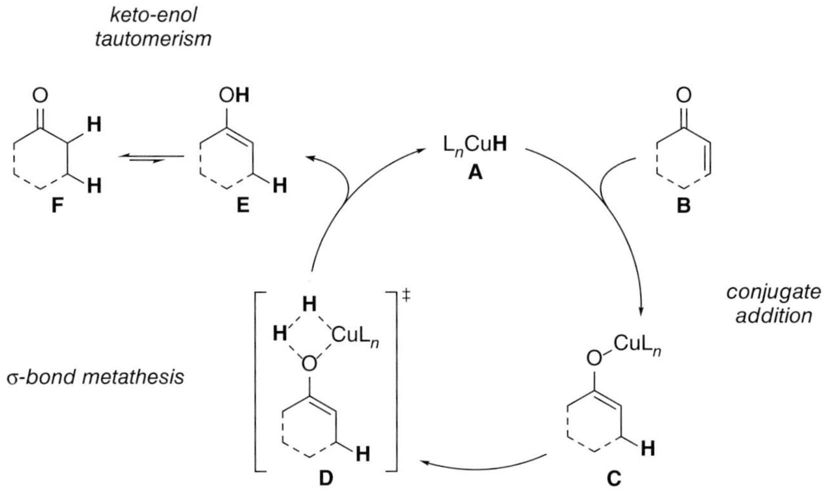

Scheme 4.3 Proposed catalytic cycle for the copper-catalyzed reduction of enones with H_2.

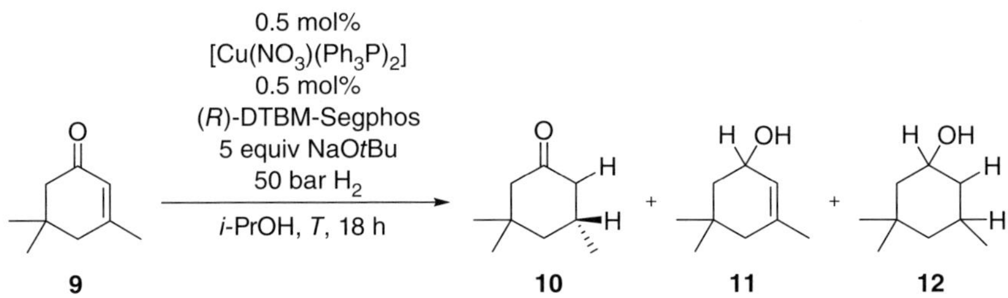

	10	11	12
30 °C: (51% conv.)	91 (e.r. 99.5 : 0.5)	2	7
30 °C: (1 mol%, 77% conv.)	84 (e.r. 99.5 : 0.5)	<1	16
50 °C: (72% conv.)	82 (e.r. 99 : 1)	4	14

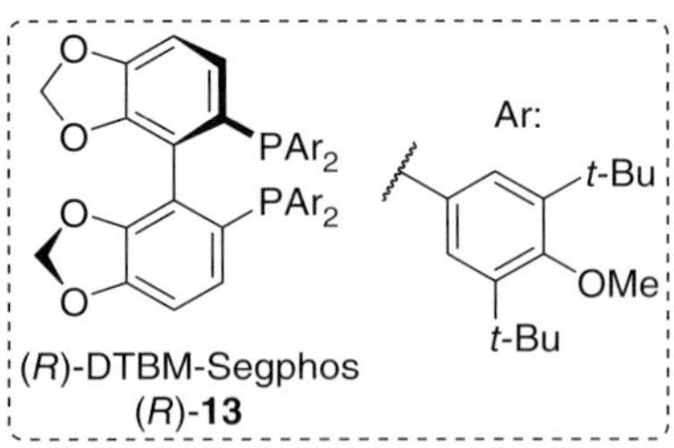

Scheme 4.4 Copper-catalyzed asymmetric conjugate reduction of isophorone (**9**).

More recently, it was demonstrated that copper(I)/NHC (NHC = *N*-heterocyclic carbene) complexes could be employed in the hydrogenation of the less reactive enoates such as **14** (Scheme 4.5) [31]. Under relatively low H_2 pressure of 10 bar, and employing catalytic amounts of the common Cu/NHC complex [SIMesCuCl] **(16)** [32], the corresponding saturated esters **15** with various substitution patterns (alkyl, aryl, heteroaryl) could be obtained with excellent yields. No reduction of the ester moieties was observed. Furthermore, in the same report it was demonstrated that chiral NHC ligands led to stereoinduction of the products **15**, although the *ee* values reached at this point were not satisfactory.

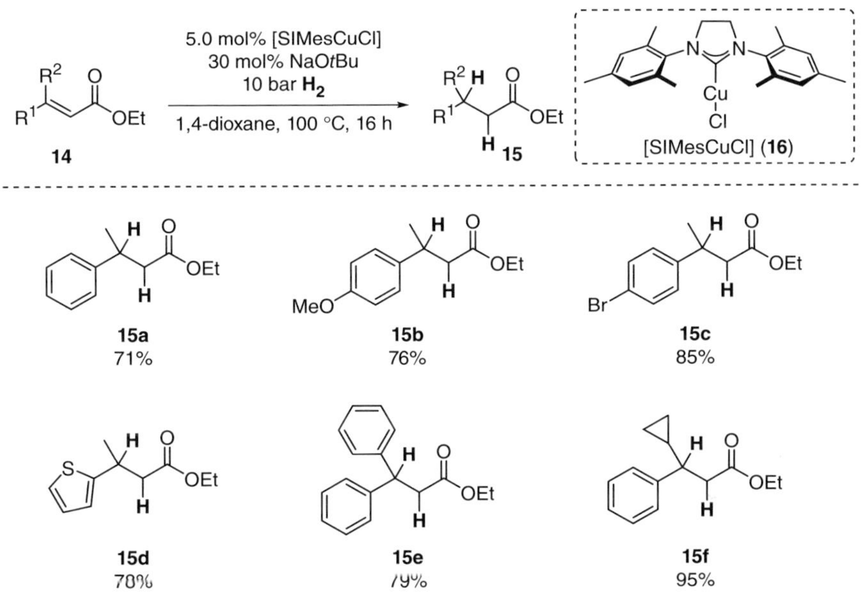

Scheme 4.5 Copper-catalyzed hydrogenation of enoates with Cu/NHC complexes.

4.2.2 1,2-Hydrogenation of α,β-Unsaturated Ketones and Aldehydes

One decade after the hydride-mediated conjugate reduction of enones with $[Ph_3PCuH]_6$ it was shown that using Me_2PhP as additional ligand and *tert*-butanol as additive, the chemoselectivity of the enone hydrogenation could be altered from the 1,4-pathway to the selective 1,2-reduction of the carbonyl moiety [33]. Thus, α,β-unsaturated ketones and aldehydes **17** were selectively reduced to give the synthetically valuable allylic alcohols **18** at lower H_2 pressure (reaching pressures as low as 5 bar, Scheme 4.6). As before, simple isolated carbon–carbon double bonds were not hydrogenated. This protocol could also be used for the hydrogenation of simple (non-conjugated) ketones under mild reaction conditions (1 bar H_2, RT) [33, 34]. Following up on these results, a chemoselective 1,2-hydrogenation of enals and simple aldehydes by employing $[Cu(NO_3)(PPh_3)_2]$ and additional 1,4-bis(diphenylphosphino)butane (dppb) as ligand was also reported [35].

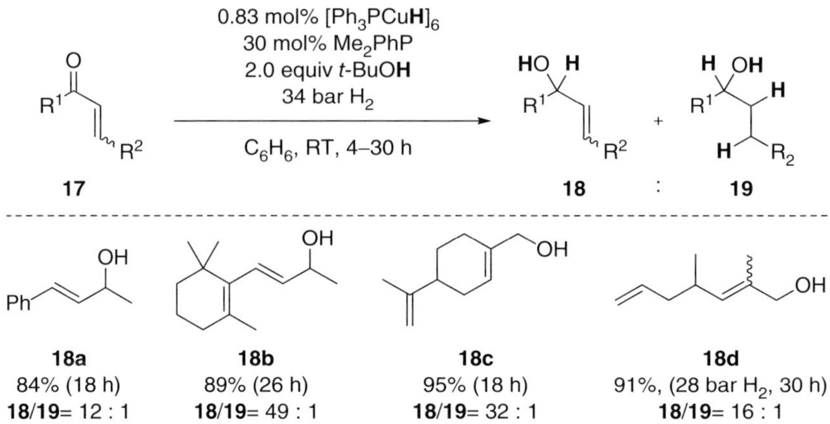

Scheme 4.6 Copper-catalyzed 1,2-hydrogenation of α,β-unsaturated ketones and aldehydes.

4.2.3 Asymmetric 1,2-Hydrogenation of Simple (Nonconjugated) Ketones and Aldehydes

As outlined in previous Section 4.2.2, the selectivity of the hydride transfer from copper hydride complexes to conjugated carbonyl and/or carboxyl compounds can be altered to the reduction of the C=O double bond instead of the C=C double bond by choice of conditions and/or ligands. Several groups have reported on the copper-catalyzed asymmetric hydrogenation of acetophenone derivatives. The first copper-catalyzed enantioselective hydrogenation of aryl ketones **20** was presented in 2007 [36] and was followed up by similar reports [37, 38] (Scheme 4.7). In all cases, an alkoxide base was employed for catalyst activation and *iso*-propanol was used as solvent (10–30 °C, 20–50 bar H_2). In the earliest report, commercially available diphosphine (*S*,*S*)-**22** [(*S*,*S*)-BDPP] (BDPP = 2,4-Bis(diphenylphosphino)pentane) led to high enantioselectivities for *ortho*-substituted aryl ketones (**20**, up to *e.r.* 95.5 : 4.5), whereas *meta*- and *para*-substituted aryl ketones gave lower enantioselectivities (up to *e.r.* 82 : 18) [36]. Addition of P(3,5-xylyl)$_3$ ensured high catalytic activity. This protocol was later extended to heteroaromatic ketones [30].

In a subsequent study, a catalyst consisting of an arguably better available copper(II) precursor [Cu(OAc)$_2$] and binaphthophosphepine (*S*)-**23** was shown to catalyze the hydrogenation of ketones at temperatures as low as 10 °C and low catalyst loading [37]. Overall, excellent yields were obtained including heteroaromatic ketones as substrates. Nevertheless, no breakthrough in terms of enantioselectivity was achieved (*e.r.* 74 : 26–94.5 : 5.5, *e.r.* 56 : 44 for an alkyl, alkyl ketone). In contrast to the earlier report, this catalyst does not require the addition of extra phosphine ligands.

Finally, a high-throughput screening with a variety of commercially available chiral diphosphine ligands was applied in order to improve enantioselectivity [38]. Here, structurally modified BoPhoz ligand (*R*,*S*)-**24** resulted in high enantiomeric ratios at a dihydrogen pressure of 20 bar at 15 °C. Again, an additional phosphine ligand was needed for high catalytic activity. A variety of

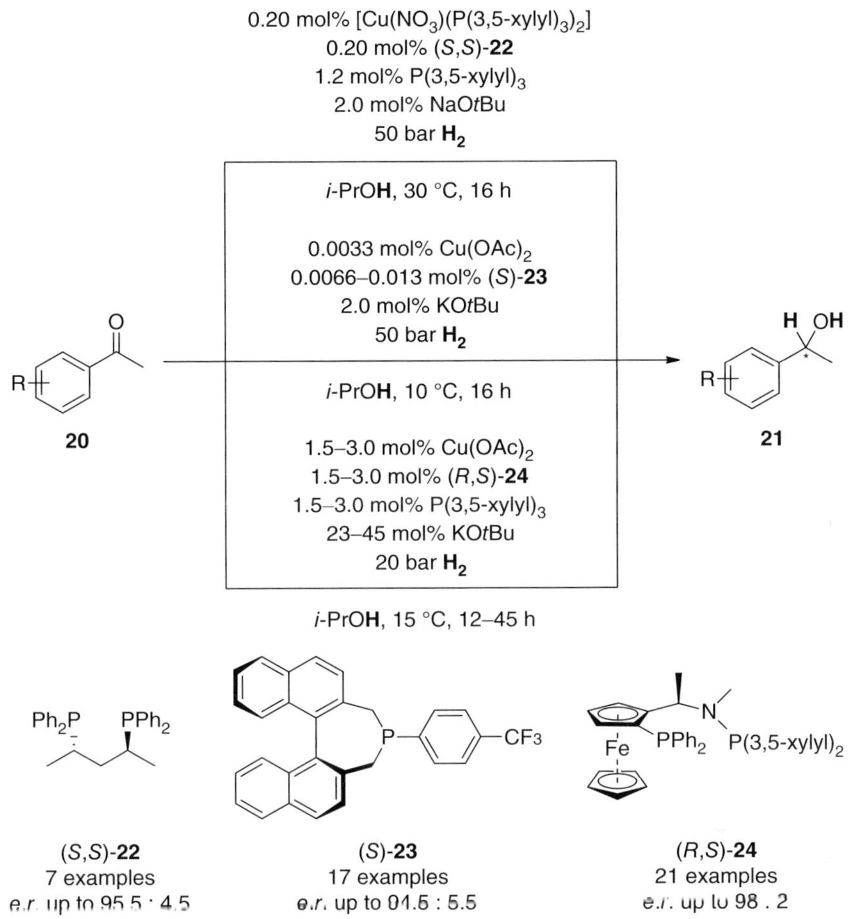

Scheme 4.7 Asymmetric copper-catalyzed hydrogenation of ketones.

aryl and heteroaryl ketones were transformed with high enantioselectivity (*e.r.* 83 : 17–98 : 2) and also an alkyl, alkyl ketone could be hydrogenated with an *e.r.* of 75 : 25.

4.3 CO_2 Reduction to Formate

The reduction of CO_2 to formate, a possible application for the valorization of carbon dioxide, by copper salts in the presence of strong organic bases was achieved in 2015 [39]. A variety of copper(I) and copper(II) salts and organic bases were investigated and under forcing conditions (60 bar H_2/CO_2 1 : 1, 100 °C) employing $Cu(OAc)_2{\cdot}H_2O$ as catalyst and 1,8-Diazabicyclo[5.4.0]undec-7-ene(25, DBU) as a trapping reagent for the formate generated in the reduction, ammonium formate **26** was isolated in 54% yield with a turnover number (TON) of 60 (Scheme 4.8). Remarkably, additional ligands (NHCs, phosphines) led to lower

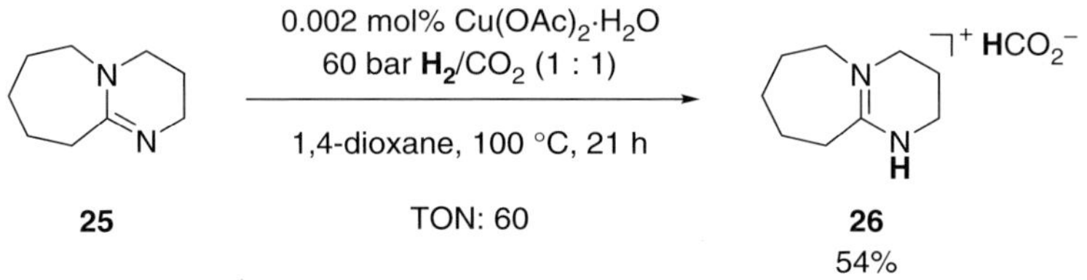

Scheme 4.8 Copper-catalyzed CO_2 reduction to formate with DBU as trapping reagent.

TON. Although the true nature of the catalyst remained unclear, a heterolytic H_2 activation and formation of a copper hydride intermediate was suggested and the crucial role of DBU was underlined by isolation of catalytically active $[CuI(dbu)_2]$. Although commonly activated by the formation of a Cu—O bond for H_2 activation, this system presumably comprises catalytically active copper(I) complexes bearing a Cu—N bond, which is rarely observed [40]. In fact, this approach to H_2 activation shows some resemblance to the earliest report of copper(I)-catalyzed hydrogenation [12, 13].

Shortly afterwards, a related study was presented which employed the well-defined triphos/copper complex **27** (Scheme 4.9) [41]. With DBU (**25**) as base, higher activity (TON: 130) at milder reaction conditions (40 bar H_2/CO_2 1 : 1, 83 °C in MeCN) was achieved; and even at pressures as low as 1.8 bar H_2/CO_2 (1 : 1) at 60 °C, some catalytic activity was observed (TON: 20). It was shown that there was no correlation between the pK_a value of the base and the

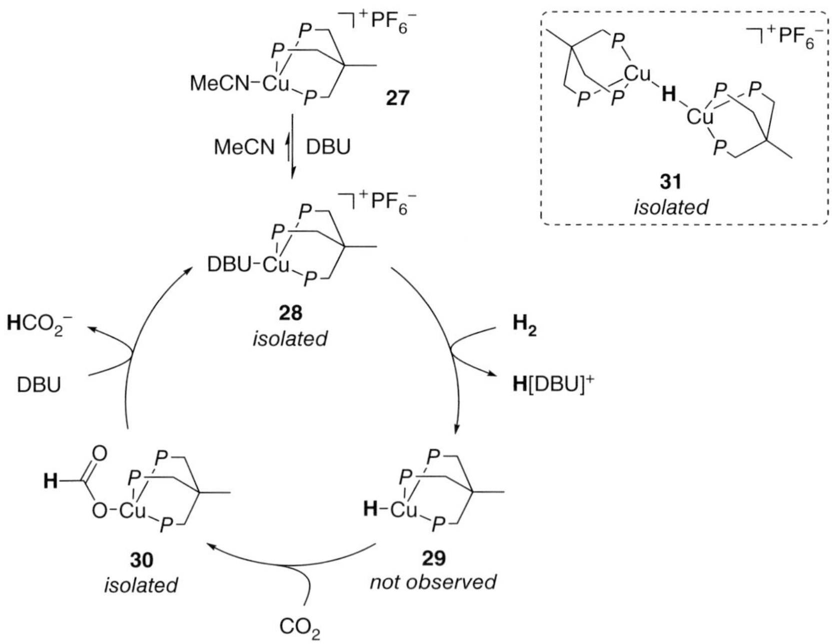

Scheme 4.9 Simplified catalytic cycle for CO_2 reduction via copper hydride intermediates. $P = Ph_3P$.

catalytic activity. Instead, the reaction rates correlated with the coordinating ability of the base to the copper atom. Thus, it was concluded that the crucial step for catalytic turnover was coordination of the base to copper. A mechanistic study later identified several catalytically active intermediates and a catalytic cycle was proposed on the basis of stoichiometric experiments (Scheme 4.9) [42]. At first, ligand exchange of acetonitrile with DBU leads to the isolable cationic triphos/copper(I) complex **28**. Heterolytic dihydrogen activation by **28** yields the proposed triphos/copper(I) hydride intermediate **29**, which in turn reduces CO_2 to give the isolated copper(I) formate complex **30**. Ligand exchange with DBU releases formate and regenerates **28**. The isolated cationic dicopper(I) complex with bridging hydride **31** showed low catalytic activity in stoichiometric reactions with CO_2. It was proposed that DBU coordinates to **31** in order to form the copper hydride monomer triphos/CuH (**29**), which serves as the actual reductant for CO_2. It remains unclear whether complex **31** is part of the catalytic cycle or represents an off-cycle resting state of the catalyst.

4.4 Allylic Substitutions with a Hydride Nucleophile Generated from H_2

Catalytic hydrogenations are characterized by the fact that two hydrogen atoms are transferred to an unsaturated substrate, independent of the actual H_2 activation mechanism [43]. The heterolytic cleavage of the H—H bond with copper complexes bearing a Cu—O bond generates a copper hydride which is nucleophilic, next to an alcohol as the proton equivalent. If the reactivity of the proton and the hydride generated in this manner could be decoupled, a catalytic and atom-economic generation of nucleophilic hydrides could be rendered feasible. To the best of our knowledge, such a reactivity is unprecedented in a homo geneous catalysis setting. At the same time, the catalytic generation of nucleophilic hydrides from H_2 could potentially open up a route to the replacement of commonly used complex hydrides (such as borohydrides, aluminum hydrides, or hydrosilanes, to name a few), which are associated with the stoichiometric generation of waste products.

An allylic reduction, the catalytic transfer of a hydride nucleophile to an allylic acceptor **32** (Scheme 4.10), was chosen as a model reaction for such a process. This reaction would generate terminal alkenes **33** as products, which again underscores the need for chemoselective catalysts, as many transition metal catalysts would simply hydrogenate the terminal double bond in **33** [43].

An allylic reduction with H_2 was achieved by employing the air-stable and well-defined NHC/copper(I) complex **36**, in combination with NaO*t*Bu as activator (Scheme 4.11) [44]. In analogy with earlier studies employing hydrosilanes as stoichiometric hydride sources [45], the regioselectivity (S_N2 vs. S_N2', α vs. γ substitution) was highly dependent on the utilized NHC/copper(I) complex. The double bond geometry of allylic acceptors **34** had a great impact on the reaction outcome: While *E*-allylic chlorides led to almost complete loss of regioselectivity ($\gamma/\alpha \sim 2:1$), the corresponding *Z*-allylic chlorides gave the terminal alkenes

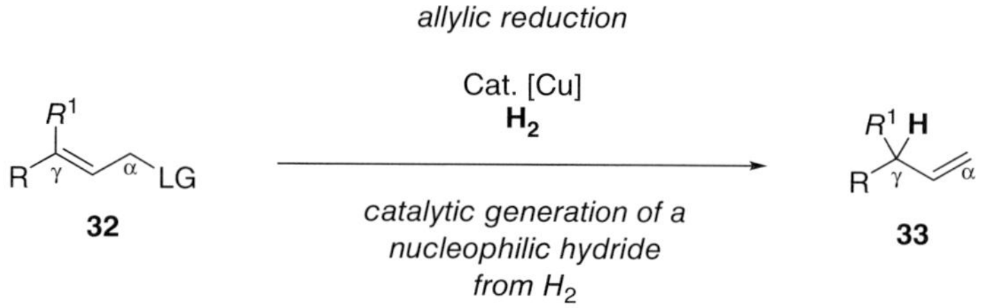

Scheme 4.10 Allylic reduction with hydrides catalytically generated from H_2.

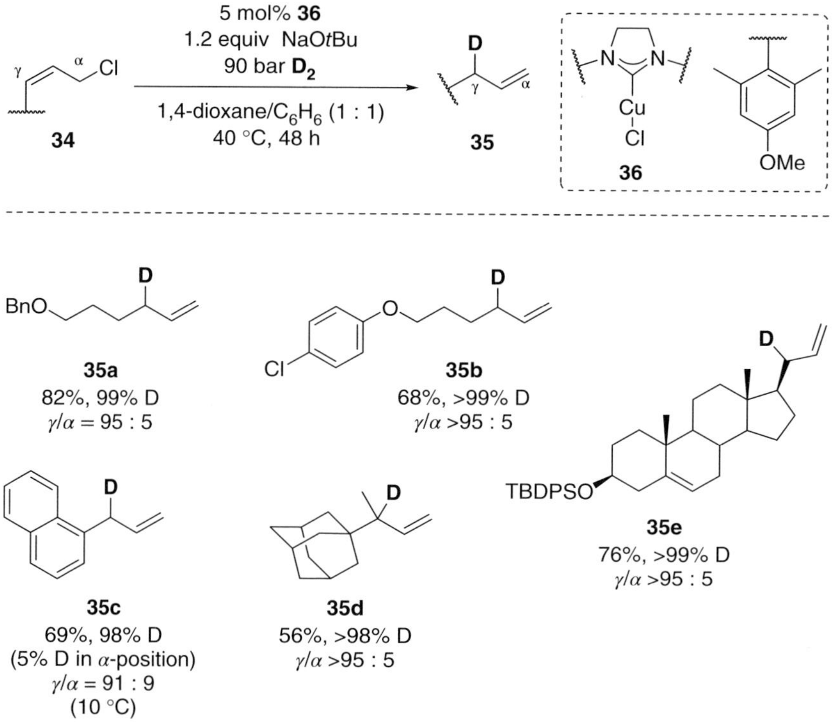

Scheme 4.11 Copper-catalyzed deuteride transfer from D_2.

with high γ selectivity ($\gamma/\alpha > 20:1$). To probe the mechanistic hypothesis and to generate value-added products, a selective monodeuteration from D_2 was carried out. In this manner, monodeuterated compounds **35** could be obtained with very high deuterium incorporations exceeding 95%. Such a process could gain importance with foresight to applications where selective isotope labeling is of prime interest (such as elucidation of (bio)organic reaction mechanisms or analytics of drug metabolites). In this manner, the selective synthesis of γ-deuterated terminal alkenes **35** by employment of D_2 (90 bar) as terminal reductant at slightly elevated temperatures was achieved (40 °C, Scheme 4.11). While aryl-substituted *Z*-allylic chlorides (**34c**) led to lower regioselectivities ($\gamma/\alpha = 91:9$, 10 °C), a variety of alkyl-substituted *Z*-allylic chlorides could be transformed into the corresponding terminal alkenes **35** selectively ($\gamma/\alpha > 20:1$). Next to the remarkable chemoselectivity of the catalysts leaving the terminal C=C double bond intact, a benzyl ether, as in **35a**, was also not hydrogenolytically cleaved, again underscoring the fact that the copper catalyst does indeed activate H_2, leading to nucleophilic copper hydrides, but does not behave like a hydrogenation catalyst.

To showcase this difference in reactivity, a copper(I)-catalyzed alkyne semihydrogenation (see Section 4.5) with tethered copper(I) complexes and the described copper(I)-catalyzed allylic hydride transfer were combined to a new deuteration strategy: At first, propargylic silyl ether **37** was deuterated by an alkyne semihydrogenation utilizing a copper complex derived from NHC precursor **39** in combination with D_2. The obtained product, an allylic silyl ether (not shown), could subsequently be deprotected and chlorinated in order to give dideuterated *Z*-allylic chloride **38**-d_2 (Scheme 4.12, top). The

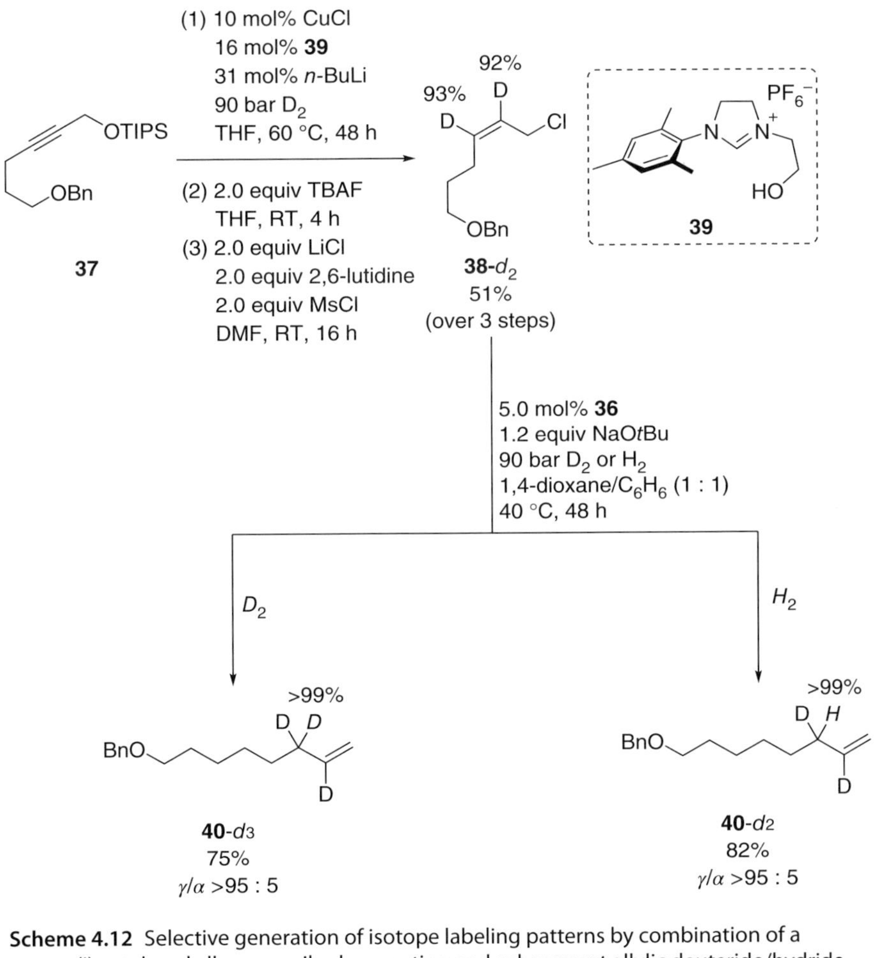

Scheme 4.12 Selective generation of isotope labeling patterns by combination of a copper(I)-catalyzed alkyne semihydrogenation and subsequent allylic deuteride/hydride transfer.

latter one is, in turn, the substrate for the following allylic reduction with either D_2 or H_2 and gave rise to terminal alkenes **40**-d_3 or **40**-d_2, respectively (Scheme 4.12, bottom). In both cases, excellent regioselectivity (γ/α >95 : 5) as well as deuterium/hydrogen incorporations were observed (>99%).

4.5 *Z*-Selective Alkyne Semihydrogenation

The copper(I)-catalyzed alkyne semihydrogenation has served as the model reaction for the development of copper complexes with the ability to activate H_2. In 2015, the first H_2-mediated *formal* alkyne semihydrogenation was reported [46]. A catalyst comprising tetrameric phosphine-stabilized copper(I) chloride complex, LiO*t*Bu (0.5 equiv) as activating agent, *i*-PrOH (1 equiv) as protic additive, and a H_2 pressure of 5 bar in toluene at 100 °C was able to transform internal alkynes bearing diaryl, dialkyl, as well as aryl, alkyl substituents (Scheme 4.13). The corresponding internal *Z*-configured alkenes can be obtained in high yields and excellent stereoselectivities (>97 : 3) with negligible amounts of overreduction to the corresponding alkanes (<2%). The variety of functional groups tolerated by this process was limited to halogenated compounds as well as an ester moiety. Ester **42a** was obtained in moderate yield (52%) and excellent stereoselectivity (99 : 1). *Z*-acrylate **42b** was not found when the corresponding propiolate was deployed; instead, a complex mixture was obtained. When a dialkyl-substituted 1,3-diyne (not shown) was used, **42c** was isolated in good yield and excellent stereoselectivity (78%, *Z/E* > 95 : 5). A terminal alkyne and a trimethylsilyl (TMS) protected acetylene gave no conversion. Deuteration

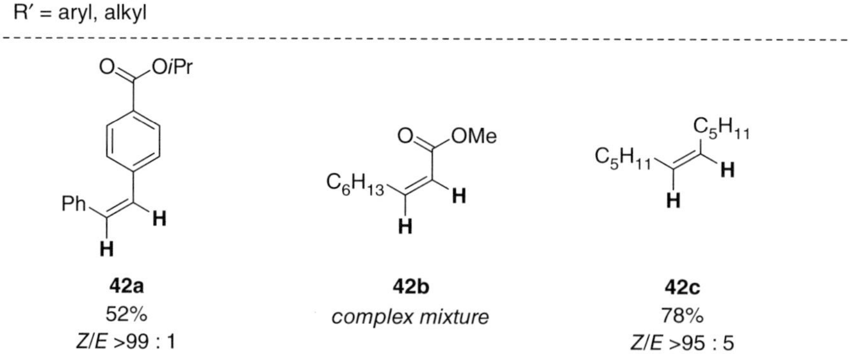

Scheme 4.13 *Formal* alkyne semihydrogenation using a phosphine/copper(I) complex.

studies showed that in this case, H_2 served as the hydride source, i.e. an intermediate copper hydride complex. *iso*-Propanol as protic additive delivered the H atom most likely through a protodecupration, resulting in an overall *formal* semihydrogenation, as the two H atoms of the alkene products **42** stem from different reagents.

Shortly after, a related study was published utilizing a tethered NHC/copper(I) alkoxide complex [47], which allowed for the transfer of both H atoms from H_2 onto internal alkynes (as displayed by deuteration experiments, Scheme 4.14). This protocol required high H_2 pressure (100 bar) and slightly elevated temperature (40 °C) in tetrahydrofuran (THF). The catalyst comprises mesitylcopper(I) and imidazolinium salt **39**. An *in situ* activation – to generate the active catalyst – requires the addition of *n*BuLi to form a Cu—O bond which is essential for reactivity. The so-prepared catalyst displayed excellent stereoselectivity in all cases ($Z/E > 95 : 5$) while giving very little overreduction to the corresponding alkane (<5%). A variety of functional groups were tolerated, including ester, nitrile, acetal, ether, ketone, as well as heteroaromatics (thiophene, pyridine). With this protocol, diaryl, dialkyl, as well as aryl, alkyl alkynes can be converted into the corresponding *Z*-alkenes in high yields. As representative examples, ketone **44a** and nitrile **44b** were obtained in good yields (84% and 70%) and

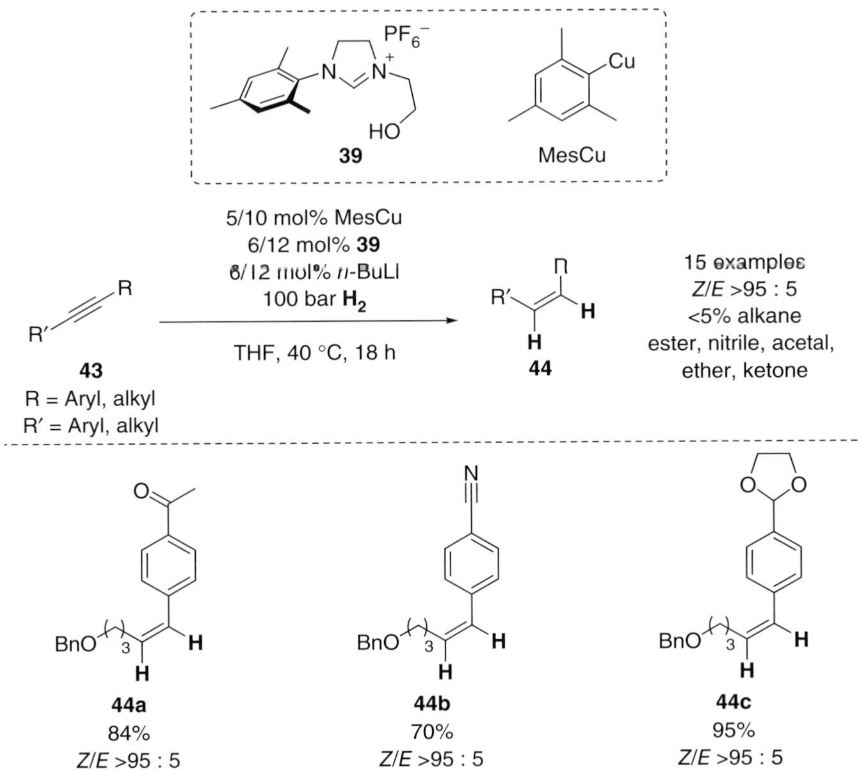

Scheme 4.14 *Z*-selective alkyne semihydrogenation using a tethered NHC/copper(I) alkoxide complex.

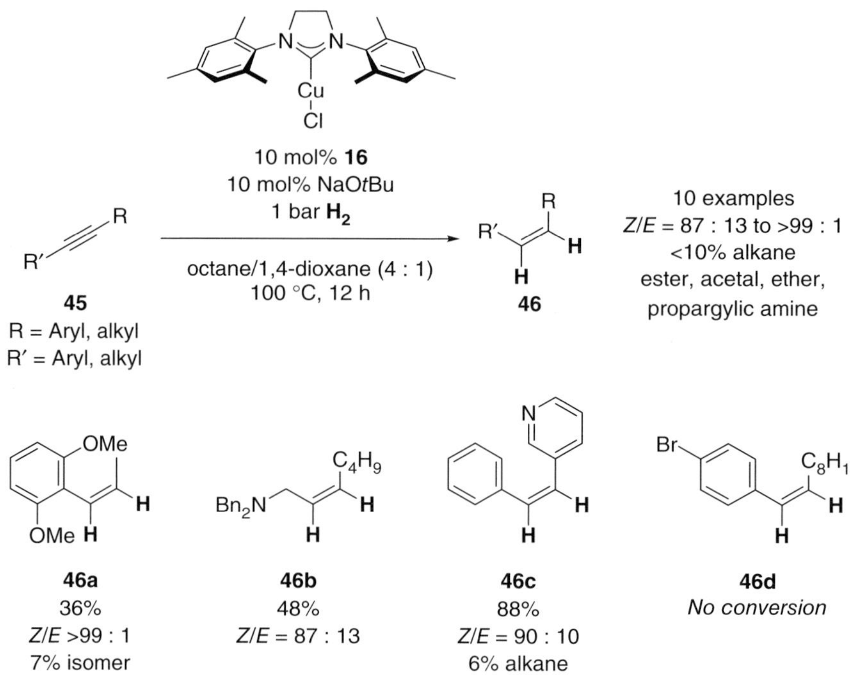

Scheme 4.15 NHC/copper(I)-catalyzed alkyne semihydrogenation at atmospheric H_2 pressure.

stereoselectivity ($Z/E > 95 : 5$). In neither case, 1,2-reduction products were observed even after prolonged reaction time (65 hours), showcasing the high chemoselectivity of this protocol. An aldehyde containing internal alkyne was not tolerated, whereas the corresponding acetal reacted smoothly to give **44c** in high yield (95%) and excellent stereoselectivity ($Z/E > 95 : 5$).

Following up on the two previously mentioned reports, a copper(I)-catalyzed alkyne semihydrogenation operative at atmospheric H_2 pressure was reported, in which a copper(I)/NHC complex was again employed (Scheme 4.15). This protocol omits the need for high-pressure equipment [48]. With [SIMesCuCl] (**16**) (10 mol%) and NaO*t*Bu (10 mol%) to generate the active catalyst and in a mixture of *n*-octane/1,4-dioxane (4 : 1) at high temperature (100 °C), the corresponding *Z*-alkenes **46** were obtained in good to excellent stereoselectivities ($Z/E = 87 : 13$ to $99 : 1$) and were isolated with up to 10% of over-reduced alkane. Therefore, while these catalysts display activity at much lower pressure than the previously mentioned ones, lower chemo- and stereoselectivities are obtained. This protocol tolerates esters, acetals, ethers, and propargylic amines. Diaryl-, dialkyl- as well as aryl, alkyl-substituted internal alkynes proved to be suitable substrates, albeit with varying stereoselectivity and alkane formation. Styrene derivative **46a** was obtained in poor yield (36%) with an excellent stereoselectivity ($Z/E > 99 : 1$). In this case, an isomerization of the double bond along the chain was observed to give 2-allyl-1,3-dimethoxybenze (7%, not shown) as side-product. The displacement of the double bond is unexpected for such copper(I)/NHC complexes, which generally show no reactivity toward alkenes. *Z*-allyl amine **46b** was obtained in moderate yield (48%) and a diminished stereoselectivity (87 : 13). Nevertheless, this example represents the first conversion of a propargylic amine in a copper-catalyzed hydrogenation reaction. Pyridine **46c** was obtained in good yield (88%) with a good stereoselectivity (*Z/E* = 90 : 10) favoring the *Z*-alkene. Bromo-substituted styrene **46d** gave no conversion under the reaction conditions.

For all abovementioned catalysts for alkyne semihydrogenation, an activation step with an alkoxide base (directly or by generation of an alkoxide within the ligand framework) is required to afford the crucial Cu—O-bond for H_2 activation [46–48]. In 2010, an air- and moisture-stable copper(I) hydroxide/NHC complex, [IPrCuOH] (**48**), which has previously been described in the literature [50], was identified as to be directly reactive with H_2 to presumably generate copper(I) hydride intermediates. Therefore, this complex omits the need for an activation step prior to catalysis. [IPrCuOH] (**48**) was shown to be applicable in the *Z*-selective alkyne semihydrogenation with H_2, representing a drastic simplification of the catalytic protocol (Scheme 4.16) [49]. Under reaction conditions mostly similar to the catalysts employed earlier (80 bar H_2 and 40 °C, compare Scheme 4.14), internal alkynes were transformed into the corresponding *Z*-alkenes in good yields and excellent stereoselectivity (*Z/E* > 95 : 5). This was accompanied by negligible amounts of overreduction to the alkane in all cases (Scheme 4.16). Aryl, alkyl-substituted internal alkynes bearing chloride, bromide, trifluoromethyl, methoxy, and methyl ester moieties were tolerated. Thiophene-substituted internal alkene **49a**, which could not be hydrogenated with the earlier system (compare Scheme 4.13), was obtained in high yield (94%) and excellent stereoselectivity (*Z/E* > 95 : 5) at 100 bar H_2

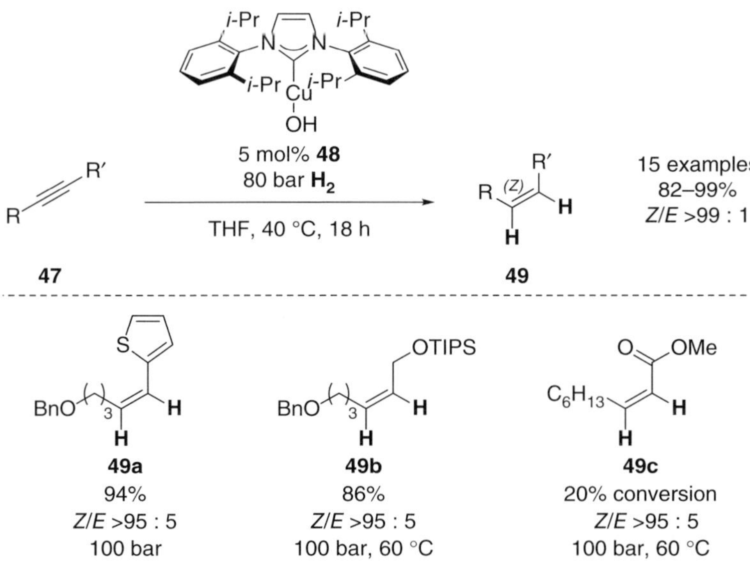

Scheme 4.16 Alkyne semihydrogenation using [IPrCuOH] (**48**).

pressure. A TIPS-protected propargylic alcohol reacted smoothly at 100 bar H_2 pressure and 60 °C to give the synthetically valuable *Z*-allyl silylether **49b** in good yield (86%) and excellent stereoselectivity. When subjecting a propiolate to the reaction conditions, low conversion (20%) was observed giving the corresponding *Z*-acrylate **49c** in excellent stereoselectivity (*Z/E* > 95 : 5) and as sole product.

Later, the application of [IPrCuOH] (**48**) was expanded to the hydrogenation of 1,3-diynes and 1,3-enynes (Scheme 4.17) [51]. Remarkably, the stereoselectivity of the diyne hydrogenation depended, to a high degree, on the nature of the substituents: *E,E*-1,3-dienes **51** were obtained in excellent stereoselectivity (*E,E/Z,Z* > 95 : 5) from diaryl-substituted 1,3-diynes **50**, whereas the corresponding dialkyl-substituted 1,3-diynes **52** gave rise to *Z,Z*-1,3-dienes **53** in excellent stereoselectivities (*Z,Z/E,E* = 95 : 5) under otherwise identical reaction conditions. Especially the formation of the *E,E*-diene is somewhat unexpected, as all of the previously mentioned copper(I)-based protocols have a strong preference for the formation of the *Z*-configured alkenes.

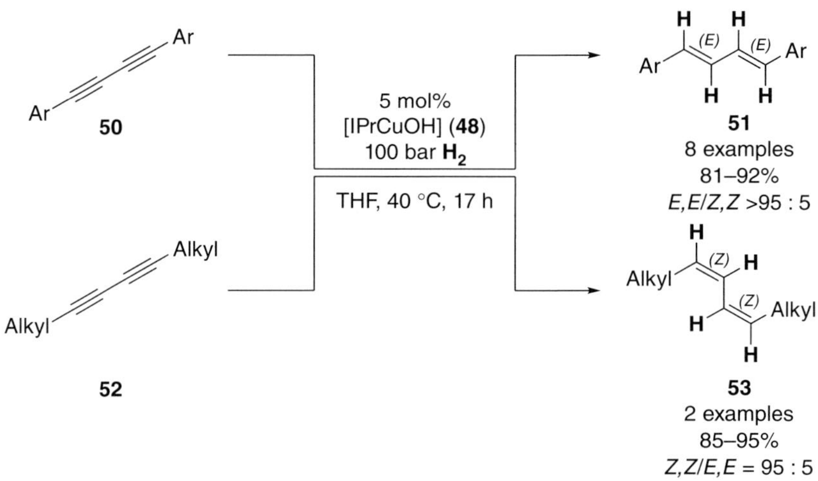

Scheme 4.17 Stereoselective hydrogenation of 1,3-diynes with [IPrCuOH] (**48**), influence of the substituents on stereoselectivity.

To shed light on this unusual selectivity, 1,3-enynes, hypothesized intermediates halfway between the diynes and dienes, were also investigated and found to be viable substrates for the copper(I)-catalyzed hydrogenation protocol with [IPrCuOH] (**48**, Scheme 4.18). Under otherwise identical reaction conditions, 1,3-dienes were obtained from 1,3-enynes. In these cases, again, the configuration of the enynes played a major role in the stereochemical outcome of the overall hydrogenation: When using *Z*-**54** at 80 bar H_2-pressure and 40 °C for 24 h, *E,E*-**55** was obtained in good yield (88%) and with excellent stereoselectivity (*E,E/Z,E* > 95 : 5). Therefore, during the alkyne hydrogenation process, a concomitant *Z* to *E* isomerization of the alkene moiety of *Z*-**54** must have taken

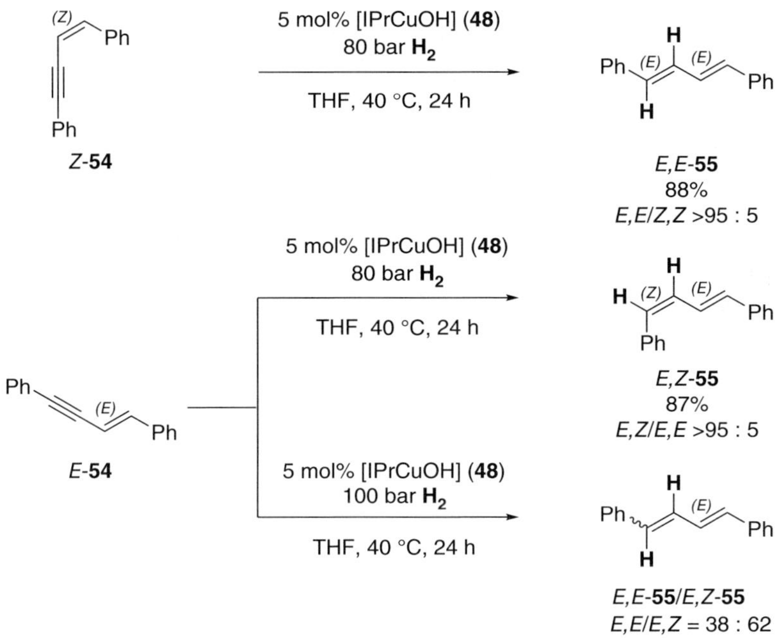

Scheme 4.18 Influence of the double bond geometry on the hydrogenation of enynes.

place. On the other hand, *E*-**54** gave the opposite stereoisomer of the 1,3-diene, namely, *E,Z*-**55** in 87% yield and an excellent stereoselectivity (*E,Z/E,E* > 95 : 5). However, when the H_2 pressure was increased to 100 bar, *E*-**54** led to a mixture of *E,E*/*E,Z*-**55** with a ratio of 38 : 62. These results show that both *Z*- and *E*-enynes are indeed possible reaction intermediates. However, from these data it can be concluded that the *E*-enyne reacted in the manner of a single alkyne, leading to the *Z*-alkene moiety (as part of the 1,3-diene). On the other hand, the *Z*-enyne led to isomerization, supporting the notion that this stereoisomer behaved like a common entity, and not like the combination of an isolated alkene and an alkyne.

This marked and somewhat surprising difference in selectivity of the obtained 1,3-dienes, in combination with further deuteration studies, led to the proposal for a catalytic cycle, in which the 1,3-diynes are first converted to a *Z*-1,3-enyne, which itself is transformed to the corresponding 1,3-dienes (Scheme 4.19). The *Z*-enyne *Z*-**57**, generated after the first *Z*-selective semihydrogenation of 1,3-diyne **56**, undergoes a hydrocupration by a copper(I) hydride (**A**), generated through heterolytic H_2-cleavage, to give vinyl copper intermediate *E*-**B**. An isomerization was suggested to give rise to the allene intermediate **C** with loss of stereoinformation. In a second isomerization process, vinyl copper intermediate *Z*-**B** could be reformed with the thermodynamically more stable *E,E*-configuration, which after protodemetalation liberates *E,E*-**58** and regenerates the catalyst. However, an aryl substituent seems to be crucial for this isomerization (B to C) to occur, since alkyl-substituted 1,3-diynes do not

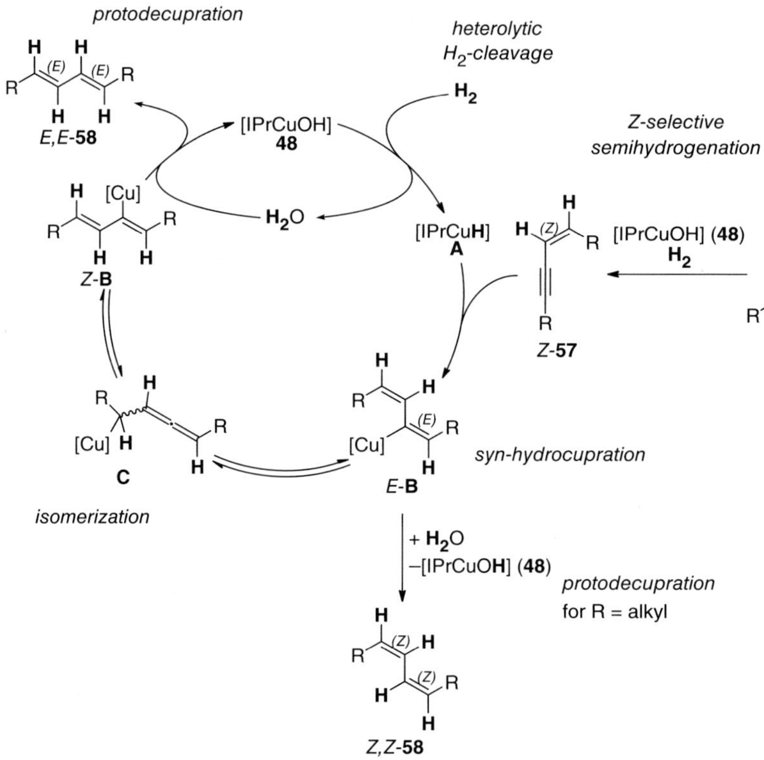

Scheme 4.19 Suggested mechanism for the hydrogenation of 1,3-diynes.

undergo this isomerization and instead directly liberate *Z,Z*-**58** most probably through protodemetalation of vinyl copper intermediate *E*-**B**.

4.6 Alkyne Transfer Semihydrogenation and Transfer Conjugate Reduction with Ammonia Borane

As a general feature, many copper(I)-catalyzed hydrogenation processes require the use of elevated pressure, and therefore the need for high-pressure equipment, rendering the overall protocols less practical. To circumvent this need, copper(I)-catalyzed *transfer* hydrogenations have also been developed. It was found that under reaction conditions almost identical to that of the alkyne semihydrogenation employing [IPrCuOH] (**48**) as catalyst, similar results can be obtained when ammonia borane ($H_3N{\cdot}BH_3$) was used as the formal hydrogen source (Scheme 4.20) [52]. In order to circumvent the catalytic generation of H_2 from $H_3N{\cdot}BH_3$, the so-called hydrodecoupling, by [IPrCuOH] (**48**), ammonia borane had to be added slowly. The transfer hydrogenation of internal alkynes **59** gave the corresponding *Z*-alkenes **60** in moderate to good yields and with excellent stereoselectivities ($Z/E \geq 95:5$). With this protocol, a variety of functional groups including ethers, halogens, esters, nitriles, and propargyl silyl

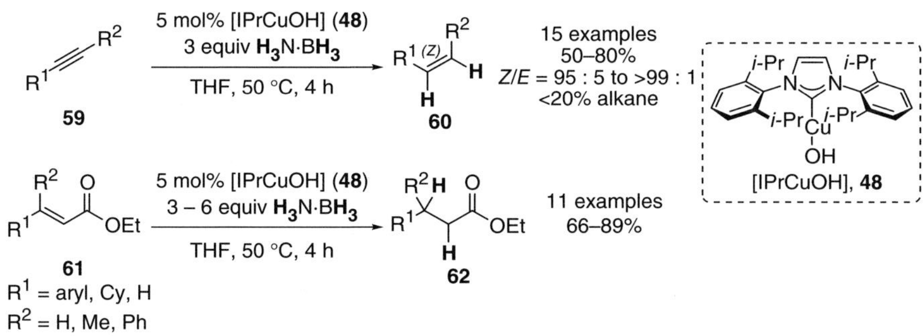

Scheme 4.20 Cu-catalyzed alkyne transfer semihydrogenation and conjugate transfer hydrogenation of α,β-unsaturated esters using ammonia borane.

ethers were tolerated. In most cases, overreduction to the alkane occurs only in negligible amounts. However, for trifluoromethyl, ester- and nitrile-substituted internal alkynes undesired alkane was observed in up to 20%.

In a similar manner, the conjugate transfer reduction of di- and trisubstituted *E*-α,β-unsaturated esters **61** could be effected giving esters **62**, although an increased amount of ammonia borane (6 equiv) was used for the latter. Employing this protocol, α,β-unsaturated esters **61** containing ethers, thiophene, halogens, and trifluoromethyl groups were tolerated. These results show that for the hallmark reaction of copper hydride complexes, namely, the conjugate reduction of α,β-unsaturated carbonyl and carboxyl compounds, transfer hydrogenation conditions can also be employed.

4.7 Dihydrogen-Mediated Cross-Coupling of Internal Alkynes and Aryl iodides

Recently, a hydrogenative cross-coupling based on a copper(I) catalyst in combination with H_2 was published [53]. Hydrogenative coupling reactions have so far only been realized with precious metal catalysts and are highly desirable transformations to effect C—C bond formation [54]. The key feature of these processes is the circumvention of stoichiometric organometallic reagents commonly used for coupling reactions. The hydrogenative cross-coupling of internal alkynes **63** with aryl iodides **64** utilizes an NHC/copper(I) complex **65** activated by LiO*t*Bu and a phosphine/Pd complex in a cooperative catalysis approach (Scheme 4.21). At 5 bar H_2 pressure in toluene at 100 °C, trisubstituted alkenes **66** were obtained in moderate to good yields (38–73%) and generally good stereoselectivity (*E*/*Z* = 86 : 14 to 92 : 8). For example, aniline **66a** was obtained in 70% yield and good stereoselectivity (*E*/*Z* = 92 : 8). Stilbene **66b** was obtained with similar results (66%, *E*/*Z* = 90 : 10), whereas chloride **66c** was obtained in moderate yield (48%) and slightly diminished *E*/*Z* ratio (86 : 14) at 3 bar H_2 pressure and 120 °C. The reported procedure was mainly limited to symmetric internal alkynes since an unsymmetrically substituted internal aryl, alkyl-alkyne resulted in 59 : 41 mixture of the regioisomers.

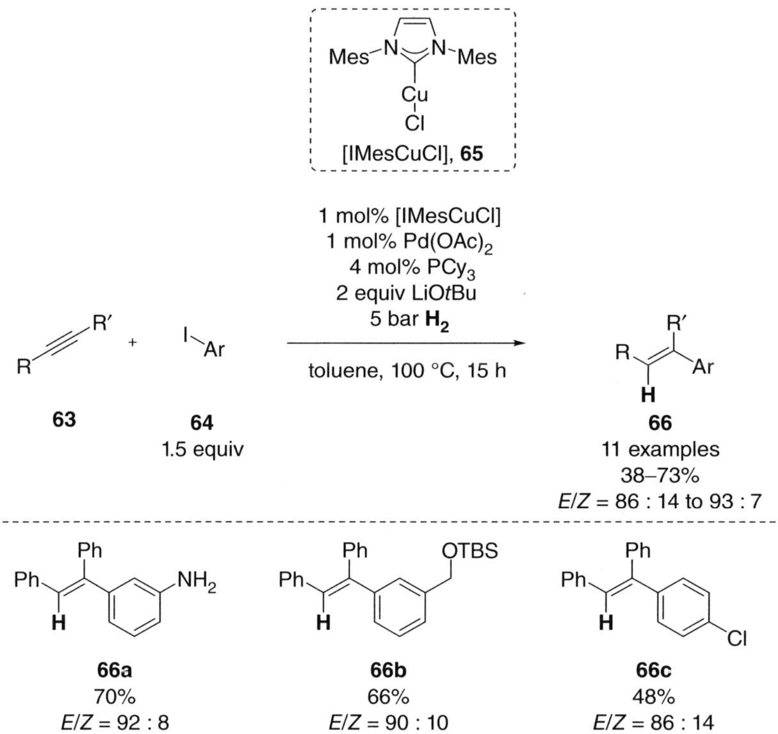

Scheme 4.21 Hydrogenative cross-coupling of internal alkynes and aryl iodides by Pd/Cu cooperative catalysis.

4.8 Conclusions and Perspectives

Even though copper complexes were among the first to be studied for homogeneous hydrogenation in the historical perspective, generally they can be considered underexplored. Even though the past decade has seen substantial progress, the use of dihydrogen to access copper hydride intermediates clearly lags behind when compared to the well-established protocols relying on hydrosilanes as terminal reducing agents. However, a variety of functional groups have been found to be reactive in hydrogenation reactions with copper catalysts based on phosphine or NHC ligands. Clearly, a major drawback at the moment is the high H_2 pressure required for most of the reactions. However, it was shown that asymmetric processes are, in principle, viable, which outlines that copper catalysts can be used for these transformations in the future as well. Furthermore, first examples of transfer hydrogenation protocols to circumvent the need for high-pressure equipment have also been developed.

One clear advantage of copper-based catalysts for hydrogenation reactions is the possibility of carrying out highly chemoselective transformations. For example, C=O double bonds or allylic chlorides can be reduced in the presence of simple C=C double bonds. From both a synthetic as well as an atom-economic perspective, the catalytic generation of nucleophilic copper hydrides from H_2 without carrying out a hydrogenation reaction of an alkene is a prime example for the future use of copper catalysts for homogeneous hydrogenation reactions. Concluding, one could state that the major groundwork has been achieved, with future research projects being able to draw upon the trends worked out in the recent studies, which paves the way for more development in the area of homogeneous hydrogenations with copper catalysts.

References

1 A.J. Jordan, G. Lalic, J.P. Sadighi, *Chem. Rev.* 2016, 116, 8318–8372.
2 C. Deutsch, N. Krause, *Chem. Rev.* 2008, 108, 2916–2927.
3 S. Rendler, M. Oestreich, *Angew. Chem. Int. Ed.* 2007, 46, 498–504.
4 B.H. Lipshutz, Copper(I)-mediated 1,2- and 1,4-reductions. In: *Modern Organocopper Chemistry* (ed. N. Krause) Weinheim: Wiley-VCH, 2002, 167–187.
5 (a) For the more recently studied hydrofunctionalizations of alkenes and alkynes utilizing catalytic copper hydride intermediates, see: M.T. Pirnot, Y.-M. Wang, S.L. Buchwald, *Angew. Chem. Int. Ed.* 2016, 55, 48–57; (b) J. Mohr, M. Oestreich, *Angew. Chem. Int. Ed.* 2016, 55, 12148–12149.
6 For a representative example, see: B.H. Lipshutz, P. Papa, *Angew. Chem. Int. Ed.* 2002, 41, 4580–4582.
7 For representative examples, see: (a) L.T. Leung, S.K. Leung, P. Chiu, *Org. Lett.* 2005, 7, 5249–5252; (b) B.H. Lipshutz, C.S. Ung, S. Sengupta, *Synlett* 1989, 64–66.
8 For the seminal study discovering hydrosilanes as hydride source, see: H. Brunner, W. Miehling, *J. Organomet. Chem.* 1984, 275, c17–c21.
9 B.M. Trost, *Angew. Chem. Int. Ed.* 1995, 34, 259–281.
10 R.A. Sheldon, *Chem. Soc. Rev.* 2012, 41, 1437–1451.
11 A. Wurtz, *Ann. Chim. Phys.* 1844, 11, 250–253.
12 M. Calvin, *J. Am. Chem. Soc.* 1939, 61, 2230–2234.
13 M. Calvin, *Trans. Faraday Soc.* 1938, 34, 1181–1191.
14 (a) S. Weller, G.A. Mills, *J. Am. Chem. Soc.* 1953, 75, 769–777; (b) W.K. Wilmarth, M.K. Barsh, *J. Am. Chem. Soc.* 1953, 75, 2237–2242; (c) L.W. Wright, S. Weller, *J. Am. Chem. Soc.* 1954, 76, 3345–3346; (d) L. Wright, S. Weller, G.A. Mills, *J. Phys. Chem.* 1955, 59, 1060–1064; (e) M. Calvin, W.K. Wilmarth, *J. Am. Chem. Soc.* 1956, 78, 1301–1305; (f) W.K. Wilmarth, M.K. Barsh, *J. Am. Chem. Soc.* 1956, 78, 1305–1307.
15 (a) R.G. Dakers, J. Halpern, *Can. J. Chem.* 1954, 32, 969–978; (b) J. Halpern, R.G. Dakers, *J. Chem. Phys.* 1954, 22, 1272–1273; (c) E. Peters, J. Halpern, *J. Phys. Chem.* 1955, 59, 793–796; (d) E. Peters, J. Halpern, *Can. J. Chem.* 1955, 33, 356–364.
16 (a) A.J. Chalk, J. Halpern, *J. Am. Chem. Soc.* 1959, 81, 5846–5852; (b) E.A. von Hahn, E. Peters, *J. Phys. Chem.* 1965, 69, 547–552.
17 A.J. Chalk, J. Halpern, *J. Am. Chem. Soc.* 1959, 81, 5852–5854.
18 J. Halpern, *J. Phys. Chem.* 1959, 63, 398–403.

19 G.V. Goeden, K.G. Caulton, *J. Am. Chem. Soc.* 1981, 103, 7354–7355.

20 T.M. Koenig, J.F. Daeuble, D.M. Brestensky, J.M. Stryker, *Tetrahedron Lett.* 1990, 31, 3237–3240.

21 J.F. Daeuble, C. McGettigan, J.M. Stryker, *Tetrahedron Lett.* 1990, 31, 2397–2400.

22 D.M. Brestensky, J.M. Stryker, *Tetrahedron Lett.* 1989, 30, 5677–5680.

23 For the first synthesis of [Ph3PCuH]6, see, S.A. Bezman, M.R. Churchill, J.A. Osborn, J. Wormald, *J. Am. Chem. Soc.* 1971, 93, 2063–2065.

24 D.M. Brestensky, D.E. Huseland, C. McGettigan, J.M. Stryker, *Tetrahedron Lett.* 1988, 29, 3749–3752.

25 (a) For a review on σ-bond metathesis with hydrosilanes, see, M. Oestreich, *Synlett* 2007, 1629–1643; (b) T. Vergote, F. Nahra, A. Merschaert, O. Riant, D. Peeters, T. Leyssens, *Organometallics* 2014, 33, 1953–1963; (c) T. Gathy, D. Peeters, T. Leyssens, *J. Organomet. Chem.* 2009, 694, 3943–3950; (d) S. Rendler, O. Plefka, B. Karatas, G. Auer, R. Fröhlich, C. Mück-Lichtenfeld, S. Grimme, M. Oestreich, *Chem. Eur. J.* 2008, 14, 11512–11528.

26 S.J. Blanksby, G.B. Ellison, *Acc. Chem. Res.* 2003, 36, 255–263.

27 W.S. Mahoney, J.M. Stryker, *J. Am. Chem. Soc.* 1989, 111, 8818–8823.

28 (a) J.A. Widegren, R.G. Finke, *J. Mol. Catal. A, Chem.* 2003, 198, 317–341; (b) R.H. Crabtree, *Chem. Rev.* 2012, 112, 1536–1554.

29 For the synthesis of copper-*C*- and copper-*O*-enolates, see, J. Plotzitzka, C. Kleeberg, *Organometallics* 2014, 33, 6915–6926.

30 H. Shimizu, T. Ohshima, K. Mashima, T. Nagano, N. Sayo, T. Saito, *Synlett* 2009, 3143–3146.

31 B.M. Zimmermann, S.C.K. Kobosil, J.F. Teichert, *Chem. Commun.* 2019, 55, 2293–2296.

32 (a) For reviews about NHC-complexes of copper, see, F. Lazreg, F. Nahra, C.S.J. Cazin, *Coord. Chem. Rev.* 2015, 293–294, 48–79; (b) J.D. Egbert, C.S.J. Cazin, S.P. Nolan, *Catal. Sci. Technol.* 2013, 3, 912–926.

33 J.-X. Chen, J.F. Daeuble, D.M. Brestensky, J.M. Stryker, *Tetrahedron* 2000, 56, 2153–2166.

34 J.-X. Chen, J.F. Daeuble, J.M. Stryker, *Tetrahedron* 2000, 56, 2789–2798.

35 H. Shimizu, N. Sayo, T. Saito, *Synlett* 2009, 1295–1298.

36 H. Shimizu, D. Igarashi, W. Kuriyama, Y. Yusa, N. Sayo, T. Saito, *Org. Lett.* 2007, 9, 1655–1657.

37 K. Junge, B. Wendt, D. Addis, S. Zhou, S. Das, S. Fleischer, M. Beller, *Chem. Eur. J.* 2011, 17, 101–105.

38 S.W. Krabbe, M.A. Hatcher, R.K. Bowman, M.B. Mitchell, M.S. McClure, J.S. Johnson, *Org. Lett.* 2013, 15, 4560–4563.

39 R. Watari, Y. Kayaki, S.-I. Hirano, N. Matsumoto, T. Ikariya, *Adv. Synth. Catal.* 2015, 357, 1369–1373.

40 F. Pape, J.F. Teichert, *Synthesis* 2017, 49, 2470–2482.

41 C.M. Zall, J.C. Linehan, A.M. Appel, *ACS Catal.* 2015, 5, 5301–5305.

42 C.M. Zall, J.C. Linehan, A.M. Appel, *J. Am. Chem. Soc.* 2016, 138, 9968–9977.

43 J.G. de Vries, C.J. Elsevier, *The Handbook of Homogeneous Hydrogenation.* Weinheim: Wiley-VCH, 2006.

44 F. Pape, L.T. Brechmann, J.F. Teichert, *Chem. Eur. J.* 2019, 25, 985–988.

45 T.N.T. Nguyen, N.O. Thiel, F. Pape, J.F. Teichert, *Org. Lett.* 2016, 18, 2455–2458.
46 K. Semba, R. Kameyama, Y. Nakao, *Synlett* 2015, 26, 318–322.
47 F. Pape, N.O. Thiel, J.F. Teichert, *Chem. Eur. J.* 2015, 21, 15934–15938.
48 T. Wakamatsu, K. Nagao, H. Ohmiya, M. Sawamura, *Organometallics* 2016, 35, 1354–1357.
49 N.O. Thiel, J.F. Teichert, *Org. Biomol. Chem.* 2016, 14, 10660–10666.
50 G.C. Fortman, A.M.Z. Slawin, S.P. Nolan, *Organometallics* 2010, 29, 3966–3972.
51 N.O. Thiel, S. Kemper, J.F. Teichert, *Tetrahedron* 2017, 73, 5023–5028.
52 E. Korytiaková, N.O. Thiel, F. Pape, J.F. Teichert, *Chem. Commun.* 2017, 53, 732–735.
53 K. Semba, R. Kameyama, Y. Nakao, *Chem. Lett.* 2018, 47, 213–216.
54 For reviews, see: (a) J.F. Bower, I.S. Kim, R.L. Patman, M.J. Krische, *Angew. Chem. Int. Ed.* 2009, 48, 34–46; (b) H.Y. Jang, M.J. Krische, *Acc. Chem. Res.* 2004, 37, 653–661.

5

Hydrogenation Reactions Using Group III to Group VII Transition Metals

Matthew L. Clarke and Magnus B. Widegren

University of St Andrews, School of Chemistry, Purdie Building, North Haugh, KY16 9ST, St Andrews, Fife, UK

5.1 Introduction

The best examples of catalytic hydrogenation using precious metal catalysts display many desirable characteristics. The contribution to process cost from the catalysts is very low, since catalyst loadings as low as 0.0001 mol% are used. These reactions are characterized by being inexpensive to run, with simple purifications and excellent atom economy. While there are dozens of multi-kilogram processes that can be optimized to this level of performance with a precious metal catalyst, the majority of catalytic hydrogenations require significantly higher catalyst loadings. With a catalyst that is expensive on a per-gram basis, this leads to high process costs, which are further increased by the purification becoming more complex, since for most products <5 ppm of Ir, Rh, or Ir impurities are required. Earth-abundant metal catalysts could be several times cheaper to produce than precious metal systems, and hence offer opportunities to operate a broader range of catalytic hydrogenations. In addition, a higher tolerance for some of these metal impurities in products is possible (e.g. 250 ppm for Mn) [1]. This reduces the complexity of purification, adding another desirable characteristic. Finally, it would be desirable if the manufacture of chemicals only made use of indefinitely sustainable sources, which are not too energy intensive to produce. This is not the case for trace elements, and indeed for some of the early transition metals such as Re. However, within groups III–VII, Ti, Cr, Zr, and Mn metal are highly abundant, and Ti, Zr, and Mn compounds are also of limited concern from a toxicity perspective.

This chapter focuses on the hydrogenation reactions using group III to group VII metal catalysts and is broken down into sections for each group in the series.

5.2 Group III Metals: Scandium and Yttrium

Neither scandium or yttrium catalysts have received a great deal of attention in the field of hydrogenation and remain an open field for novel catalysts. Scandium and yttrium are most often used as Lewis acids in a variety of reactions.

Homogeneous Hydrogenation with Non-Precious Catalysts, First Edition. Edited by Johannes F. Teichert.

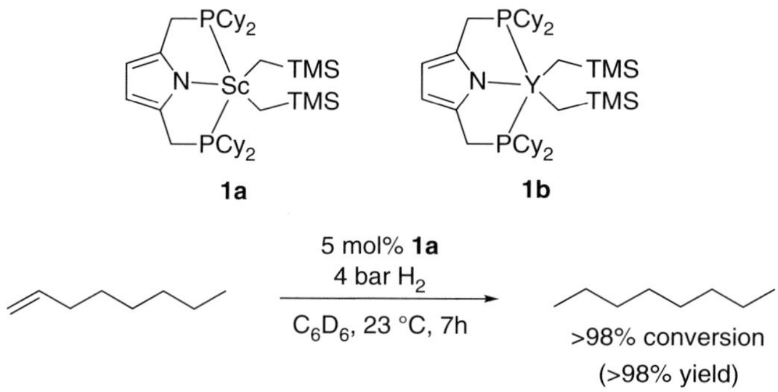

Figure 5.1 Scandium and yttrium PNP complexes [2].

A recent publication by Andersen and co-workers studies Sc catalyzed hydrogenations [2]. Building on recent findings that scandium hydrides can insert into double bonds [3], they investigated whether the catalysts (**1a** and **1b**, Figure 5.1) could be used in the hydrogenation of alkenes.

The authors reported that catalyst **1a** was more reactive than **1b** in the reduction of 1-octene. At 5 mol% catalyst loading, 4 atm H_2 and at 23 °C, **1a** fully reduced 1-octene in 7 hours, while **1b** needed 48 hours for complete conversion. Similar trends were observed with cyclohexene and 3-hexyne (to *cis*-3-hexene). This initial report is a useful proof of concept, and merits further study of scandium and yttrium catalysts in hydrogenation reactions.

5.3 Group IV Metals: Titanium, Zirconium, and Hafnium

Of the metals in group IV, titanium dominates the field of catalytic hydrogenation. Titanium, zirconium, and hafnium hydrogenation catalysts most often take the form of cyclopentadienyl (Cp) complexes with variable substitution patterns. The dicyclopentadienyl metal species (metallocenes) often show better activity than $CpML_n$. Low-valent titanium species are prone to form dimers, which are catalytically inactive (Scheme 5.1) [4]. Substituents on the cyclopentadienyl ring can alleviate this problem [5].

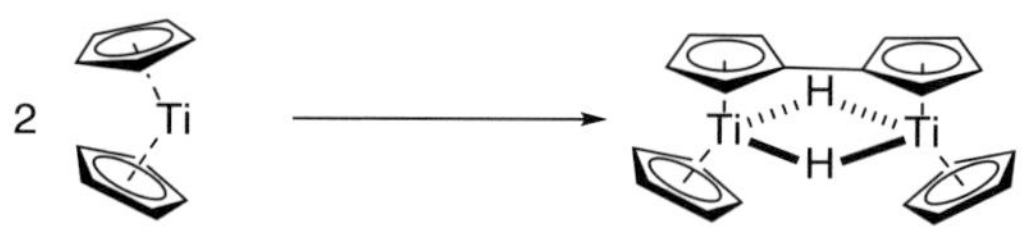

Scheme 5.1 Dimerization of low-valent titanocene species.

In many cases, the pre-catalysts need to be activated using highly reactive reagents such as *n*-butyllithium (typically 2 equiv with respect to the catalyst) [6], sometimes in combination with hydrosilanes [6a, 6b, 7] and in some cases trialkylaluminium [8]. Elemental magnesium was successfully used to activate Cp_2TiCl_2 in tetrahydrofuran (THF) and found to be an efficient way to reduce olefins and acetylenes at room temperature (2.5 mol% catalyst, 1.5 equiv (to substrate) Mg, 1 bar H_2) [9]. Initially, the active catalyst failed to completely reduce

1-hexyne and gave a mixture starting material and 1-hexene and 2-hexene. The authors found that by the addition of triphenylphosphine to the system, the catalyst could reduce both 1-hexyne (to a mixture of hexane and 1-hexene in a ratio of 5/1) and the ester methyl oleate. Sodium hydride has also been used to activate titanium-based catalysts in the reduction of 1-hexene [10] and styrene [11]. With titanium fluorides (Cp_2TiF_2), milder reducing agents such as phenylsilane have been successfully used [12].

One of the earliest examples of homogeneous catalyzed olefin hydrogenation using titanium and zirconium catalysts was reported by Sloan et al. in 1963 [13]. By treating the pre-catalyst species (either $Ti(O\textit{i}Pr)_4$, titanocene, or zircocene) with an excess of either trialkylaluminium or *n*-butyllithium, they formed an active catalyst capable of reducing olefins at room temperature and low dihydrogen pressure. The use of reactive alkyl metals is believed to initially displace a halide to form a titanium or zirconium alkyl species, and then transformed to the hydride in the presence of hydrogen gas. This was shown by Bercaw and Brintzinger who generated the catalytically active Cp_2TiH_2 by hydrogenating $Cp_2Ti(CH_3)_2$ [14].

Brubaker and coworkers synthesized $CpTiCl_2$, $CpTiCl_3$, $CpZrCl_3$, and $CpHfCl_3$ tethered to styrene-divinylbenzene copolymers (Scheme 5.2) and investigated the formed complexes in the hydrogenation of olefins [15] using *n*-butyllithium to activate the catalysts.

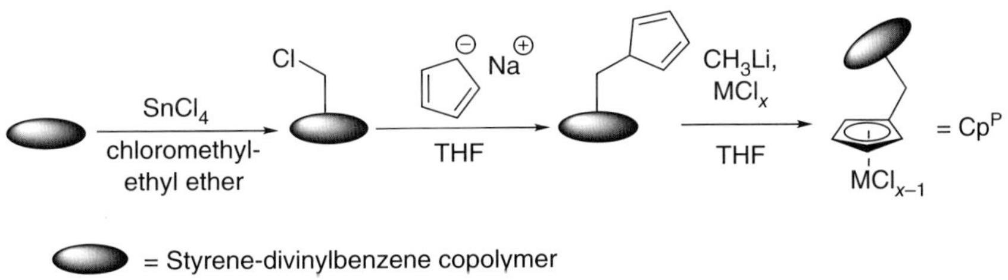

Scheme 5.2 Synthesis of styrene-divinylbenzene copolymer tethered cyclopentadienyl metal species.

They found minimal activity for the Cp^PZr and Cp^PHf complexes, but the Cp^PTi complexes showed good activity in the reduction of 1-hexene and cyclohexene, with Cp^PTiCl_3 being the better of the two. Comparison with the untethered CpTi species revealed that the tethered catalysts exhibited higher activity. The higher activity of the tethered complexes was attributed to the smaller propensity of the tethered catalysts to dimerize and become inactive. Both tethered Cp^PTiCl_3 and Cp^PTiCl_2 complexes showed poor performance in the reduction of cyclooctene. A study of Cp_2TiCl_2 tethered to silica showed that linker length was important for activity, with increasing linker length leading to lower activity [16].

Royo and coworkers performed an extensive study on titanocene and zirconocene derivatives for the hydrogenation of olefins by comparing a number of catalysts including *ansa* (bridged) cyclopentadienyl ligands with either chloro or alkyl ligands on the metal [17]. Titanium catalysts were found to be more active than zirconium catalysts, and *ansa*-type ligands were found to be more active than Cp_2ML_n species in general. Interestingly, when L = alkyl or aryl, a slowdown of the activity was found when reactions were run in the presence of

$MePh_2P$, which indicated a competitive interaction of the phosphine with the alkene for the binding site of the catalyst. However, at the same time with certain catalysts (with dichloro ligands), the absence of $MePh_2P$ in the activation step (using Mg) was found to be detrimental to the activity; however, the presence of the phosphine did not affect the catalytic activity. It is likely the phosphine favorably interacts with the pre-catalyst (where L = Cl) during the activation with magnesium but plays no further role in the actual catalytic cycle. This could be an indication that the active catalyst is different when L = Cl than when L = alkyl or aryl.

The compound $Cp_2Zr(H)Cl$ (Schwartz's reagent) [18] has often been used as a stoichiometric reagent for the selective synthesis involving alkenes and alkynes, but it has seen limited use as a catalytic reagent. The compound $[Cp_2ZrH(CH_2PPh_2)]_n$, on the other hand, has been shown to be an active catalyst for the reduction of alkenes and furthermore does not need any activating reagent [19]. It was found that branched alkenes gave higher reactivity, although excessively bulky alkenes slowed down the reactions (due to difficulty forming the alkyl-Zr intermediate). Competitive isomerization (with dienes) was found to slow down the hydrogenation rate. The proposed simplified mechanism is detailed in Scheme 5.3.

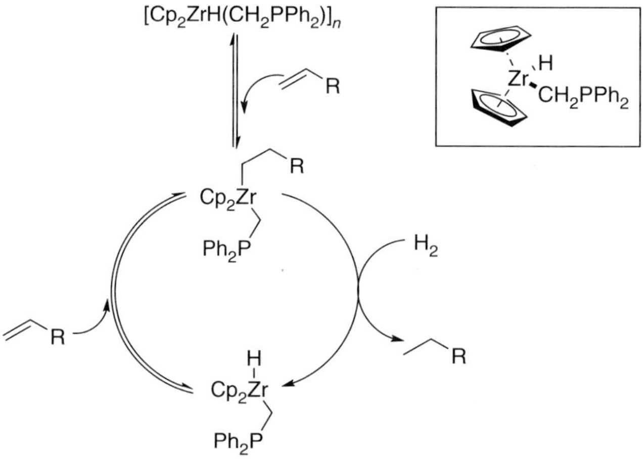

Scheme 5.3 Proposed simplified mechanism for the reduction of alkenes using $Cp_2ZrH(CH_2PPh_2)$ catalyst [19a]. Typical conditions: substrate to catalyst ratio: 100–300 : 1, 40 bar H_2, THF, 80 °C, 1–16 hours.

In the first step, $[Cp_2ZrH(CH_2PPh_2)]_n$ reacts with the olefin via addition of the hydride to the β-carbon forming a linear alkyl-zirconium species. This species then reacts with dihydrogen gas to generate the alkane and a monomeric catalytic Zr species that continues to react with olefins. The role of the $-CH_2PPh_2$ ligand is unclear, but it is believed to stabilize the zirconium species involved in the catalytic cycle and is presumably far less prone to hydrogenolysis than the alkyl ligands generated from the alkene substrate.

Compared to titanium catalysts that need activation with corrosive and flammable activators, this zirconium-based catalyst could very well be a more promising catalyst for the reduction of olefins but has not received much attention since the initial findings were published. The benefits of being able to use a pre-catalyst that does not need to be activated prior to the catalysis are significant. The catalytic loading was also reported to be lower than that for many titanium-based catalysts (typically in the region of 2–5 mol%) and added benefit if the cost of the catalyst is important.

When investigating reduction of SBS (styrene–butadiene–styrene copolymer), Tsai and Chiang found that from catalysts of the type investigated Cp_nMCl_x (where $n = 1$ or 2, M = Ti, Zr or Hf, $x = 2$–3) in the presence of trialkylaluminium as activators and Cp_2TiCl_2 was found to be the most active [8]. The catalysts could function without the alkylaluminium as activator. The presence of an activator is believed to increase catalyst stability at higher temperatures and might influence the nature of the catalytically active species. Triisobutylaluminium was found to be a better activator than triethylaluminium, trioctylaluminium, or diethylaluminium chloride. The pre-catalysts still needed to be activated by *n*-butyllithium (4 equiv) regardless of the presence or absence of the trialkylaluminium activator.

5.3.1 Asymmetric Hydrogenation Using Titanium and Zirconium Catalysts

One of the potential benefits of using catalytic systems is the option for the asymmetric synthesis of chiral chemicals. Using a chiral ligand, enantiomeric induction can often be achieved where one enantiomer (or diastereomer) is preferentially formed over the others resulting in an enantiomerically enriched material. The use of asymmetric hydrogenation to synthesize enantiomerically pure chemicals is not only a cost-saving process but is also less taxing on the environment as less waste is produced compared to the more classical resolution methods for separating enantiomers and diastereomers.

Kagan and coworkers were one of the first to use a chiral titanium catalyst to reduce alkenes [20]. Using a titanium catalyst based on (+)-menthol or (−)-neomenthol (**2a** and **2b** in Figure 5.2), the group successfully reduced 2-phenyl-1-butene using Red-Al ($[LiH_2Al(OCH_2CH_2OCH_3)_2]$) as activator. The enantiomeric excess was low (14.9%), but it showed the potential of using chiral titanium catalysts to generate chiral compounds. A few years later, Vollhardt and coworkers published their results on the asymmetric hydrogenation of 2-phenyl-1-butene using a more sterically hindered titanocene (**3**, Figure 5.2) [21]. The catalyst showed a clear temperature dependence on the enantiomeric excess, with a lower temperature giving higher ee (up to 96% at −75 °C). In a second article, the authors synthesized and compared the activity and enantioselectivity of several titanocenes derived from the chiral pool [22]. Of all the catalysts tested, only one stood out (**4**, Figure 5.2). Although it only showed moderate enantioselectivity in the reduction of 2-phenyl-1-butene (25% at 0 °C), the reaction was finished within 20 minutes at a low (1 mol%) catalyst loading.

Buchwald's group published a number of papers using their titanocene catalyst (**5**, Figure 5.2) for the reduction of imines [6a, 23]. Catalyst **5** was found to be active in the reduction of both acyclic and cyclic imines using 2–10 mol%

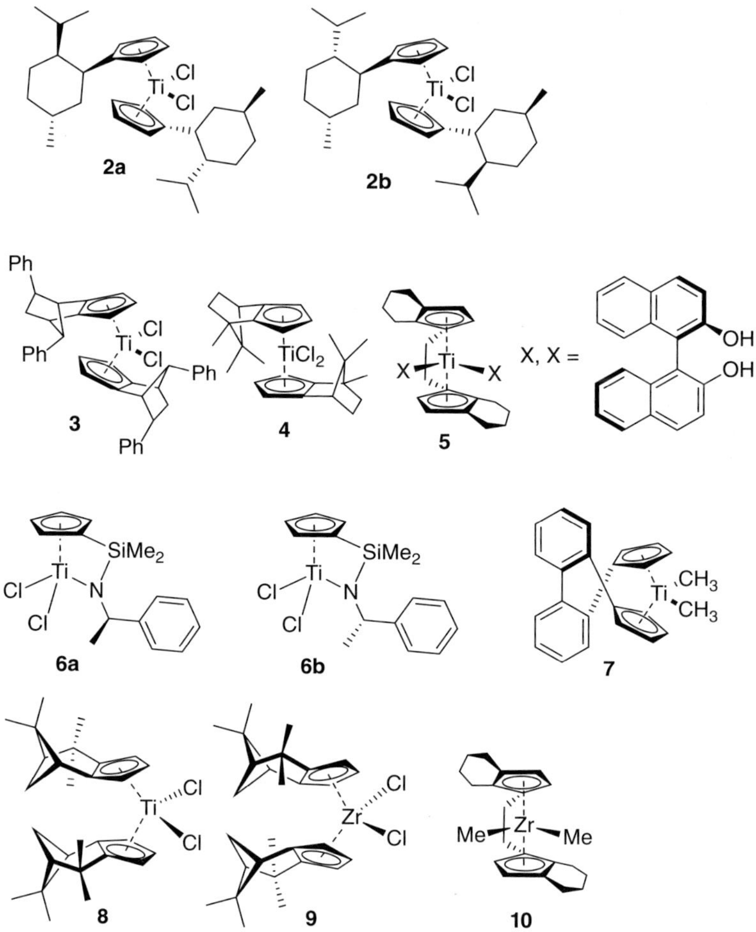

Figure 5.2 Titanium and zirconium catalysts used in asymmetric hydrogenation.

catalyst. The catalyst was activated using a combination of *n*-butyllithium (1.0 equiv relative to **5**) and phenylsilane (1.5 equiv relative to **5**) prior to hydrogenation. In general, the acyclic imines showed moderate to good ee (up to 86%), while for cyclic imines the ee was excellent (up to 99%). The reactions were run at high pressure (138 bar) and at room temperature. The proposed mechanism for the imine reduction can be seen in Figure 5.3. The pre-catalyst **5** is first reduced using *n*-butyllithium and phenylsilane to the Ti (III) hydride (**X**). This is the catalytically active species which reacts with the imine to generate a titanoceneamide. This intermediate reacts with dihydrogen to reform the Ti–H catalyst and releases the formed amine. Buchwald and coworkers also investigated the mechanism with regard to the enantioselectivity and found that for acyclic substrates the pressure had an impact on the ee, with higher pressures leading to higher enantioselectivity. They explained this by postulating four different transition states depending on the *E* or *Z* conformation of the

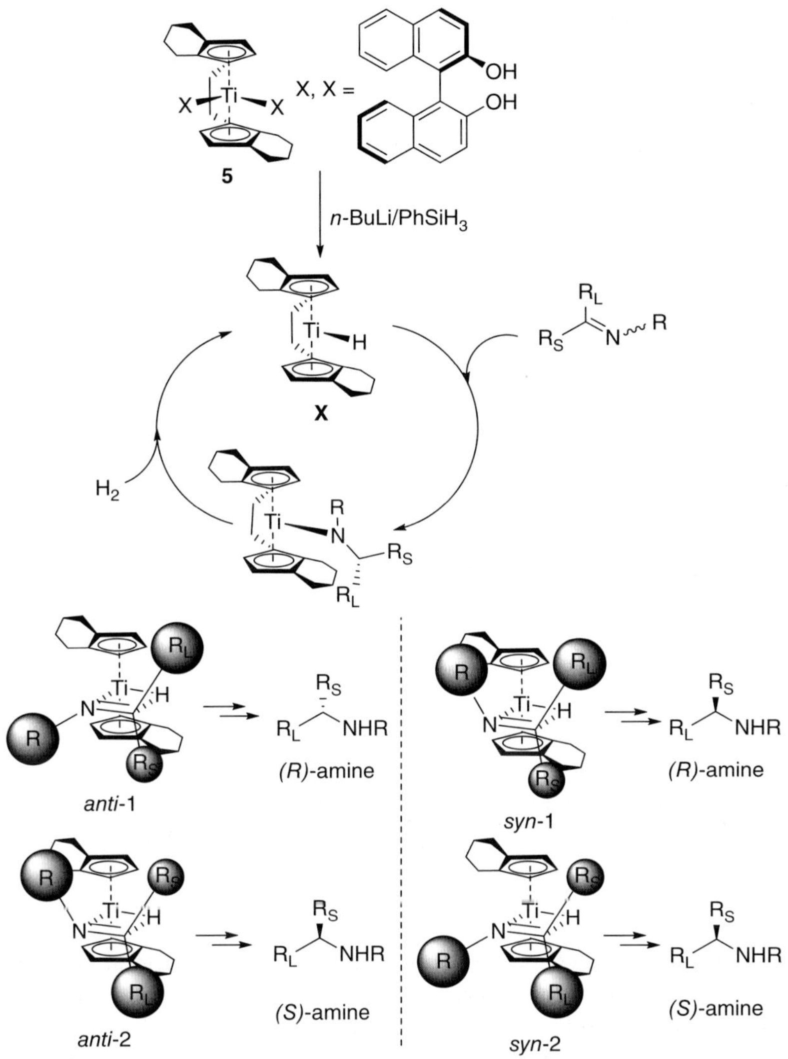

Figure 5.3 Proposed mechanism for the reduction of imines using catalyst **5**. Source: Adapted from Willoughby and Buchwald 1994 [23b].

imine and the position of the substituents on the carbonyl carbon (*anti*-1/2 and *syn*-1/2 in Figure 5.3).

They found that the *E*-isomer imine gave the (*R*) enantiomer (via *anti*-1 transition state) and the *Z*-isomer the (*S*) enantiomer (via *syn*-2 transition state). At lower pressure, the *Z*-imine isomer would react faster than the *anti*-imine isomer. For imines containing an *N*-methyl group, the enantioselectivity was found to be dependent on the size of the α-substituents on the carbonyl carbon atom (R_S and R_L in Figure 5.3). With cyclic imines, the problem of *E*/*Z*-isomerization was not an issue and the authors found that excellent enantioselectivity could be achieved as well as good functional group tolerance.

Catalyst **5** has also been used in the reduction of enamines [7]. Using methodology similar to that with imines, the Buchwald group successfully reduced a number of acyclic enamines prepared from pyrrolidine, morpholine, diethylamine, and methylbenzylamine (Scheme 5.4). The enantioselectivity varied from 89% to 98%, and lower pressure (1–5.5 bar of dihydrogen) could be used relative to imine reduction. No pressure dependence on enantiomeric excess was observed. Good enantiomeric excess was observed for all substrates used in the study.

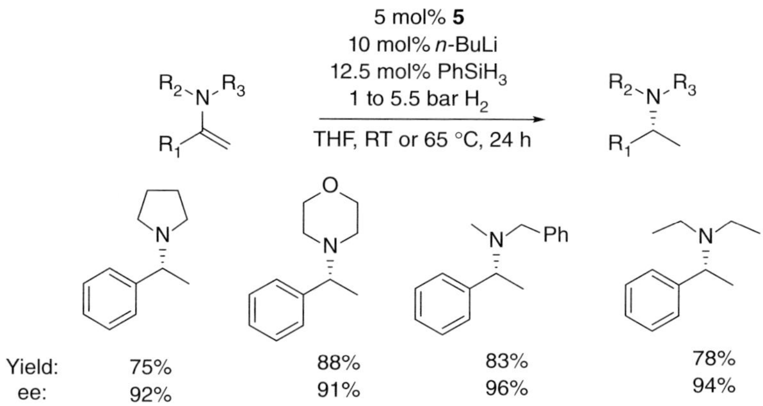

Scheme 5.4 Asymmetric reduction of enamines using catalyst **5**.

The same group also reported the use of catalyst **5** in the reduction of unfunctionalized trisubstituted alkenes [6b]. The reactions were slower compared to those of imine reduction (days compared to hours) using 5 mol% of catalyst **5**, but the enantioselectivity was excellent (up to 99%), albeit dependent on the substrate structure.

Titanium compounds such as **6a** and **6b** (Figure 5.2) containing an amine-tethered cyclopentadienyl ligand were shown to be active in the reduction of imines by Okuda and coworkers [24]. The enantioselectivity was low (12% ee). In a follow-up article, the authors expanded on **6a** and **6b** and incorporated other substituents α to the amine instead of the methyl group [6c], although enantioselectivity never exceeded 24%.

Brintzinger and coworkers reported the use of a biphenyl-bridged titanocene catalyst (**7**, Figure 5.2) [25]. The zirconium variant of the same catalyst showed no activity, but the titanocene was shown to reduce both acyclic and cyclic imines at low catalyst loading (0.1 mol%) with moderate to excellent enantioselectivity (76% ee for acyclic and 98% ee for cyclic imines) at 150 bar dihydrogen pressure and 80 °C for these reactions (Scheme 5.5).

Paquette and coworkers explored a number of titanocene and zircocene catalysts for the enantioselective reduction of 1,1-disubstituted alkenes [26]. From their investigation, two catalysts stood out (**8** and **9**, Figure 5.2). Catalyst **8** showed moderate enantioselectivity in the reduction of alkenes (up to 61%) at low pressure (1.9 bar) and temperature (−18 to 22 °C) at 1–5 mol%

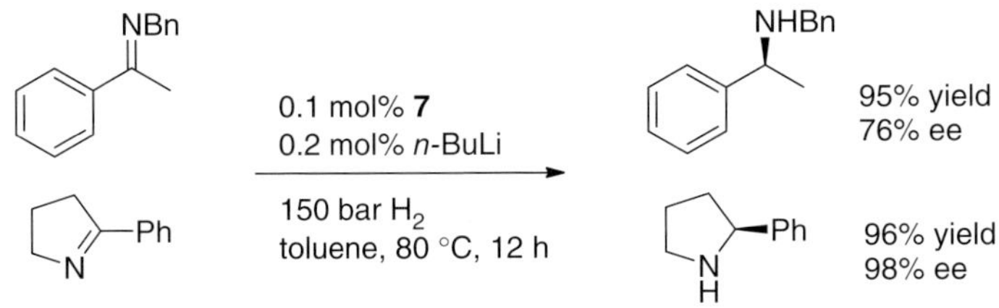

Scheme 5.5 Imine reduction using catalyst **7**.

catalyst using *n*-butyllithium (10–20 mol%) as activator. Catalyst **9** showed low enantioselectivity (14% ee) under the same conditions.

Buchwald and coworkers reported the use of catalyst **10** (Figure 5.2) in the reduction of unfunctionalized tetrasubstituted olefins [27]. This catalyst was activated using 1 equiv of $[PhMe_2NH][B(C_6F_5)_4]$ and found to reduce a number of different tetrasubstituted alkenes under the reaction conditions (8 mol% **10**, 8 mol% $[PhMe_2NH][B(C_6F_5)_4]$, room temperature, either 5.5 bar or 69–138 bar dihydrogen pressure) with enantiomeric excess of the products in the range of 5–99% depending on the substitution and ring size (in the case of cycloalkenes).

In group VI metals, titanium dominates in the literature over zirconium and hafnium. Zirconium- and hafnium-based catalysts do not tend to show the same activity or selectivity compared to titanium, but there are exceptions such as $[Cp_2ZrH(CH_2PPh_2)]_n$, which does not need an activator and gives activity similar to that of titanium catalysts in the reduction of alkenes. A notable feature has been the ability to prepare enantiomerically pure titanium catalysts, which are highly competent at reducing various unsaturated compounds with enantioselectivity. A disadvantage to the titanium-based systems is the need for quite large amounts of catalyst generally 1–5 mol% or higher. While titanium itself is inexpensive, taking the synthesis of the ligand into consideration, the actual total cost of the catalysts can be high. Successful broader implementation would probably require catalyst loading to be reduced by a factor of 100. A second disadvantage is the use of significant amounts of expensive and reactive activators such as *n*-butyllithium. It would also be desirable to operate at dihydrogen pressures below 10 bar. Since both titanium and zirconium are earth abundant and of, generally, low toxicity, further research would be welcome [28].

5.4 Group V Metals: Vanadium, Niobium, and Tantalum

Catalysts based on group V metals have not been used in hydrogenation as much as group IV metals. Vanadium is more known for its use in catalytic oxidation reactions than in catalytic reductions using dihydrogen gas. Arnold, Toste, and coworkers reported the use of catalyst **11** (Figure 5.4) in the *Z*-selective reduction of alkynes to alkenes [29]. The authors investigated the reaction mechanism and found the active species to be the cationic **11a** formed by a formal 1,2-addition of hydrogen over the V=N double bond. Arnold, Bergman, and coworkers later reported the use of the niobium catalyst **12** (Figure 5.4) in the *Z*-selective reduction of alkynes [30]. Both catalysts were found to be reactive only with alkynes and

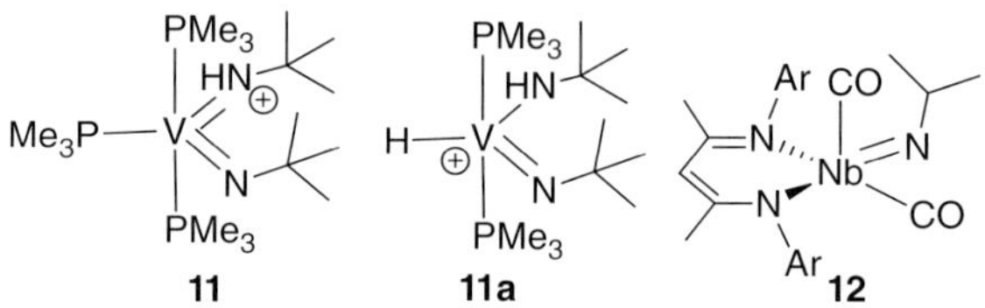

Figure 5.4 Vanadium and niobium catalysts for alkyne hydrogenations.

at high catalyst loading (20 mol%) limiting their usefulness. Only alkynes were studied in the article. Catalyst **12** was found to be most reactive in the presence of CO which the authors speculated plays a role in the mechanism.

Rothwell and coworkers published reports on the use of niobium and tantalum aryloxide catalysts (**13** and **14**, Scheme 5.6) in the reduction of aromatic compounds at relatively moderate temperatures (80–100 °C, 5 mol% catalyst, 83 bar dihydrogen gas, cyclohexane) [31]. Catalyst **13** was found to reduce alkyldiarylphosphines to alkyldicycloalkylphosphines (using 5 mol% catalyst, 83 bar dihydrogen and 60 °C, Scheme 5.6); here, *n*-butyllithium was needed as activator (3 equiv) [32].

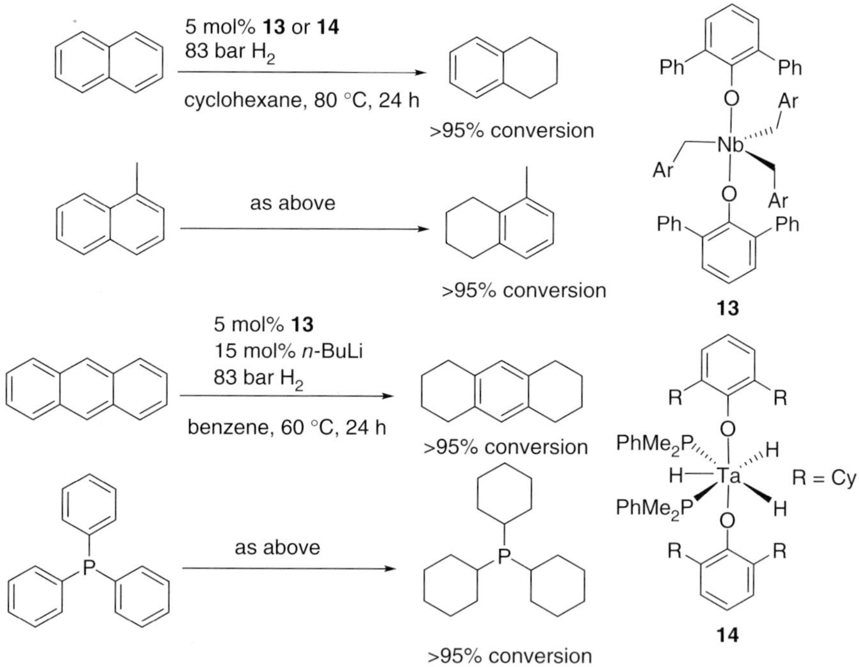

Scheme 5.6 Reduction of aromatic compounds and phosphines using catalyst **13** or **14**.

Rothwell and coworkers later published a more detailed account on the mechanism of these reductions [33]. Very recently, a computational study on the feasibility of vanadium, niobium, and tantalum in the hydrogenation of carbonyl groups, nitriles, and alkenes using pincer-type ligands was published and could be an indication that more is to come using these metals [34].

Thus far, V, Nb, and Ta have been sparingly used, although the reactivity observed is interesting: Vanadium is less common in the earth's crust than chromium or manganese and on par with tungsten, but more abundant than either copper or zinc. Both niobium and tantalum are less common in the earth's crust than vanadium and less common than either copper and zinc. There has been little research into the toxicity of these metals [28] and their salts, but caution should always be advised when handling metals.

5.5 Group VI Metals: Chromium, Molybdenum, and Tungsten

As with group V metals, the metal catalysts from group VI are more known for reactions other than reductions. That said, there are several examples of group VI metals being used in hydrogenation, as detailed here.

Darensbourg and coworkers first reported the use of $[HM(CO)_5]^-$ and *cis*-$[HM(CO)_4P(OMe)_3]^-$ (M = Cr, W) as stoichiometric reagents in the reduction of ketones and aldehydes [35]. The addition of a Brønsted acid was necessary for the reduction to be achieved. The group followed up with a paper on the use of catalytic amounts (5 mol%) of $PPN^+[M(CO)_5OAc]^-$ (PPN^+ = bis(triphenylphosphine)iminium, M = Cr, Mo, W) in the hydrogenation of ketones and aldehydes [36] at 125 °C and at 41.4 bar dihydrogen. Based on the evidence gathered, the authors postulated the following mechanism (Figure 5.5).

The pentacarbonyl metal acetate (**15a**) dissociates a carbon monoxide ligand to form the tetracarbonyl metal acetate (**15b**), which reacts with dihydrogen and carbon monoxide and dissociates acetic acid to form the pentacarbonyl metal hydride (**15c**). The hydride reacts with a ketone or an aldehyde to generate the alkoxide intermediate (**15d**), which then reacts with acetic acid to release the alcohol and regenerate the acetate (**15a**). Studying $Cr(CO)_6$, Marko and Nagymagos reported the successful catalytic (5 mol%) hydrogenation of ketones

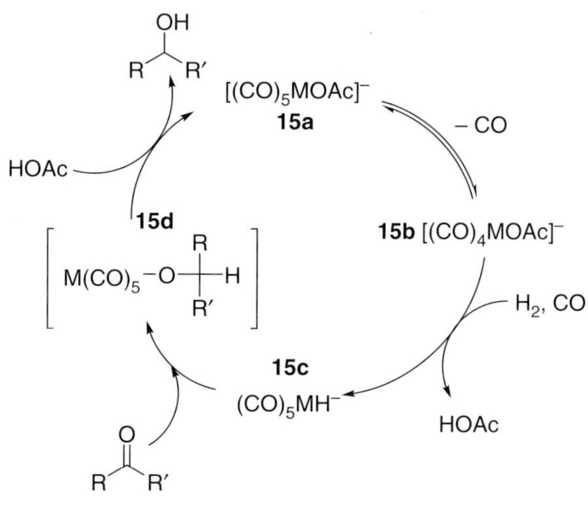

Figure 5.5 Proposed mechanism for the reduction of ketones and aldehydes using $PPN^+M(CO)_5{}^-OAc$. Source: Adapted from Tooley et al. 1986 [36].

at 100–120 °C and 100 bar H_2 in the presence of either sodium methoxide in methanol or in neat triethylamine (at 160 °C) [37]. Their proposed mechanism involved the $HCr(CO)_5$ species that formed the dimeric $Cr_2(CO)_{10}^-$ after reducing the ketone to the alcohol. This dimer then reacted with dihydrogen gas forming the active hydride species. There was a distinct rate dependence found for both dihydrogen (positive) and carbon monoxide (negative) pressure. In the same article, the authors reported that $Mo(CO)_6$ was active in the hydrogenation of ketones at lower reaction temperature (5 mol% catalyst loading, 100 bar dihydrogen gas, 70–80 °C), but $W(CO)_6$ reacted sluggishly. Several ketones and aldehydes were successfully reduced using these systems. Operating the reaction in neat triethylamine, *n*-butyraldehyde could be quantitatively reduced, while in methanol the reaction failed to yield any product.

Brunet and coworkers reported the use of catalytic $KHCr(CO)_5$ in the transfer hydrogenation of ketones using formic acid/triethylamine as the donor system [38]. Using 20 mol% of the metal and a 1 : 1 : 1 ketone/formic acid/trimethylamine mixture at room temperature in THF gave >95% conversion of cyclohexanone after 24 hours. Other ketones were also reduced, albeit at lower conversion after 24 hours.

$[Cp_2Mo(\mu{-}OH)_2MoCp_2](OTs)_2$ (**16**) has been shown to catalyze the transfer hydrogenation of 2-butanone [39] and acetophenone [40]. Initially believed to operate via the Meerwein–Ponndorf–Verley reduction mechanism, Kuo and coworkers showed that a Mo–H is actually the catalytic species in the reaction [40b]. The proposed mechanism is shown in Figure 5.6.

According to the proposed mechanism, the active monomeric species **16a** exchanges water for isopropanol to generate an alkoxide species (**16b**). The alkoxide is oxidized to the ketone and molybdene hydride complex **16c** is formed. Water then displaces the formed acetone to generate the active catalyst **16d**. In the following step, acetophenone displaces the water to generate **16e** and gets reduced to the alkoxide to generate **16f** (with a water molecule filling the open site, and in the final step water displaces the 1-phenylethanol to regenerate **16a**).

It appears that the ability of ligands to modulate reactivity has been less explored for Cr, Mo, and W catalysts relative to Ti. The activity and selectivity observed are certainly far worse than observed with titanium, although for those metals reduction of carbonyl compounds is not stereoselective. This can be contrasted with titanium catalysts in alkene and enamine reductions. The activity observed so far for carbonyl reduction using a relatively slim range of catalysts are far below most precious metal systems or the best earth-abundant metal catalysts. Molybdenum and tungsten catalysts are also known to operate through the ionic hydrogenation mechanism [41].

5.6 Group VII Metals: Manganese and Rhenium

Until recently, most of the chemistry utilizing manganese and rhenium was in the field of oxidation. Rhenium complexes have been used to hydrogenate imines and ketones using either ionic hydrogenation or hydrosilanes as reagents.

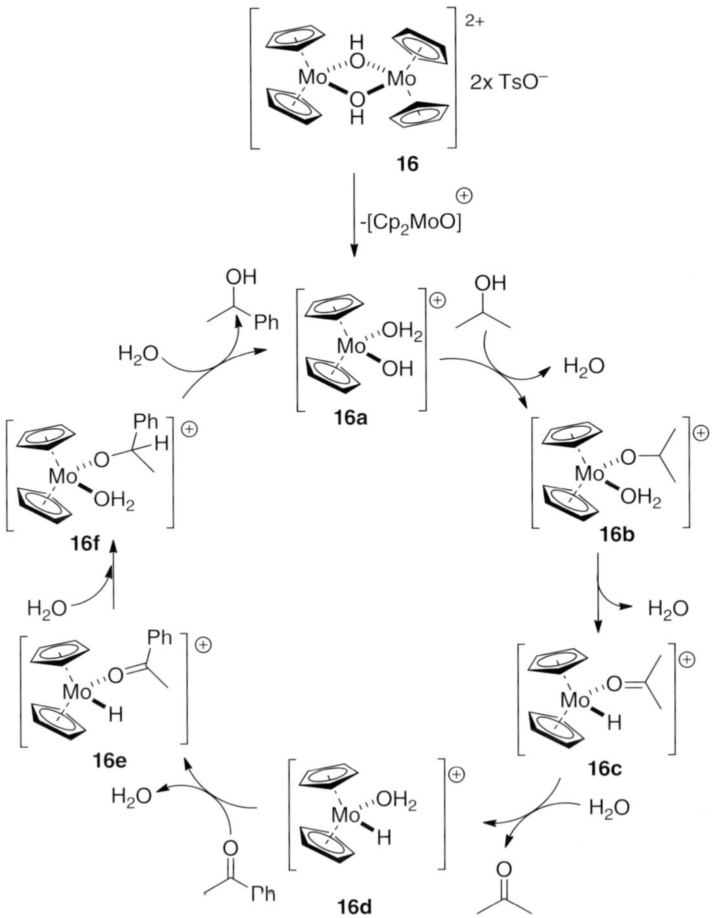

Figure 5.6 Transfer hydrogenation of ketones using $[Cp_2Mo(\mu\text{-}OH)_2MoCp_2](OTs)_2$ [40b, 37].

The first reported use of a manganese pincer complex in the hydrogenation of aldehydes, ketones, and nitriles was made by Beller and coworkers using catalysts **17a** and **17b** (Figure 5.7) [42]. They found that complex **17a** was more active than **17b** in their investigation.

For nitriles, the catalyst loading was set to 3 mol% and 50 bar dihydrogen gas was used, although a loading as low as 2 mol% and a pressure of 30 bar could be used to achieve good conversion (42–94%). The catalyst could reduce both aromatic and aliphatic nitriles with good to excellent yields. With ketones and aldehydes, the catalyst loading could be lowered (1 mol%) and a pressure of hydrogen gas between 10 and 30 bar dihydrogen was found to be sufficient (Scheme 5.7).

Kempe and coworkers reported the use of catalyst **18** in the reduction of ketones [43]. They showed that by varying the R′ substituent they could tune the activity of the catalyst with a reactivity order of R′ = cyclopropyl > Ph ~ CH_3. It is quite possible that the effect is not on a step on the catalytic cycle but rather on catalyst stability, activation, or solubility. Using as low a catalyst loading as

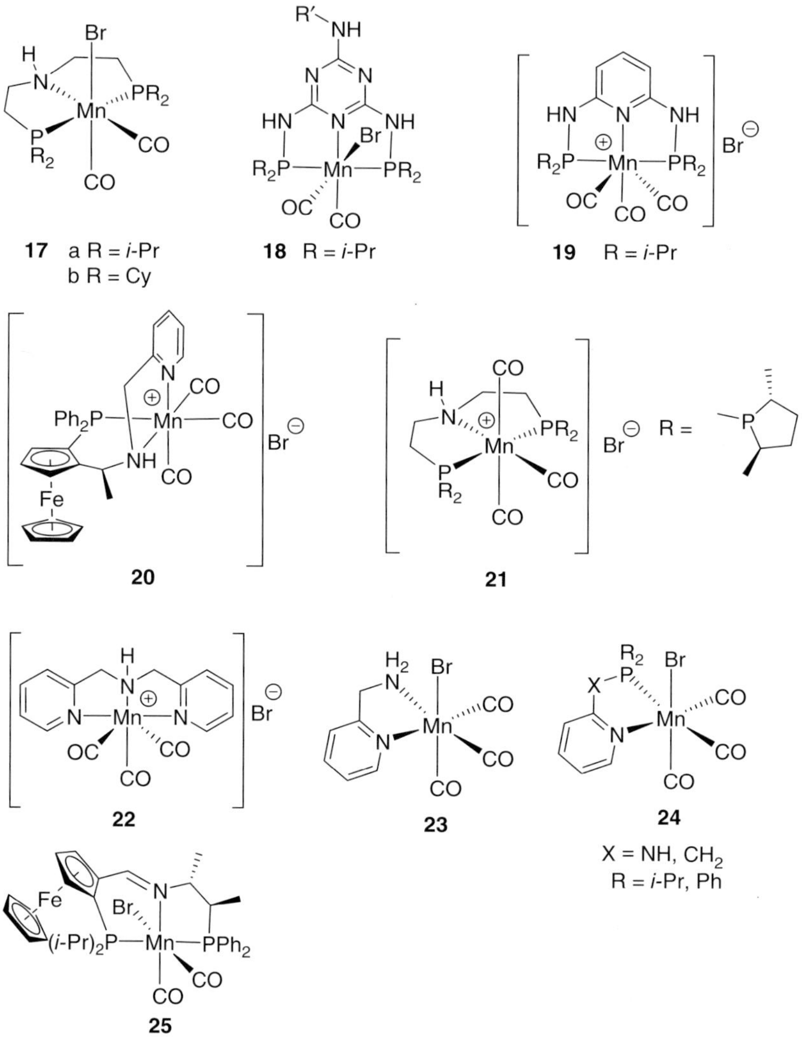

Figure 5.7 Manganese-based catalysts used in hydrogenation of ketones.

0.1 mol% (at 80 °C in toluene and 20 bar H_2), catalyst **18** was able to reduce both aliphatic and aromatic ketones and showed good substrate tolerance including pyridines, chloro, fluoro, cyano, and alkene substituents. A similar catalyst (**19**, Figure 5.7) was reported by Sortais and coworkers [44]. This catalyst was less active than the previously reported catalyst **18** (5 mol% loading and 130 °C). Sortais and coworkers recently reported the use of **24**, a manganese catalyst derived from a bidentate in the reduction of ketones using as low as 0.5 mol% catalyst loading under the conditions used (2 mol% KHMDS, 50 bar H_2, 50 °C in toluene, 20 hours) for a number of aliphatic and aromatic ketones [45].

The first enantioselective hydrogenations of ketones with manganese catalysts were reported by Clarke, Widegren, and coworkers using catalyst **20**

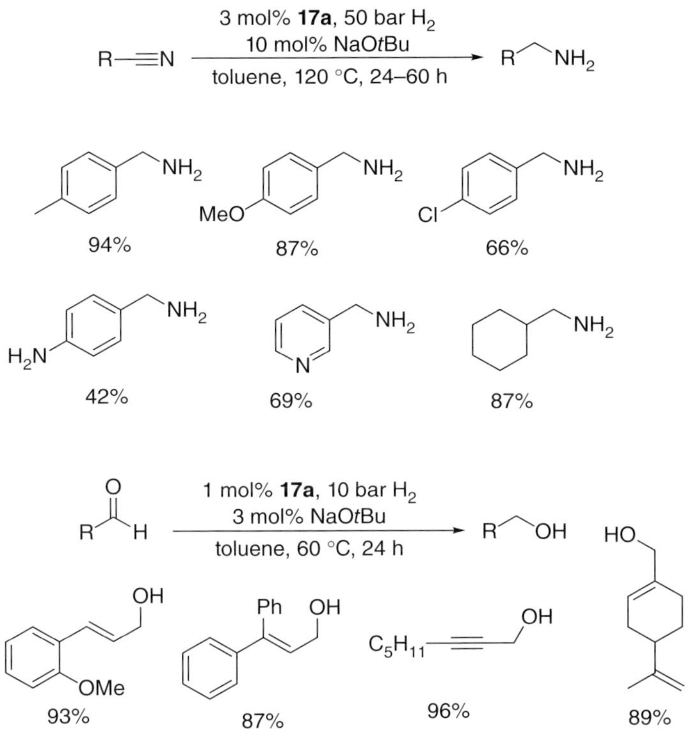

Scheme 5.7 Hydrogenation of nitriles (upper) and aldehydes (lower) using catalyst **17a**.

[46]. Although the catalyst produced low to moderate enantioselectivity with acetophenone derivatives (20–23%), branched aromatic ketones and *ortho*-substituted aromatic ketones gave good to excellent enantioselectivities (58–96%, Scheme 5.8). The best solvent was found to be ethanol. The catalyst loading was generally 1 mol%, but it was shown that it could be lowered to 0.1 mol%. Although potassium *t*-butoxide was used as base (10 mol%), the catalyst was found to be as active with inorganic bases such as potassium carbonate and potassium phosphate.

Beller and coworkers reported the use of the chiral catalyst **21** (Figure 5.7) to reduce both aliphatic and aromatic ketones under mild conditions [47] (Scheme 5.9). With catalyst **21**, *tert*-amyl alcohol was found to be the best solvent for aliphatic ketones, but 1,4-dioxane was found to be the best for aromatic ketones. The enantioselectivity varied from 84% ee (for aliphatic ketones) to as low as 18–20% ee (aromatic ketones).

Transfer hydrogenation of ketones and imines is well established using transition metals and one of the first reports using manganese catalyst was published by Xu and coworkers in 2011 [48]. Using 20 mol% manganese (II) oxide in neat benzyl alcohol at 100 °C in the presence of potassium carbonate (20 mol%) for eight hours, *N*-benzylidenebenzenesulfonamide was selectively reduced to *N*-benzylbenzenesulfonamide. Beller and coworkers published a paper in early 2017 on transfer hydrogenation using catalyst **22** (Figure 5.7)

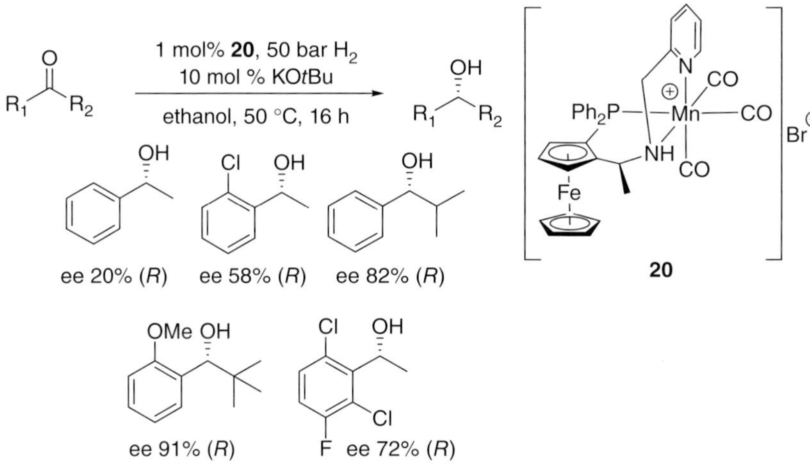

Scheme 5.8 Examples of asymmetric hydrogenation of ketones using manganese catalyst **20**.

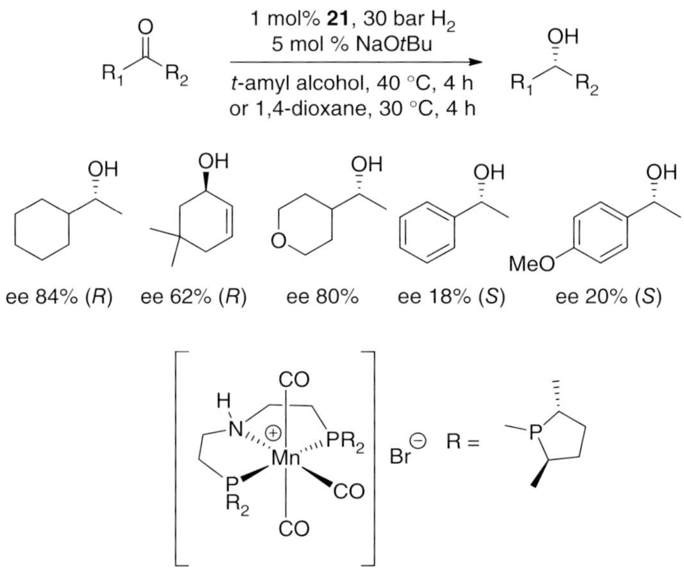

Scheme 5.9 Examples of asymmetric hydrogenation of ketones using manganese catalyst **21**.

[49]. Several aromatic and aliphatic ketones were reduced under the conditions (1 mol% **22**, 2 mol% KO*t*Bu, 70 °C in isopropanol). The catalyst was shown to be able to tolerate various functional groups including thioethers, nitriles, and esters. Complex **23** (Figure 5.7), derived from a bidentate ligand (the commercially available 2-aminomethylpyridine), was shown to be very active in the transfer hydrogenation of ketones with a catalyst loading as low as 0.1 mol% [50]. Isopropanol was used as solvent and reductant and potassium *t*-butoxide (2 equiv to catalyst) as base at temperatures between 30 and 80 °C. Several

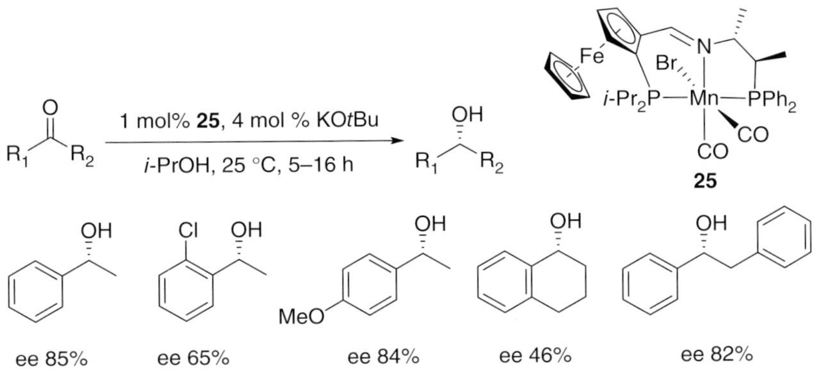

Scheme 5.10 Examples of asymmetric transfer hydrogenation of ketones using manganese catalyst **25**.

aliphatic and aromatic ketones were readily reduced to the corresponding alcohols under the reaction conditions. The catalyst was found to be inhibited by pyridine-based ketones, especially the 2-acetylpyridine, which could act as a competitive ligand. An example of asymmetric transfer hydrogenation was reported by Kirchner and coworkers using catalyst **25** (Scheme 5.10) [51]. Using very mild conditions (room temperature), the catalyst successfully reduced substituted acetophenones with moderate to very good enantioselectivities (46–85% ee).

Sortais and coworkers published an article on an *in situ* formed catalytically active manganese complex from $Mn(CO)_5Br$ and the chiral (1*R*, 2*R*)-*N*,*N*′-dimethyl-1,2-diphenylethane-1,2-diamine [52]. The catalytic system was able to reduce both aliphatic and aromatic ketones using isopropanol as the solvent and reductant (1 mol% metal precursor, 1 mol% ligand, 2 mol% KO*t*Bu, 80 °C, three hours). Enantioselectivities as high as 84% ee were reported. Manganese-based catalysts have also been used in the hydrosilylation of ketones and esters [53].

The homogeneous catalytic hydrogenation of esters is an area of research that has matured significantly over the past decade. Most publications make use of ruthenium or iron catalysis. Milstein was the first to report the use of manganese catalyst in the reduction of esters using catalyst **26** (Figure 5.8) [54].

The group later followed up with a second paper using catalyst **27** (Figure 5.8). Both catalysts **26** and **27** are based on the phosphine-pyridine ligand type that had been explored by Milstein and coworker over the years [55]. Catalyst **27** was found to be more active than **26** in the hydrogenation of esters and could be used with lower catalyst loading (1 mol% for **27** vs. 2 mol% for **26**). Using KH as the base catalyst, **27** was found to be able to reduce different esters to alcohols under at 100 °C and 20 bar H_2 with good conversions, being obtained after 50 hours using toluene as the solvent (Table 5.1 entry 1).

Beller and coworkers found that catalyst **28** could be used in the hydrogenation of esters [56]. While structurally similar to **17a** and **17b** (Figure 5.7), neither **17a** nor **17b** were found to be competent catalysts for ester hydrogenation. Catalyst **28** was found to exist in two forms; the main structure being *facially* coordinating

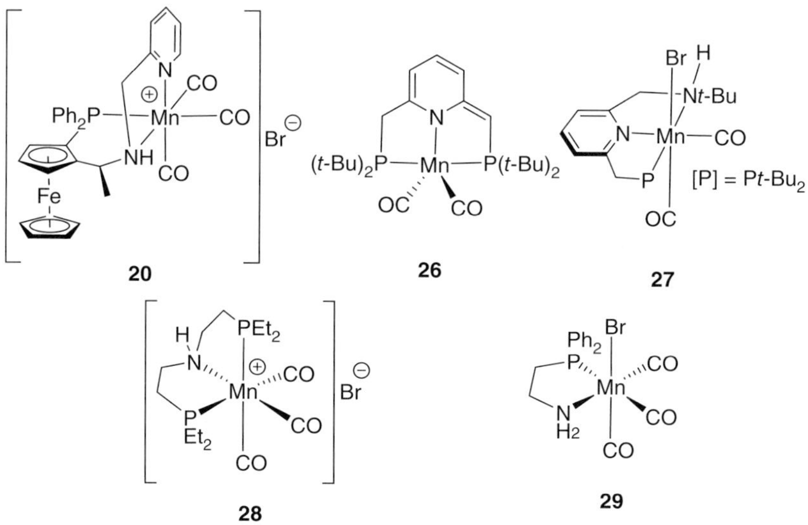

Figure 5.8 Manganese catalysts used in the reduction of esters.

to the manganese (as depicted in Figure 5.8). The minor form showed a *meridional* coordination mode (same as **17a**/**17b**, Figure 5.7). The authors showed that **28** could be turned into the minor form by extended heating. Both forms behaved similarly under the reaction conditions, which pointed to a common intermediate (**28i**, Figure 5.9) [53]. Using **28**, both aromatic and aliphatic esters as well as lactones could be reduced under the reaction conditions (2 mol% catalyst, 30 bar H_2, 110 °C in 1,4-dioxane, Table 5.1, entry 2). The catalyst showed excellent functional group tolerance. A proposed mechanism for the ester hydrogenation can be seen in Figure 5.9.

Complex **28** is deprotonated to form the imido complex **28I**. This complex reacts reversibly with hydrogen gas to generate the catalytically active hydride species (**28H**). The active catalyst reduces the ester via an outer-sphere mechanism to first generate the hemi-acetal species that decomposes to the aldehyde and methanol, while **28H** reverts to **28I**. The benzaldehyde then gets further reduced by **28H** to generate benzyl alcohol.

A manganese complex derived from two bidentate ligands was reported by the groups of Pidko, Beller, and coworkers (**29**, Figure 5.8 and Table 5.1, entry 3). This catalyst was able to reduce esters with as low a loading as 0.2 mol% but needed higher amounts of base (75 mol% KO*t*Bu) to achieve good conversion [57]. The reason behind the high loading of the base was believed to be due to manganese catalyst inhibition of the alkoxide formed in the final step of the catalytic cycle. The catalyst was found to be most active using 1,4-dioxane as solvent and required 80–120 °C and 50 bar dihydrogen pressure to achieve good conversions.

Catalyst **20** (Figure 5.7) was reported by Clarke, Widegren, and coworkers to be able to reduce esters initially using isopropanol as reaction solvent at

Table 5.1 Examples of ester hydrogenation using different manganese catalysts.

Entry	Ester	Catalyst (Figure 5.8)	Catalyst loading (mol%)	Base loading (mol%)	Base	Temperature (°C)	Pressure (bar)	Reaction time (h)	Solvent	Conversion (yield, %)
1	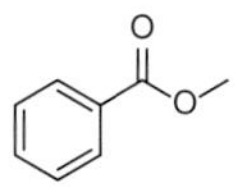	**27**	1	2	KH	100	20	50	toluene	99 (98)
2	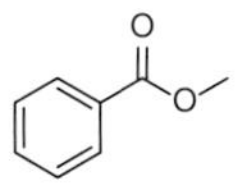	**28**	2	10	KO*t*Bu	110	30	24	1,4-dioxane	99 (97)
3	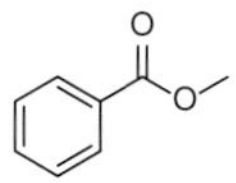	**29**	0.2	75	KO*t*Bu	100	50	26	1,4-dioxane	95 (87)
4	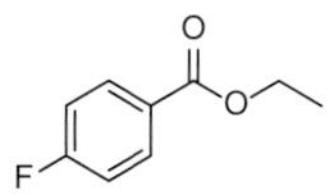	**20**	0.1	10	K_2CO_3	90	20	16	ethanol	99 (76)

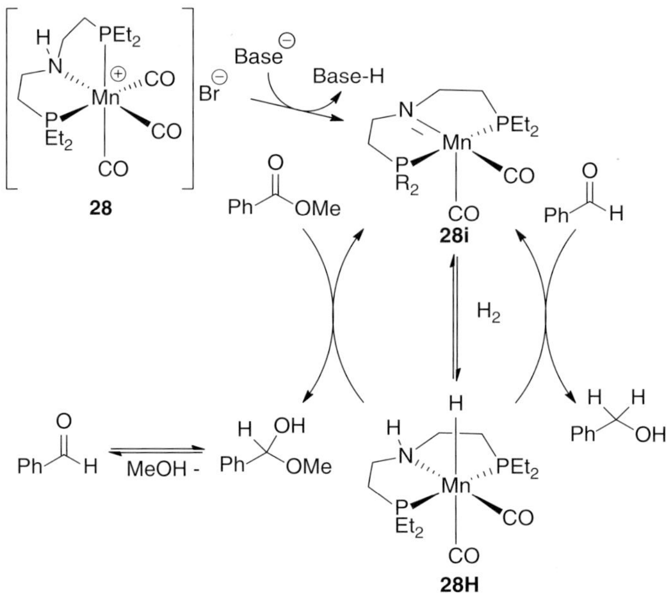

Figure 5.9 Proposed mechanism for the hydrogenation of esters using catalyst **28** [56].

75 °C and 50 bar of dihydrogen pressure with 1 mol% catalyst loading and 10 mol% base (KO*t*Bu) [46]. One example of operating at 0.1 mol% of catalyst was reported. Later it was found that the catalyst could be used in ethanol at 0.1 mol% loading using 20 bar H_2 and potassium carbonate as base (Table 5.1, entry 4) [55].

More examples of ester and lactone hydrogenation can be found in Table 5.2. A range of esters and lactones have been successfully reduced to the corresponding alcohols using manganese catalysts. The catalysts show excellent substrate tolerance capable of selectively reducing esters in the presence of nitriles (Table 5.2, entry 3) or capable of reducing esters also bearing free amino groups (Table 5.2, entry 4). Halides such as bromide and chloride can be tolerated (Table 5.2, entries 5–7). More recently, further investigations have revealed that catalyst **20** can operate at low catalyst loadings using both industrially desirable conditions (0.1 mol% catalyst loading, 20 bar H_2) and using an inorganic base in ethanol (Table 5.1, entry 4 and Table 5.2 entries 4 and 10). Of special interest is the compound Sclareolide (Table 5.2, entry 11), a precursor to Ambrox® which is used in the perfume industry. Furthermore, the catalyst was found to be able to reduce esters bearing stereogenic centers in the α-position with full retention of stereochemistry using potassium carbonate and with isopropanol as solvent (Table 5.3) [58]. Reductions of α-chiral esters are known to be difficult without racemization, but it has been done using ruthenium catalysts [59]. Both α-amino acid esters and α-alkyl esters were successfully hydrogenated under the reaction conditions with minimal racemization. To minimize racemization,

Table 5.2 Further examples of ester hydrogenation using different manganese catalysts.

Entry	Ester	Catalyst[a]	Catalyst loading (mol%)	Base loading (mol%)	Base	Temperature (°C)	Pressure (bar)	Reaction time (h)	Solvent	Conversion[b]
1	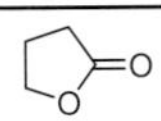	**28**	2	10	KO*t*Bu	110	50	24	1,4-dioxane	99 (88)
2	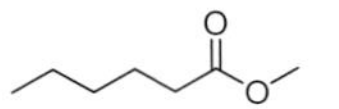	**27**	1	2	KH	100	20	28	toluene	98 (94)
3	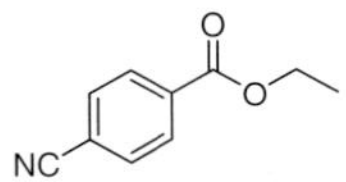	**27**	3	6	KH	100	20	60	toluene	75 (60)
4	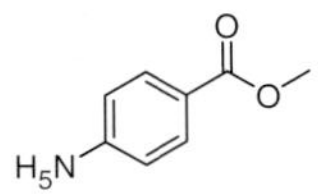	**20**	0.1	10	K_2CO_3	90	50	16	isopropanol	>99 (88)
5	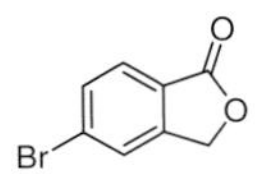	**28**	2	10	KO*t*Bu	110	50	24	1,4-dioxane	79 (57)
6	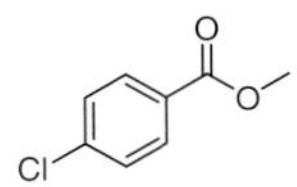	**29**	0.2	75	KO*t*Bu	100	50	16	1,4-dioxane	89 (75)
7	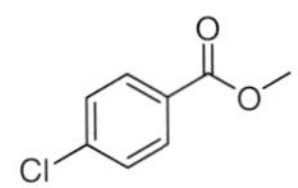	**28**	2	10	KO*t*Bu	110	30	24	1,4-dioxane	>99 (91)
8	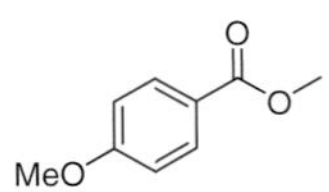	**29**	0.2	75	KO*t*Bu	100	50	16	1,4-dioxane	89 (81)
9	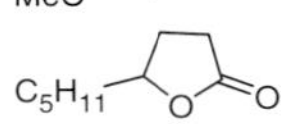	**28**	2	10	KO*t*Bu	110	30	24	1,4-dioxane	>99 (82)
10	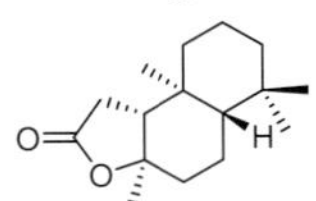	**20**	0.1	10	K_2CO_3	90	50	20	ethanol	>99 (75)

a) See Figure 5.8.
b) Conversion to alcohol in percent (isolated yield in brackets).

Table 5.3 Hydrogenation of chiral esters to chiral alcohols using catalyst **20**.[a)]

Entry	Ester	ee (Starting material) (%)	ee (Product) (%)	Conversion (%)	Yield (%)
1	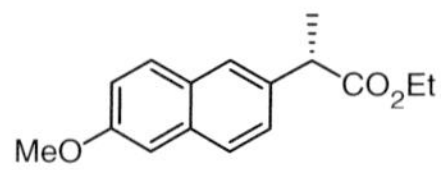	> 99	97.5	> 99	74
2	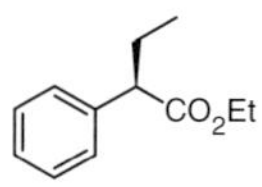	> 99	97.5	> 99	90
3	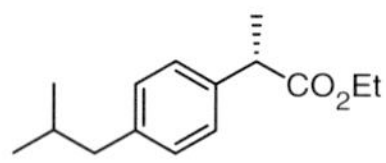	> 99	98.5	> 99	96
4	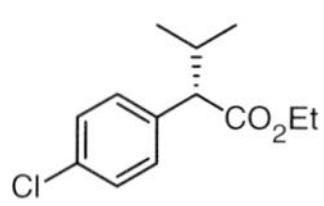	> 99	98.0	> 99	91
5[b)]	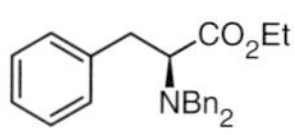	> 99	>99	> 99	66
6	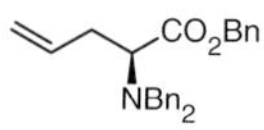	98	98	> 99	76

a) Typical conditions: 1.00 mmol substrate, 0.01 mmol **20**, 0.10 mmol K_2CO_3, 50 bar H_2 (g), isopropanol, 50 °C, 16 hours.
b) Reaction run at 110 °C.

higher catalyst loading (1 mol%) had to be used for the reaction rate to be high enough at the lower reaction temperature (50 °C).

Due to their stability, amides are one of the harder functional groups to hydrogenate. Reduction of amides either gives the substituted amine (path A, Scheme 5.11) or breaks the amide bond to generate an amine and an alcohol (path B, Scheme 5.11).

Catalyst **30** (Scheme 5.11) was found to be very active in the reduction of amides and gave products that were formed from path B [60]. Cyclohexane was found to be the best solvent for the amide reduction. The catalyst was able to reduce primary, secondary, and tertiary amides in moderate to excellent yield.

Carbon dioxide is a desirable renewable chemical (if exhaust gases are used). The potential of using carbon dioxide as a reusable source to generate liquid fuels (formic acid and methanol) via reduction using hydrogen gas is therefore being actively researched.

Catalysts **17a** [61] and **31** [62] (Figure 5.10) have both been used to investigate the reduction of carbon dioxide to methanol. The protocol used is a two-step

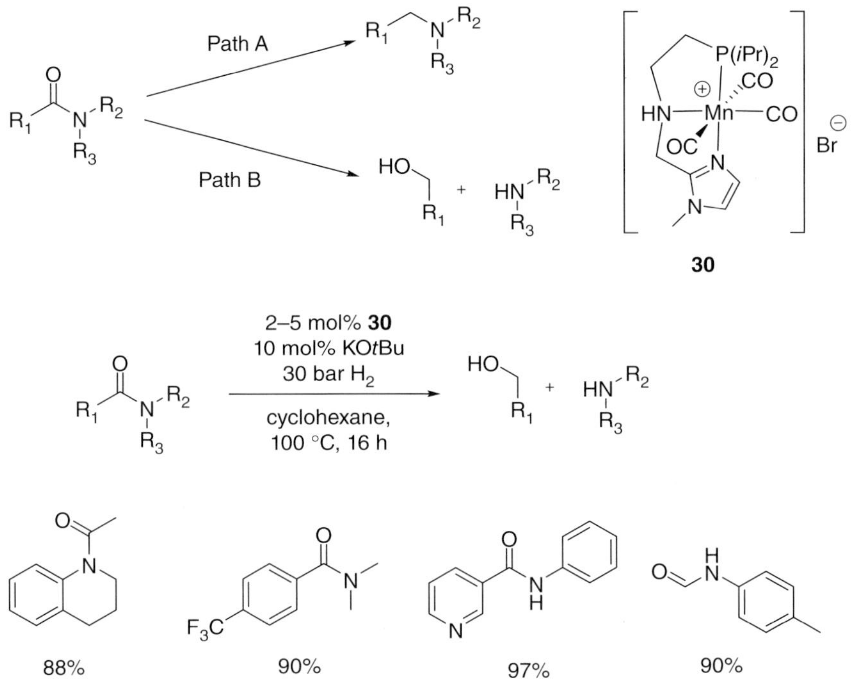

Scheme 5.11 Amide hydrogenation using manganese catalyst **30** [60]. Values under the structures are yields in percent of the products (calculated using GC and hexadecane as external standard).

Figure 5.10 Manganese-based catalysts used in the reduction of carbon dioxide.

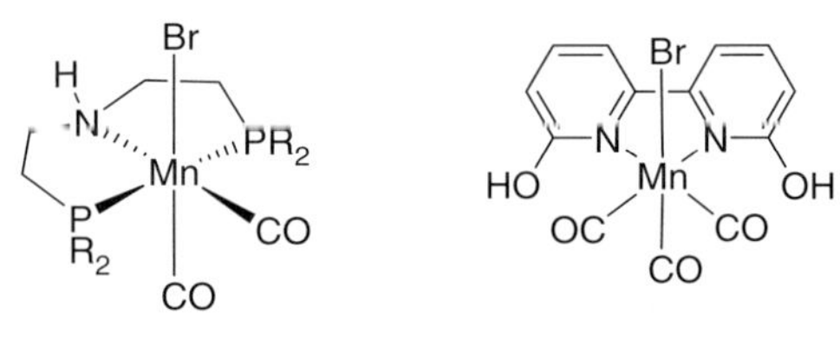

17a (R = *i*-Pr) **31**

procedure involving the formation of a formamide in the first step, followed by the reduction of the formamide to methanol and amine (Scheme 5.12).

Catalyst **17a** was found to catalyze both the reductions of carbon dioxide to formamides and the formamides reduced to methanol. Interestingly, the authors found that the reduction of carbon dioxide to the formamide of morpholine was achievable in the presence of potassium phosphate in THF. The reduction of carbon dioxide and the formed formamides both needed high pressure (60 bar of a 1 : 1 mixture of H_2 and CO_2) and temperature (110 °C) in THF to be achieved at reasonable rates.

Catalyst **31** reduced carbon dioxide to either formate (using 1,8-DiazaBicylo-7-Undecene [DBU] as base) or formamides (when using diethylamine) under carbon dioxide and dihydrogen gas (30 bar each) at 65–80 °C in acetonitrile with good turnover number (TON) (up to 6250 reported) over 24 hours. The authors

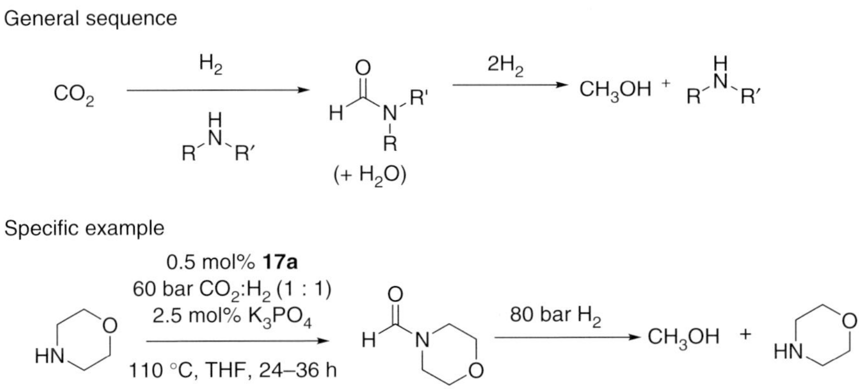

Scheme 5.12 Reduction of carbon dioxide to methanol [61].

saw a sharp effect of having the *ortho*-hydroxy groups on the bipyridyl moiety with considerably rate enhancement compared to other substituents in the *ortho* position. Computational studies on the reduction of carbon dioxide have been reported for either **17a** and **31** [63] looking at the mechanism and base-free reactions.

Although rhenium is not an earth-abundant metal, it has been included in this chapter as a matter of consistency. Rhenium complexes have been reported in the hydrogenation of olefins over the years [64]. Berke and coworkers reported the use of rhenium nitrosyl complexes to reduce olefins using hydrogen gas [64c, 64e, 64f, 65]. The group found that adding a Lewis acid ($HSiR_3/B(C_6F_5)_3$) increased the activity of the catalyst, and believed this was due to a change in the coordination mode of the nitrosyl ligand from linear to bent (Figure 5.11).

In short, rhenium complex **32** can exchange water for hydrogen to form **32a**. This complex reacts with a Lewis acid to form **32b**. The Lewis acid causes the nitrosyl to go from a linear binding mode to a bent mode (**32c**). This activates the PCy_3 ligand which acts as a base and deprotonates the coordinated dihydrogen to give the hydride **32d**. This complex has an open coordination site in which an alkene can coordinate (**32e**). A migratory insertion of the rhenium-hydride occurs to form a rhenium alkyl species, which undergoes a reductive elimination to deliver the alkane product [66]. The authors found that the oxophilic silyl cation (formed from the reaction of a hydrosilane with a borane) gave the best results. Interestingly, the catalyst was found to be active in the absence of a cocatalyst, giving a turnover frequency (TOF) of 5000 without a cocatalyst, but an impressive TOF of 49 000 in the presence of the cocatalyst for the hydrogenation of 1-hexene. The group later published a study on different kinds of Lewis acids used as a cocatalyst for the reduction of alkenes using dinitrosyl complexes [64f].

A study on asymmetric hydrogenation of α,β-unsaturated acids using rhenium complexes of Josiphos and Walphos showed a modest enantioselectivity (8–57%) but reasonable conversion under the reaction conditions (1 mol% catalyst, 50 bar H_2, 100 °C, in ethanol/toluene 1 : 1, 24 hours) [67].

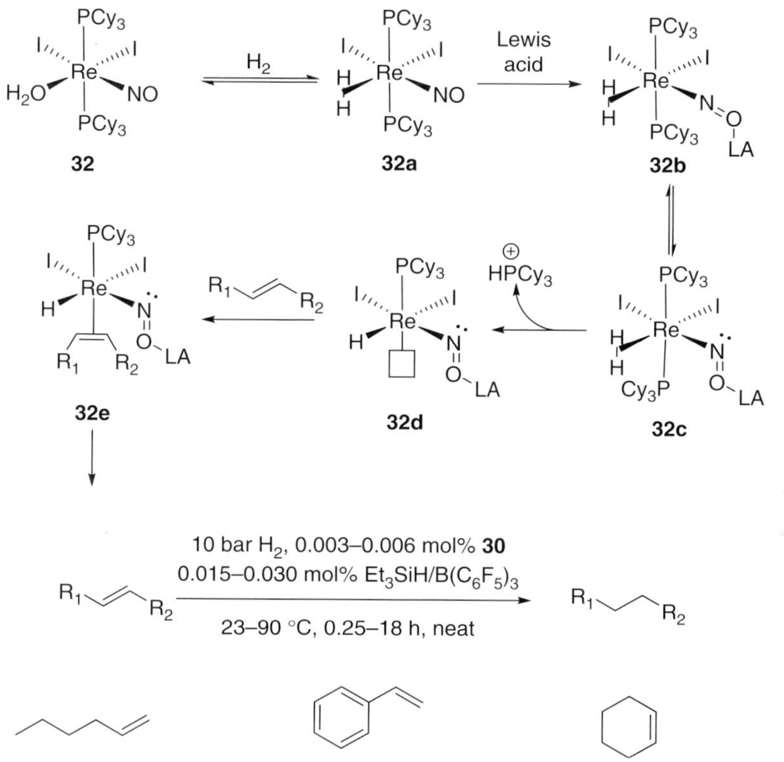

TOF 5×10^3 (w/o Lewis acid) TOF 3816 (w/Lewis acid) TOF 5.2×10^4 (w/Lewis acid)
TOF 4.9×10^4 (w/Lewis acid)

Figure 5.11 Proposed action of a Lewis acid on the rhenium catalyst **32** [65].

Transfer hydrogenation of alkenes using a nitrosyl rhenium catalyst was reported using dimethylamine-borane as the hydrogen source [64b, 64d]. Here, the authors found that the rhenium complex (of the type $Br_2Re(H_2)(NO)(P\textit{i}Pr_3)_2$, similar in structure to compound **32** where the iodides have been replaced by bromides and the water molecule by dihydrogen) could effectively dehydrogenate the dimethylamine-borane and use that to reduce olefins in good yield under the reaction conditions (1 mol% catalyst, 1 equiv of dimethylamine-borane to alkene, 1,4-dioxane, 85 °C, 1–4 hours). The authors believe that the dimethylamine-borane reacts with the rhenium complex to form an unsaturated rhenium hydride complex that then reacts with the alkene [64d]. Another example of transfer hydrogenation using a rhenium catalyst was reported for the reduction of ketones and imines (Figure 5.12) [68].

Catalyst **33** (Figure 5.12) is a bifunctional catalyst similar to Shvo's ruthenium catalyst [69] in that it can shuttle between two oxidation states (**33** and **34** in Figure 5.12). Rhenium catalyst **33** reacts with a ketone and gets oxidized to **34** while reducing the ketone to the alcohol. Catalyst **34** can then be reduced back to **33** using an alcohol (here isopropanol) as the hydrogen source. Both aromatic

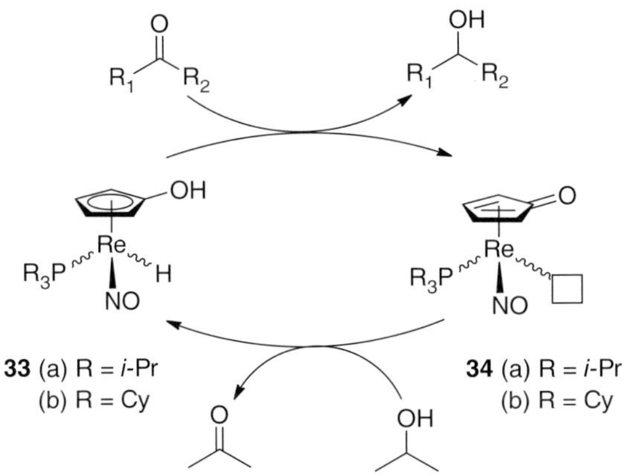

Figure 5.12 Transfer hydrogenation using catalyst **33**. Source: Adapted from Landwehr et al. 2012 [68].

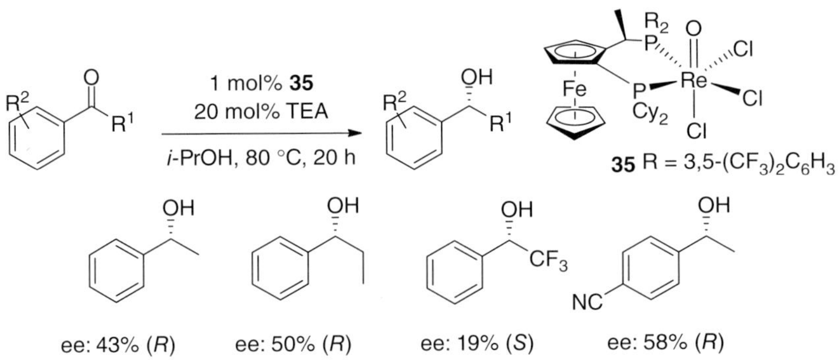

Figure 5.13 Asymmetric transfer hydrogenation of ketones using rhenium catalyst **35** [73].

and aliphatic ketones as well as aromatic imines were found to be readily reduced under the reaction conditions (0.5 mol% **33**, isopropanol, 120 °C). These conditions are relatively forcing compared to similar systems using either Shvo's ruthenium catalyst [70] (0.33 mol% catalyst, 70 °C, 1–4 hours) or iron-based catalysts [71] (5 mol% catalyst, 80 °C, 18 hours), both of which are active at lower temperatures.

A final example of transfer hydrogenation using rhenium catalyst **35** based on the Josiphos ligand family [72] was reported by Togni and coworkers (Figure 5.13) [73].

Using 1 mol% catalyst (prepared *in situ*) and 20 mol% triethylamine in isopropanol at reflux, complex **35** was able to reduce aromatic ketones with moderate enantioselectivities in the range of 19–58% but with excellent conversion and isolated yields.

Manganese is the third most common transition metal found in the earth's crust, while rhenium is as rare as the noble metals palladium and iridium and should not be considered an earth-abundant metal. While the chemistry of the rhenium is fascinating and worth exploring further, manganese being more readily available shows more promise for industrial applications.

5.7 Summary and Conclusions

Metal catalysts from group III to VII have been used in the hydrogenation of a multitude of functional groups such as alkenes, alkynes, imines, enamines, nitriles, amides, ketones, aldehydes, and esters. Titanium has been extensively researched and shown to be able to successfully reduce not only alkenes but also imines and enamines with good enantioselectivity. In most cases, titanium and zirconium catalysts need to be activated using reactive reagents such as alkyllithium and often need to be used at a relatively high catalyst loading, which limits their use on scale. Other metals such as vanadium, molybdenum, and chromium have been reported in the hydrogenation (and transfer hydrogenation) of functional groups, and recently even scandium and yttrium catalysts have been found to be active in hydrogenation reactions. A very recent development has been the use of manganese catalysts in the hydrogenation of carbonyl groups such as ketones and esters. Even α-chiral esters have been reduced with minimal racemization. Although the enantioselectivity in the reduction of ketones are good, there is still some way to go before the manganese catalysts can fully compete with other metal catalysts based on ruthenium and iridium in terms of not only enantioselectivity but also in catalyst loading.

With the increasing focus on more sustainable chemistry, the research into earth-abundant metal catalysts will most likely continue and perhaps intensify even more. Research into the use of earth-abundant metals in groups III–VII is still an open field and there is ample opportunity to explore the chemistry and catalysis of these metals in hydrogenation.

References

1 Guidelines from the European Medicines Agency can be found here, http://www.ema.europa.eu/docs/en_GB/document_library/Scientific_guideline/2009/09/WC500003586.pdf; in effect since 1 September 2008; see also *Elemental Impurities in Drug Products,* Food and Drug Administration draft guidance; http://www.fda.gov/ucm/groups/fdagov-public/@fdagov-drugs-gen/documents/document/ucm509432.pdf.

2 D.S. Levine, T.D. Tilley, R.A. Andersen, *Chem. Commun.* 2017, 53, 11881–11884.

3 (a) D.S. Levine, T.D. Tilley, R.A. Andersen, *Organometallics* 2015, 34, 4647–4655; (b) D.S. Levine, T.D. Tilley, R.A. Andersen, *Organometallics* 2017, 36, 80–88.

4 W.D. Bonds, C.H. Brubaker, E.S. Chandrasekaran, C. Gibbons, R.H. Grubbs, L.C. Kroll, *J. Am. Chem. Soc.* 1975, 97, 2128–2132.

5 Y.P. Zhang, S.J. Liao, Y. Xu, S.S. Chen, *J. Organomet. Chem.* 1990, 382, 69–76.

6 (a) C.A. Willoughby, S.L. Buchwald, *J. Am. Chem. Soc.* 1992, 114, 7562–7564; (b) R.D. Broene, S.L. Buchwald, *J. Am. Chem. Soc.* 1993, 115, 12569–12570; (c) J. Okuda, S. Verch, R. Sturmer, T.P. Spaniol, *J. Organomet. Chem.* 2000, 605, 55–67.

7 N.E. Lee, S.L. Buchwald, *J. Am. Chem. Soc.* 1994, 116, 5985–5986.
8 C.H. Chiang, J.C. Tsai, *J. Polym. Sci. Polym. Chem.* 2017, 55, 2141–2149.
9 F. Scott, H.G. Raubenheimer, G. Pretorius, A.M. Hamese, *J. Organomet. Chem.* 1990, 384, C17–C20.
10 Y.P. Zhang, S.J. Liao, Y. Xu, *J. Mol. Cat.* 1993, 84, 211–221.
11 M. Zhang, A.-Q. Zhang, R. Zhang, S.-H Yang, Y.-J Hong, *Catal. Lett.* 2008, 124, 146–149.
12 X. Verdaguer, U.E.W. Lange, M.T. Reding, S.L. Buchwald, *J. Am. Chem. Soc.* 1996, 118, 6784–6785.
13 M.F. Sloan, A.S. Matlack, D.S. Breslow, *J. Am. Chem. Soc.* 1963, 85, 4014–4018.
14 J.E. Bercaw, H.H. Brintzinger, *J. Am. Chem. Soc.* 1971, 93, 2046–2048.
15 E.S. Chandrasekaran, R.H. Grubbs, C.H. Brubaker, *J. Organomet. Chem.* 1976, 120, 49–63.
16 M. Capka, A. Reissova, *Collect. Czech. Chem. Commun.* 1989, 54, 1760–1769.
17 T. Cuenca, J.C. Flores, P. Royo, *J. Organomet. Chem.* 1993, 462, 191–201.
18 (a) D.W. Hart, J. Schwartz, *J. Am. Chem. Soc.* 1974, 96, 8115–8116; (b) S.L. Buchwald, S.J. LaMaire, R.B. Nielsen, B.T. Watson, S.M. King, *Org. Synth.* 1993, 71, 77.
19 (a) Y. Raoult, R. Choukroun, M. Bassobert, D. Gervais, *J. Mol. Cat.* 1992, 72, 47–58; (b) R.B.-B. Choukroun, M. Bassobert, D. Gervais, *J. Chem. Soc. Chem. Commun.* 1986, 1317–1319.
20 E. Cesarotti, R. Ugo, H.B. Kagan, *Angew. Chem.* 1979, 91, 842–843.
21 R.L. Halterman, K.P.C. Vollhardt, M.E. Welker, D. Blaeser, R. Boese, *J. Am. Chem. Soc.* 1987, 109, 8105–8107.
22 R.L. Halterman, K.P.C. Vollhardt, *Organometallics* 1988, 7, 883–892.
23 (a) A. Viso, N.E. Lee, S.L. Buchwald, *J. Am. Chem. Soc.* 1994, 116, 9373–9374; (b) C.A. Willoughby, S.L. Buchwald, *J. Am. Chem. Soc.* 1994, 116, 8952–8965.
24 J. Okuda, S. Verch, T.P. Spaniol, R. Sturmer, *Chem. Ber.* 1996, 129, 1429–1431.
25 M. Ringwald, R. Sturmer, H.H. Brintzinger, *J. Am. Chem. Soc.* 1999, 121, 1524–1527.
26 L.A. Paquette, M.R. Sivik, E.I. Bzowej, K.J. Stanton, *Organometallics* 1995, 14, 4865–4878.
27 M.V. Troutman, D.H. Appella, S.L. Buchwald, *J. Am. Chem. Soc.* 1999, 121, 4916–4917.
28 K.S. Egorova, V.P. Ananikov, *Organometallics* 2017, 36, 4071–4090.
29 H.S. La Pierre, J. Arnold, F.D. Toste, *Angew. Chem. Int. Ed. Engl.* 2011, 50, 3900–3903.
30 T.L. Gianetti, N.C. Tomson, J. Arnold, R.G. Bergman, *J. Am. Chem. Soc.* 2011, 133, 14904–14907.
31 J.S. Yu, B.C. Ankianiec, I.P. Rothwell, M.T. Nguyen, *J. Am. Chem. Soc.* 1992, 114, 1927–1929.
32 I.P. Rothwell, *Chem. Commun.* 1997, 1331–1338.
33 D.R. Mulford, J.R. Clark, S.W. Schweiger, P.E. Fanwick, I.P. Rothwell, *Organometallics* 1999, 18, 4448–4458.
34 Z. Wei, K. Junge, M. Beller, H. Jiao, *C. R. Chim.* 2018, 21, 303–309.

35 P.L. Gaus, S.C. Kao, K. Youngdahl, M.Y. Darensbourg, *J. Am. Chem. Soc.* 1985, 107, 2428–2434.

36 P.A. Tooley, C. Ovalles, S.C. Kao, D.J. Darensbourg, M.Y. Darensbourg, *J. Am. Chem. Soc.* 1986, 108, 5465–5470.

37 L. Marko, Z. Nagymagos, *J. Organomet. Chem.* 1985, 285, 193–203.

38 J.J. Brunet, R. Chauvin, P. Leglaye, *Eur. Inorg. Chem.* 1999, 713–716.

39 (a) C. Balzarek, D.R. Tyler, *Angew. Chem. Int. Ed. Engl.* 1999, 38, 2406–2408; (b) C. Balzarek, T.J.R. Weakley, L.Y. Kuo, D.R. Tyler, *Organometallics* 2000, 19, 2927–2931; (c) C. Balzarek, T.J.R. Weakley, D.R. Tyler, *J. Am. Chem. Soc.* 2000, 122, 9427–9434.

40 (a) L.Y. Kuo, T.J.R. Weakley, K. Awana, C. Hsia, *Organometallics* 2001, 20, 4969–4972; (b) L.Y. Kuo, D.M. Finigan, N.N. Tadros, *Organometallics* 2003, 22, 2422–2425.

41 (a) M.H. Voges, R.M. Bullock, *J. Chem. Soc. Dalton* 2002, 759–770; (b) B.F.M. Kimmich, P.J. Fagan, E. Hauptman, W.J. Marshall, R.M. Bullock, *Organometallics* 2005, 24, 6220–6229; (c) R.M. Bullock (ed.) *Catalysis Without Precious Metals.* Weinheim, Wiley-VCH, 2010, 51–81; (d) A. Dybov, O. Blacque, H. Berke, *Eur. Inorg. Chem.* 2011, 2011, 652–659; (e) S. Chakraborty, O. Blacque, T. Fox, H. Berke, *Chem. Asian J.* 2014, 9, 2896–2907.

42 S. Elangovan, C. Topf, S. Fischer, H. Jiao, A. Spannenberg, W. Baumann, R. Ludwig, K. Junge, M. Beller, *J. Am. Chem. Soc.* 2016, 138, 8809–8814.

43 F. Kallmeier, T. Irrgang, T. Dietel, R. Kempe, *Angew. Chem. Int. Ed. Engl.* 2016, 55, 11806–11809.

44 A. Bruneau-Voisine, D. Wang, T. Roisnel, C. Darcel, J.-B Sortais, *Catal. Commun.* 2017, 92, 1–4.

45 D. Wei, A. Bruneau-Voisine, T. Chauvin, V. Dorcet, T. Roisnel, D.A. Valyaev, N. Lugan, J.-B. Sortais, *Adv. Synth. Catal.* 2017, 1–4.

46 M.B. Widegren, G.J. Harkness, A.M.Z. Slawin, D.B. Cordes, M.L. Clarke, *Angew. Chem. Int. Ed. Engl.* 2017, 56, 5825–5828.

47 M. Garbe, K. Junge, S. Walker, Z. Wei, H. Jiao, A. Spannenberg, S. Bachmann, M. Scalone, M. Beller, *Angew. Chem. Int. Ed. Engl.* 2017, 56, 11237–11241.

48 X. Yu, C. Liu, L. Jiang, Q. Xu, *Org. Lett.* 2011, 13, 6184–6187.

49 M. Perez, S. Elangovan, A. Spannenberg, K. Junge, M. Beller, *ChemSusChem* 2017, 10, 83–86.

50 A. Bruneau-Voisine, D. Wang, V. Dorcet, T. Roisnel, C. Darcel, J.-B. Sortais, *Org. Lett.* 2017, 19, 3656–3659.

51 A. Zirakzadeh, S.R.M.M. de Aguiar, B. Stöger, M. Widhalm, K. Kirchner, *ChemCatChem* 2017, 9, 1744–1748.

52 D. Wang, A. Bruneau-Voisine, J.B. Sortais, *Catal. Commun.* 2018, 105, 31–36.

53 T.K. Mukhopadhyay, M. Flores, T.L. Groy, R.J. Trovitch, *J. Am. Chem. Soc.* 2014, 136, 882–885.

54 N.A. Espinosa-Jalapa, A. Nerush, L.J.W. Shimon, G. Leitus, L. Avram, Y. Ben-David, D. Milstein, *Chem.-Eur. J.* 2017, 23, 5934–5938.

55 J.R. Khusnutdinova, D. Milstein, *Angew. Chem. Int. Ed. Engl.* 2015, 54, 12236–12273.

56 S. Elangovan, M. Garbe, H. Jiao, A. Spannenberg, K. Junge, M. Beller, *Angew. Chem. Int. Ed. Engl.* 2016, 55, 15364–15368.

57 R. van Putten, E.A. Uslamin, M. Garbe, C. Liu, A. Gonzalez-de-Castro, M. Lutz, K. Junge, E.J.M. Hensen, M. Beller, L. Lefort, E.A. Pidko, *Angew. Chem. Int. Ed. Engl.* 2017, 56, 7531–7534.

58 M.B. Widegren, M.L. Clarke, *Org. Lett.* 2018, 20, 2654–2658.

59 (a) Y. Ino, W. Kuriyama, O. Ogata, T. Matsumoto, *Top. Catal.* 2010, 53, 1019–1024; (b) W.M. Kuriyama, T. Matsumoto, Y. Ino, O. Ogata, Novel ruthenium carbonyl complex having a tridentate ligand and manufacturing method and usage therefor. WO/2011/048727, Takasago International Corporation. patentscope.wipo.int/search/en/detail.jsf?docId=WO2011048727&tab=PCTBIBLIO&maxRec=1000.

60 V. Papa, J.R. Cabrero-Antonino, E. Alberico, A. Spanneberg, K. Junge, H. Junge, M. Beller, *Chem. Sci.* 2017, 8, 3576–3585.

61 S. Kar, A. Goeppert, J. Kothandaraman, G.K.S. Prakash, *ACS Catal.* 2017, 7, 6347–6351.

62 A. Dubey, L. Nencini, R.R. Fayzullin, C. Nervi, J.R. Khusnutdinova, *ACS Catal.* 2017, 7, 3864–3868.

63 (a) K.S. Rawat, A. Mahata, I. Choudhuri, B. Pathak, *J. Phys. Chem. C* 2016, 120, 16478–16488; (b) K.S. Rawat, B. Pathak, *Catal. Sci. Technol.* 2017, 7, 3234–3242.

64 (a) X.Y. Liu, K. Venkatesan, H.W. Schmalle, H. Berke, *Organometallics* 2004, 23, 3153–3163; (b) Y. Jiang, H. Berke, *Chem. Comm.* 2007, 3571–3573; (c) A. Choualeb, E. Maccaroni, O. Blacque, H.W. Schmalle, H. Berke, *Organometallics* 2008, 27, 3474–3481; (d) Y.F. Jiang, O. Blacque, T. Fox, C.M. Frech, H. Berke, *Organometallics* 2009, 28, 5493–5504; (e) B. Dudle, K. Rajesh, O. Blacque, H. Berke, *J. Am. Chem. Soc.* 2011, 133, 8168–8178; (f) Y. Jiang, W. Huang, H.W. Schmalle, O. Blacque, T. Fox, H. Berke, *Organometallics* 2013, 32, 7043–7052; (g) L.J. Liu, S.W. Bi, M. Sun, X. Yuan, N. Zheng, P. Li, *J. Organomet. Chem.* 2009, 694, 3343–3348.

65 Y. Jiang, B. Schirmer, O. Blacque, T. Fox, S. Grimme, H. Berke, *J. Am. Chem. Soc.* 2013, 135, 4088–4102.

66 R. Schrock, J.A. Osborn, *J. Am. Chem. Soc.* 1976, 98, 2134–2143.

67 A.F. Abdel-Magied, M.S. Patil, A.K. Singh, M. Haukka, M. Monari, E. Nordlander, *J. Clust. Sci.* 2015, 26, 1231–1252.

68 A. Landwehr, B. Dudle, T. Fox, O. Blacque, H. Berke, *Chem. Eur. J.* 2012, 18, 5701–5714.

69 Y. Shvo, D. Czarkie, Y. Rahamim, D.F. Chodosh, *J. Am. Chem. Soc.* 1986, 108, 7400–7402.

70 J.S.M. Samec, J.-E. Bäckvall, *Chem. Eur. J.* 2002, 8, 2955–2961.

71 T.N. Plank, J.L. Drake, D.K. Kim, T.W. Funk, *Adv. Synth. Catal.* 2012, 354, 597–601.

72 H.-U. Blaser, W. Brieden, B. Pugin, F. Spindler, M. Studer, A. Togni, *Top. Catal.* 2002, 19, 3–16.

73 E. Mejía, R. Aardoom, A. Togni, *Eur. Inorg. Chem.* 2012, 2012, 5021–5032.

6

Early Main Group Metal Catalyzed Hydrogenation

Heiko Bauer and Sjoerd Harder

Universität Erlangen-Nürnberg, Department Chemie und Pharmazie, Chair für Inorganic and Organometallic Chemistry, Egerlandstraße 1, 91058 Erlangen, Germany

6.1 Introduction

Unsaturated double bonds such as C=C, C=N, or C=O consist of a strong stable σ-bond in conjunction with a much weaker π-interaction. The latter forms the basis for a rich chemistry and allows for the addition of a wide range of substrates. Hydrogenation, the addition of H_2 to a double bond, is probably one of the best investigated transformations. As the product is lower in energy than the educts, the hydrogenation of double bonds with H_2 is a thermodynamically favorable reaction. Nevertheless, a catalyst is needed for this exergonic chemical transformation. Simple frontier orbital theory demonstrates that all HOMO/LUMO combinations of the double bond containing substrate and H_2 result in an overlap integral of zero (Figure 6.1) meaning that the four-membered ring transition state for the synchronous addition of H_2 to the double bond is symmetry forbidden [1]. The high activation barrier is overcome by a catalyst promoting the heterolytic cleavage of H_2 through the interaction of orbitals of the metal with the hydrogen orbitals. As d-orbitals of the transition metals show perfect symmetry for the interaction with the HOMO and LUMO of H_2, most catalytic hydrogenation reactions are based on transition metal catalysts.

The first catalytic hydrogenation was already discovered in 1874 by Wilde, who reported the catalytic conversion of acetylene and H_2 into ethane using "platinum black" (finely divided Pt powder) [2]. This early example of heterogeneous hydrogenation was the starting point for the discovery of further heterogeneous hydrogenation catalysts. Sabatier developed around 1900 a procedure for the catalytic hydrogenation of ethylene with H_2 using finely divided nickel at surprisingly low temperatures of 30–45 °C. Further improvements on catalytic hydrogenation of acetylene with Co, Fe, and Cu catalysts were ultimately acknowledged with the 1912 Nobel Prize in chemistry [3]. At the beginning of the twentieth century, Murray Raney developed his famous Raney nickel and showed that his finely powdered nickel is an active heterogeneous catalyst for the hydrogenation of C=N double and C≡N triple bonds. Other well-known heterogeneous

Homogeneous Hydrogenation with Non-Precious Catalysts, First Edition. Edited by Johannes F. Teichert.

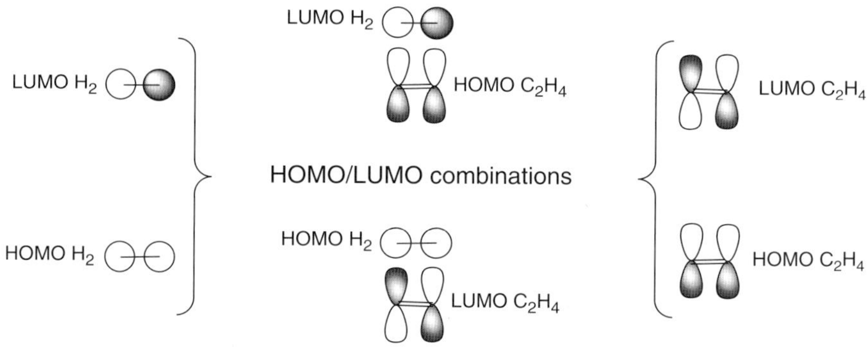

Figure 6.1 Representation of the HOMOs and LUMOs of H_2, C_2H_4, and unproductive orbital interactions for the concerted addition of H_2 to C_2H_4.

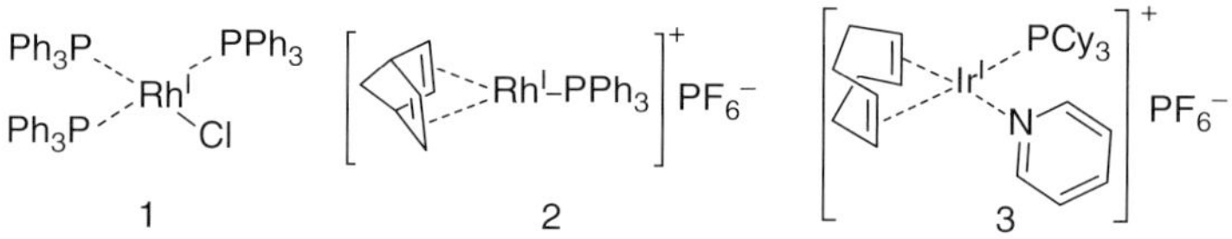

Figure 6.2 Early homogeneous transition metal hydrogenation catalysts.

catalysts, which are still used today, are, for example, platinum dioxide or Pt and Pd on carbon [4].

In conjunction with the rapidly developing field of transition metal complex chemistry in the early 1960s, also homogeneous catalysis started to develop rapidly. The very first catalysts for homogeneous hydrogenation are based on late transition metals. An early example is $RhCl(PPh_3)_3$ (**1**), introduced by Wilkinson as a precatalyst for the hydrogenation of olefins (Figure 6.2). The catalytically active species was reported to be $RhCl(PPh_3)_2$, which is able to hydrogenate olefins at room temperature and normal pressure [5]. Only a few years later in 1976, Osborn and Schrock developed a further catalyst, which is still in use. The rhodium complex $[Rh(nbd)PPh_3{}^+][PF_6{}^-]$ (nbd = norbornadiene) **(2)** (Figure 6.2) was synthesized by exchanging the chlorine atom for neutral nbd, thus creating a cationic rhodium species that led to significantly increased activities in the catalytic hydrogenation of olefins [6]. A third example of a highly active olefin hydrogenation catalyst is the cationic iridium complex $[Ir(cod)PCy_3(pyr)^+][PF_6{}^-]$ (cod = cyclooctadiene, pyr = pyridine) **(3)** (Figure 6.2) introduced by Crabtree in 1979 [7].

These pioneering hydrogenation catalysts incorporate expensive and rare transition metals. In contrast, enzymes in nature are generally based on more abundant, less noble metals. The catalytic center in hydrogenases, enzymes that regulate both the uptake and production of H_2 in microorganisms, was generally believed to be based on Fe and/or Ni. The early 1990s saw the surprising discovery of a metal-free enzyme (Hmd: H_2-forming methylene-tetrahydromethanopterin) that catalyzed hydrogenation without the need for Ni or Fe–S clusters [8]. This spectacular breakthrough in hydrogenation catalysis was later unmasked. The cofactor of Hmd was found to contain Fe as could be shown unequivocally by determination of its molecular structure [8]. Although not purely organocatalytic [9–11], it still presents an important milestone in the trend toward using less toxic and cheaper metals in hydrogenation catalysis.

Apart from using the more economical first row transition metals there is also increasing interest in using redox-inactive main group or lanthanide metals in hydrogenation catalysis. First attempts in the homogeneously catalyzed hydrogenation of C=C double bonds without transition metals started already in the 1960s. Slaugh et al. could show that $LiAlH_4$ or alkaline metal hydrides and alkaline earth metal hydrides [12, 13] are able to hydrogenate alkenes. However, the harsh conditions required (35 mol% cat, 190 °C, 80 bar H_2) and poor performance did not encourage any further research and clearly favored the transition metal catalysts. Thus, it took nearly 30 years until the development of non-transition metal catalysts started to take off. Marks and coworkers in 1985 proved, in a mechanistic study, that lanthanides are very active hydrogenation catalysts [14]. The oxidation state of the lanthanide metal does not change during this transformation and therefore a mechanism different from the oxidative addition/reductive elimination protocol, typical for transition metal catalyzed hydrogenations, was proposed. Lanthanide catalyzed alkene hydrogenation proceeds via an olefin/metal hydride insertion, followed by a metal alkyl/H_2 σ-bond metathesis. The catalytic procedure could be extended to asymmetric hydrogenation [15] or hydrogenation of C=N double bonds [16]. On a similar note, the Buchwald group reported on the asymmetric hydrogenation of imines with chiral redox-inactive titanocene catalysts [17]. These methods started a change from precious late transition metal catalysts toward cheaper, more abundant, early transition metal catalysts.

The last decade has seen the rapid development of a new concept in hydrogenation catalysis: Stephan introduced organocatalytic metal-free hydrogenation using frustrated Lewis pairs (FLPs) [18–21]. The key to this chemistry is the rupture of the H—H bond by a Lewis base/acid combination that for steric reasons does not form a Lewis pair. Starting with catalytic imine hydrogenation [18] (Figure 6.3), this concept was soon extended to the diastereoselective

Figure 6.3 Catalytic cycle for metal-free imine hydrogenation.

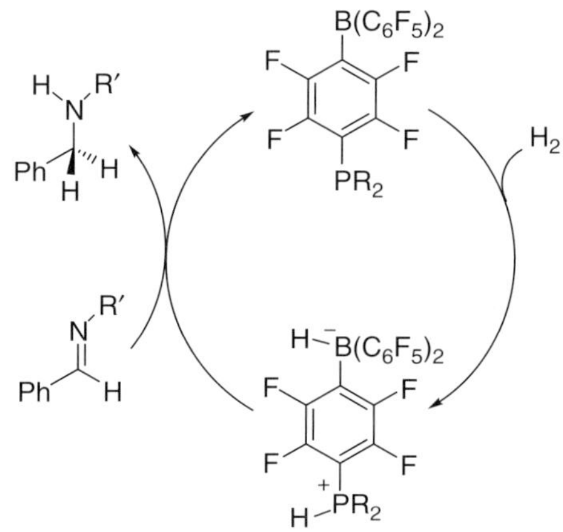

hydrogenation of imines [19] and to FLP-mediated alkene hydrogenation [20]. In imine hydrogenation the imine substrate itself may play the role of the Lewis base, thereby simplifying the catalyst to the Lewis acid $B(C_6F_5)_3$ [21]. Wang et al. introduced highly Lewis acidic hydridoboranes as catalysts for alkene hydrogenation [22]. This unusual process starts with alkene hydroboration by $HB(C_6F_5)_2$, which is followed by cleavage of the alkyl-$B(C_6F_5)_2$ bond by H_2. Although this process does not need a Lewis base, the reaction conditions (140 °C, 72–120 hours) and catalyst loadings (20 mol%) are not very attractive when compared to FLP hydrogenation catalysis.

Apart from metal-free hydrogenation catalysis, there is also a growing interest in the use of d^{10}-metals such as Zn [23] and a revival of late main group metals such as Al [24]. Major breakthroughs, however, were made by using highly earth-abundant and nontoxic alkaline earth metals. The dogma that efficient alkene hydrogenation is only possible with transition metal catalysts was for the first time broken in 2008 by the organocalcium-catalyzed hydrogenation of conjugated alkenes by the Harder group [25]. As alkaline earth metals do not dispose of partially filled d-orbitals, binding of the substrate can occur only via an electrostatic interaction in which the alkene is activated by polarization induced by the Lewis acid. This chapter describes the first steps and the rapid development of early main group metal catalyzed hydrogenation reactions, which is mainly dominated by group 2 metal catalysts. It exclusively focuses on the challenging reduction with the bulk commodity H_2. Addition of the much more polar, highly reactive, hydridoboranes ($R_2B^{\delta+}$-$H^{\delta-}$) and hydridosilanes ($R_3Si^{\delta+}$-$H^{\delta-}$) to unsaturated double bonds is not part of this chapter. This contribution zooms in on the many pitfalls but also focuses on the major successes and future challenges. This chapter is divided in three categories that describe C=C, C=N, and C=O bond hydrogenation, respectively.

6.2 Hydrogenation of C=C Double Bonds

The challenging isolation of the first Ca hydride complex (Figure 6.4 (**4**)) could be seen as the starting point in the development of alkaline earth metal catalyzed hydrogenation chemistry [26]. This complex with the constitution $[(DIPPnacnac)CaH{\cdot}(THF)]_2$ (DIPPnacnac = $HC[(CMe)N(C_6H_3\text{-}2,6\text{-}i\text{Pr})]_2$) is stabilized by a large sterically demanding β-diketiminate ligand that prevents ligand exchange reactions via the Schlenk equilibrium. The latter equilibrium, which is well known for Grignard reagents, describes the ligand exchange between heteroleptic complexes L^1–Ae–L^2 (Ae = alkaline earth metal) to give homoleptic complexes L^1–Ae–L^1 and L^2–Ae–L^2 carrying equal ligands at the metal. The isolation of a Ca hydride complex LCaH, in which L is a spectator ligand, is fully dependent on successful suppression of ligand scrambling. The Schlenk equilibrium would lead to formation of highly insoluble $(CaH_2)_\infty$ salts, thus shifting the equilibrium completely to the side of the homoleptic species. Hitherto, the factors that influence these Schlenk equilibria are not very well understood; however, it is clear that strongly coordinating multidentate ligands

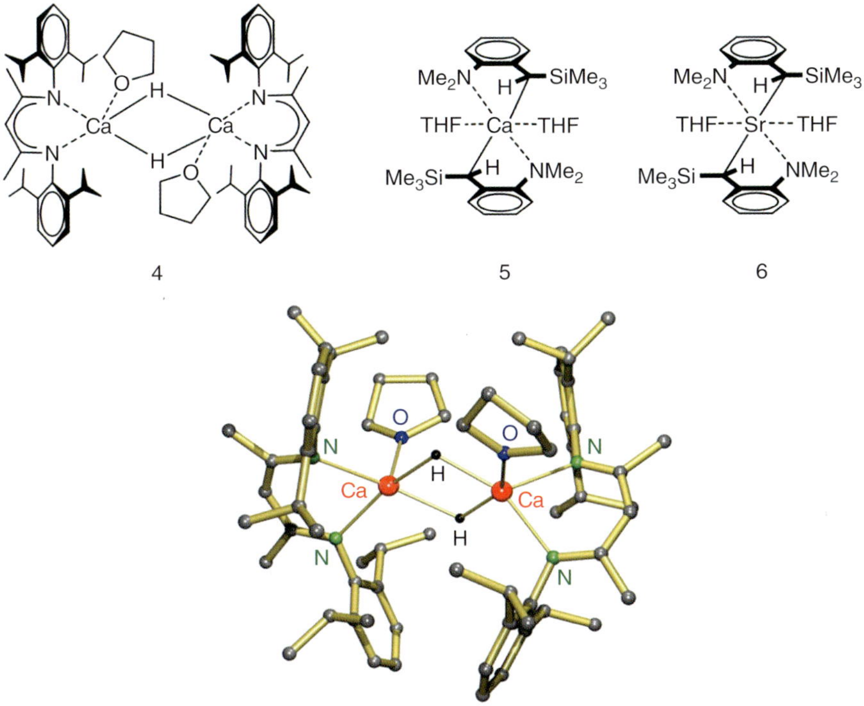

Figure 6.4 Catalysts for the hydrogenation of alkenes: [(DippNacNac)CaH·(THF)]$_2$ (**4**), (DMAT)$_2$Ca·(THF)$_2$ (**5**), and (DMAT)$_2$Sr·(THF)$_2$ (**6**), and the crystal structure of **4**.

are crucial for the isolation of Ca hydride complexes. Since Schlenk equilibria are faster for the bigger metals Sr and Ba, the recent isolation of the first Sr [27] and Ba hydride [28] complexes turned out to be even more challenging.

Just after the isolation of the first Ae hydride complex, catalytic application in the hydrogenation of alkenes was reported by Harder et al. [25]. Besides the calcium hydride complex [(DIPPnacnac)CaH·(THF)]$_2$ (**4**) also the benzyl complexes (DMAT)$_2$Ca·(THF)$_2$ (**5**) and (DMAT)$_2$Sr·(THF)$_2$ (**6**) (DMAT = 2-dimethylamino-α-trimethylsilyl-benzyl) (Figure 6.4) were introduced as (pre)catalysts for alkene hydrogenation. The calcium hydride complex **4** is able to convert styrene to ethylbenzene at room temperature with 5 mol% loading within 15 hours. The reaction proceeds at 20 bar H_2 pressure but besides the main product ethylbenzene also 19% of styrene oligomers are formed. The proposed mechanism (Figure 6.5), which was verified by stoichiometric conversion and characterization of the intermediates, shows that alkene polymerization is competitive with alkene hydrogenation. After exclusive transfer of the hydride to the terminal C of styrene a resonance-stabilized benzylcalcium complex is formed. This reaction is highly regioselective: the alternative product, a primary alkylcalcium complex, is not resonance stabilized. The benzylcalcium intermediate subsequently reacts with H_2 through a σ-bond metathesis mechanism, which is well established in lanthanide chemistry [14] but had previously not been demonstrated for group 2 metals. The latter

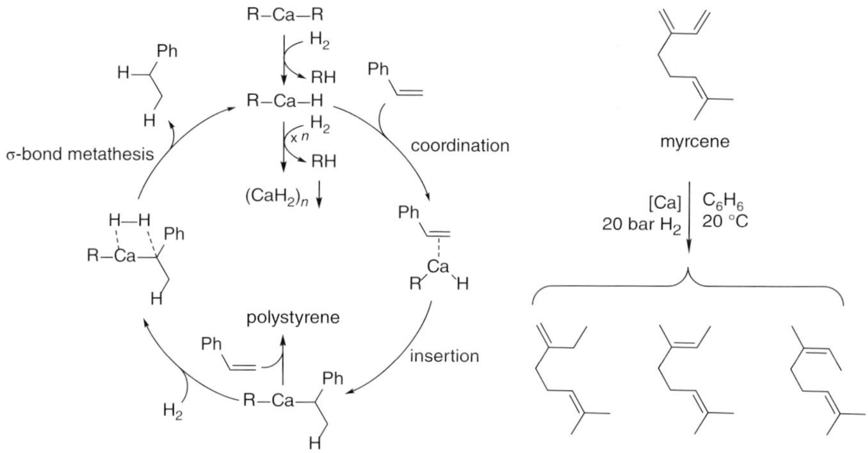

Figure 6.5 Mechanism for the calcium-catalyzed hydrogenation of styrene and the conversion of myrcene leaving the isolated C=C bond unaffected.

conversion, which formally could be viewed as a deprotonation of H_2 by a benzylcalcium reagent, results in product formation and recovery of the Ca hydride catalyst. Formation of oligomers can be explained by a competing side reaction: styrene polymerization. Instead of reaction of the benzylcalcium intermediate with H_2, the intermediate could react with the other substrate present, i.e. styrene. Given the well-established use of benzylcalcium reagents as initiators in styrene polymerization [29], it is a major challenge to prevent this side reaction. Oligomeric products can partially be prevented by using a higher H_2 pressure, which accelerates the σ-bond metathesis route. Formation of oligomers is also substrate and catalyst dependent. Catalytic hydrogenation with the Ca hydride complex **4** could also be achieved for 1,1-diphenylethylene (49% 1,1-diphenylethane) and α-methylstyrene (60% isopropylbenzene) without observation of oligomers but only incomplete conversion was observed at an elevated temperature of 60 °C. The catalytic performance could be considerably improved using the homoleptic dibenzyl calcium complex $(DMAT)_2Ca{\cdot}(THF)_2$ (Figure 6.4) (**5**). Despite the potential formation of insoluble CaH_2 salts, hydrogenation of styrene was complete after 15 hours at room temperature with only half of the catalyst loading (2.5 mol%) and decreased oligomer formation (15%). The analog strontium complex $(DMAT)_2Sr{\cdot}(THF)_2$ **(6)** gave a comparable catalytic performance. Exchanging the benzene solvent with a more polar medium such as THF accelerated the catalytic reaction and decreased the reaction time to 3.5 hours with only 8% oligomer formation. This effect could be further increased by addition of hexamethylphosphoramide (HMPA), a highly polar cosolvent. In this case, complete conversion was obtained after 1.5 hours with only 4% oligomer formation. The rate-enhancing effect of a polar reaction medium can be explained by its general ability to stabilize polar transition states. It may, however, also be related to its ability to keep in situ generated species such as nascent "CaH_2" in solution. Reactions with finely ground CaH_2 in the presence of a HMPA cosolvent were, however, not successful. Harder et al. also tested simple alkali metal catalysts [25]. While catalytic quantities of BuLi/TMEDA gave essentially no hydrogenation of 1,1-diphenylethylene (DPE), (DMAT)K converted DPE nearly quantitatively to 1,1-diphenylethane (only 3% oligomerization), but high H_2 pressures up to 100 bar were necessary. In this case, however, solid KH could also simply be used as a catalyst.

The Ca-catalyzed alkene hydrogenation was applied to the reduction of myrcene, an ingredient of many natural ethereal oils, containing two conjugated and one isolated C=C double bond. The products (Figure 6.5) showed that only one of the conjugated bonds was hydrogenated. It seemed therefore that the method is limited to the use of activated alkenes in which the C=C bond is conjugated either with a Ph ring or another C=C bond, an assumption that was found not to be true (*vide infra*). Fact is that conjugated double bonds react faster than isolated double bonds. The ease of hydrogenation of conjugated double bonds can be explained by the stability of the intermediates formed, i.e. either benzyl or allylcalcium complexes in which the negative charge is resonance stabilized. The limitation of the substrate scope to conjugated double bonds may be seen as a curse or a blessing. The Ca-catalyzed hydrogenation of cyclohexadiene gave exclusively cyclohexene as the product. Such very high selectivity for a single double bond reduction is not easily reached in transition metal catalyzed diene hydrogenation.

Ten years after the first alkene hydrogenation with group 2 metal catalysts, Hill and coworkers reported catalysis with a THF-free **4** [30]. This calcium hydride dimer is much more Lewis acidic and was able to react with isolated alkenes such as ethylene and 1-hexene. The latter is the basis for 1-hexene reduction with H_2 but the catalytic transformation has to be carried out at room temperature in order to prevent ligand exchange to homoleptic species and nucleophilic attack of the solvent, which makes this transformation very slow (C_6D_6, 10 mol% cat., 25 °C, 99%, 21 days) [30].

There are not many ligands that are able to stabilize Ca hydride complexes against ligand distribution reactions. In 2012, Okuda and coworkers isolated Ca hydride complexes using the strongly coordinating tetradentate monoanionic ligand Me_3TACD^- (1,4,7-trimethyl-1,4,7,10-tetraazacyclododecane). A cationic complex of composition $(Me_3TACD)_3Ca_3H_2^+$ was isolated and structurally characterized (**7**, Figure 6.6) [31]. This complex was shown to be an active catalyst in the hydrogenation of 1,1-diphenylethylene, DPE (Figure 6.7). With

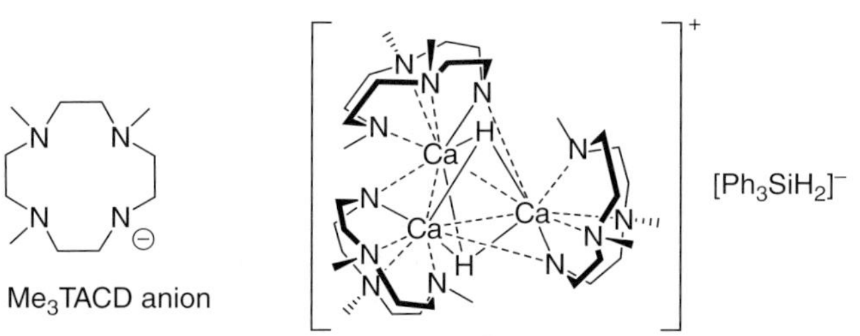

7

Figure 6.6 A cationic Me_3TACD-stabilized calcium hydride complex (**7**).

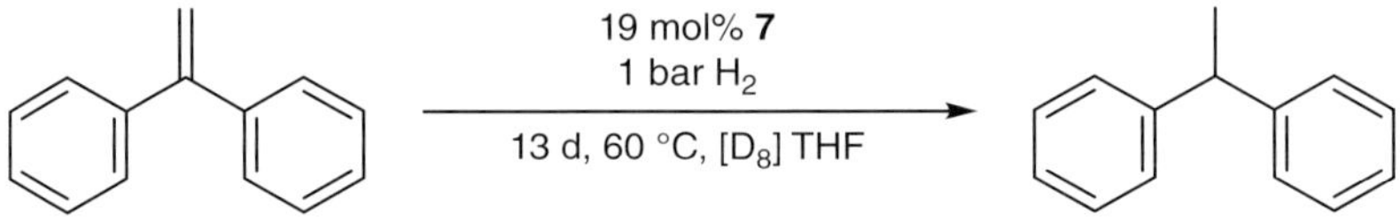

Figure 6.7 Catalytic hydrogenation of 1,1-diphenylethylene with a Me_3TACD-stabilized cationic calcium hydride complex (**7**).

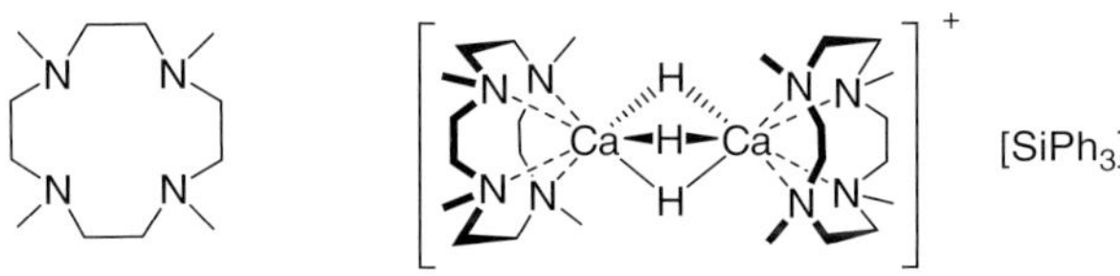

Figure 6.8 Me_4TACD-stabilized cationic calcium hydride complex (8).

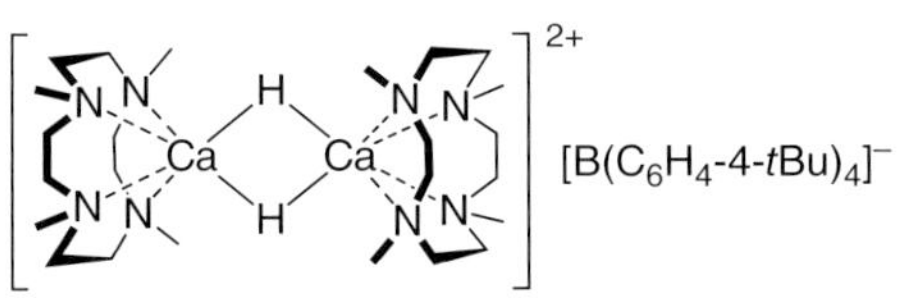

Figure 6.9 The Me_4TACD-stabilized dicationic calcium hydride complex (9).

a catalyst loading of 19 mol% full conversion to 1,1-diphenylethane could be observed at 60 °C and 1 bar H_2 pressure after 13 days. Catalysis proceeds at a surprisingly low pressure of 1 bar, and no oligomeric by-products were observed (oligomerization of DPE is known to be difficult [29]). In contrast, it suffers from a high catalyst loading and very long reaction times. It was suggested that the low activity may originate from the poor solubility of the cationic catalyst: complex **7** is quite insoluble even in THF in which the catalysis was performed.

Complexes with increased solubility could be obtained using the neutral tetradentate ligand Me_4TACD [32]. The resulting dinuclear complex of composition $(Me_4TACD)_2Ca_2H_3^+$ (**8**) formed in the reaction of $Ca(SiPh_3)_2$ with H_2 in the presence of Me_4TACD (Figure 6.8) and was found to be soluble in THF. Further development and different synthetic approaches led to the dicationic calcium hydride complex $(Me_4TACD)_2Ca_2H_2^{2+}$ that was isolated as its borate salt (**9**) (Figure 6.9). The latter dicationic complex showed a remarkable activity in the hydrogenation of alkenes, especially with less or nonactivated alkenes [33]. The activated alkene styrene could be hydrogenated at room temperature with 2.5 mol% catalyst loading at 1 bar H_2 (Table 6.1). The low reaction temperature significantly reduced formation of oligomeric by-products to a mere 5%, which is especially noteworthy considering the ease of styrene oligomerization. More importantly, it was found that the catalyst could also fully hydrogenate isolated double bonds such as that in $Me_3SiCH{=}CH_2$. While the latter alkene is still substantially polarized by the electropositive Si substituent, it was shown that also 1-hexene could be reduced. This more challenging unactivated alkene needs an increased catalyst loading of 5 mol% and a somewhat higher reaction temperature of 60 °C. Reaction times for nonactivated alkenes

Table 6.1 Cationic calcium hydride catalyzed hydrogenations of alkenes with 1 bar H_2.

Loading (mol%)	Substrate	Temperature (°C)	Time (h)	Conversion (%)
2.5	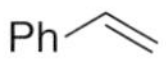	25	10	92
2.5	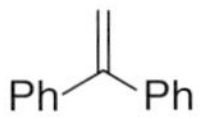	25	6	98
5	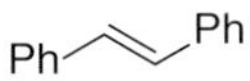	60	24	96
5	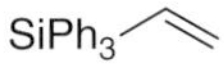	60	16	97
5	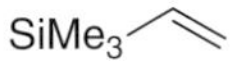	60	36	98
5	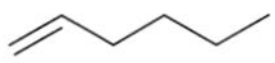	60	24	95
5	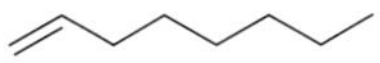	60	24	95
5	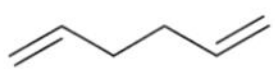	60	36	91
5	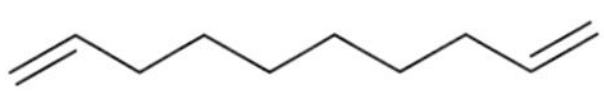	60	36	95
5	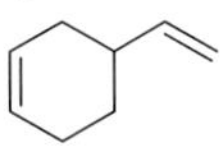	60	36	97[a)]
5		60	24	0
10	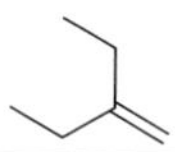	80	24	0

a) Only hydrogenation of the vinyl double bond.

such as 1-hexene, 1-octene, vinylcyclohexene, or other dienes were found to be generally longer (24–36 hours). Although some derivatives such as cyclohexene or 2-ethyl-1-butene could not be hydrogenated with the cationic calcium catalyst, the dicationic complex **9** showed an unprecedented performance. It was reasoned that the 2+ charge on the complex is critical in imparting sufficient electrophilicity to the electropositive calcium center and is crucial for the activation of isolated olefinic bonds by metal–alkene coordination. As the nucleophilicity of the hydride ligands is decreased by the 2+ charge, it seems that electrophilic substrate activation is more important than nucleophilicity. The proposed mechanism for the hydrogenation of alkenes with cationic calcium complexes follows the earlier reported cycle shown in Figure 6.5. A major difference is the reversibility of the insertion step of 1-alkenes into the Ca—H bond. Stoichiometric reactions of the Ca hydride complex with unactivated 1-alkenes showed that the intermediate Ca alkyl species could not be observed or isolated [33]. It was proposed that this Ca alkyl intermediate is unstable toward ß-hydride elimination. However, the small part of Ca alkyl reagent formed in this equilibrium is highly reactive, thus enabling catalysis.

Most recently, the Harder group demonstrated that very simple homoleptic Ae amide reagents, AeN″$_2$ (N″ = N($SiMe_3$)$_2$), can effectively be used in catalytic imine hydrogenation (*vide infra*) [34]. This rather unexpected observation meant that these easily accessible catalysts may be useful in alkene hydrogenation

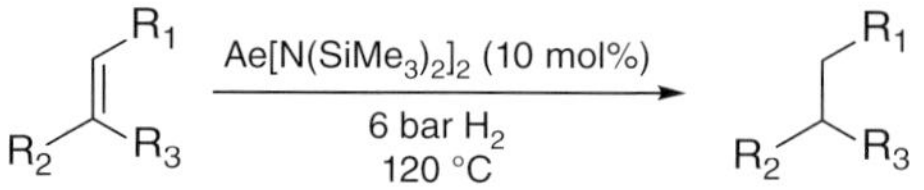

Figure 6.10 Catalytic hydrogenation of alkenes with alkaline earth metal amides AeN''_2 (Ae = Ca, Sr, Ba; N″ = $N(SiMe_3)_2$).

Table 6.2 Alkaline earth metal amide catalyzed hydrogenation of alkenes.

Catalyst	mol%	Substrate	*P* (bar)	*T* (°C)	Time (h)	Product	%
CaN''_2	10	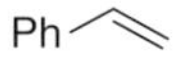	6	80	1.5	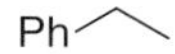	99
CaN''_2	10		6	120	0.5		99
CaN''_2	10		12	80	1		99
CaN''_2	10		1	80	24		97
CaN''_2	1		6	120	24–48		99[a)]
MgN''_2	10		6	120	24		99[a)]
SrN''_2	10		6	120	0.25		99
BaN''_2	10		6	120	0.25		99
BaN''_2	10	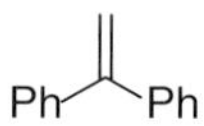	6	120	0.75	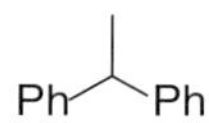	99
BaN''_2	10	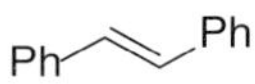	6	120	0.5	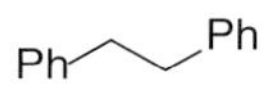	99
BaN''_2	10	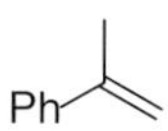	6	120	0.25	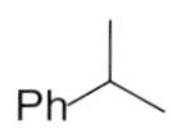	99
BaN''_2	10	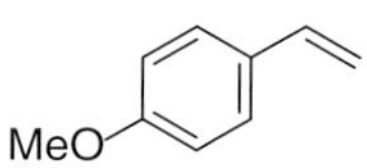	6	120	0.25	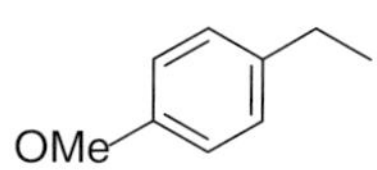	99
BaN''_2	10		6	120	0.5	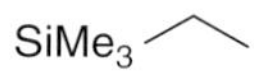	99
BaN''_2	10		6	120	24	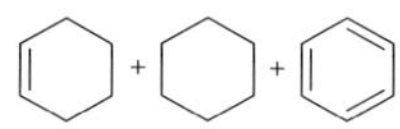 76 1 23	99
BaN''_2	10		6	120	24	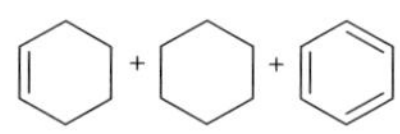 50 30 20	50
BaN''_2	10	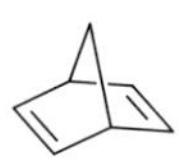	6	120	24	 29 39 32	99
BaN''_2	10	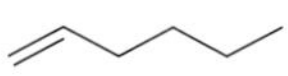	6	120	24	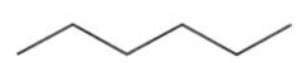	54[b)]
$(DMAT)_2Ca(THF)_2$	10		6	120	20		99

a) C. 5% oligomers formed by thermal styrene polymerization.
b) Isomerization to 2-hexene is observed as a side reaction.

as well. Indeed, a range of catalysts with Ae = Ca, Sr, and Ba smoothly hydrogenated activated alkenes such as styrene (Figure 6.10, Table 6.2) [35]. The activity increases along the row Mg < Ca < Sr < Ba. Using 10 mol% of BaN''_2, hydrogenation of styrene to ethylbenzene was completed within 15 minutes (120 °C, 6 bar H_2). The strontium amide catalyst SrN''_2 showed a similar fast conversion but CaN''_2 is slightly slower (full conversion after 30 minutes). With MgN''_2 full conversion needed 24 hours. Increase in pressure from 6 to 12 bar showed only minor acceleration. An increase in temperature from 80 to 120 °C decreased the reaction times by a factor of c. 2–3. Using only 1 mol% of catalyst is possible but this extended the reaction times to one to two days. In contrast to alkene hydrogenation using $(DMAT)_2Ae{\cdot}(THF)_2$ (Ae = Ca, Sr) [25], addition of THF inhibited the catalysis completely.

The unique feature of these Ae amide catalysts is their ability to suppress the formation of oligomeric by-products. The key to directing the catalytic cycle toward hydrogenation is the presence of small quantities of $HN(SiMe_3)_2$ formed in the catalyst initiation step (Figure 6.11). The competitive alkene polymerization, which may partially be avoided by applying high H_2 pressure, is now completely shut off by a very fast reaction of the benzyl (or alkyl) metal intermediate with the relatively Brønsted acidic $HN(SiMe_3)_2$ ($pK_a = 25.8$ [36]). This reaction is much faster than deprotonation of the much less acidic H_2 ($pK_a \approx 49$ [37]). Only for sensitive monomers (e.g. styrene) and in those cases where longer reaction times (>24 h) or higher temperature (120 °C) were needed (e.g. for the slower Mg catalyst or at low catalyst loadings) c. 5% oligomers formed by thermally induced styrene polymerization. The free amine $HN(SiMe_3)_2$ generated in the catalyst initiation step is therefore not just a by-product but essentially traps the highly reactive metal benzyl intermediates,

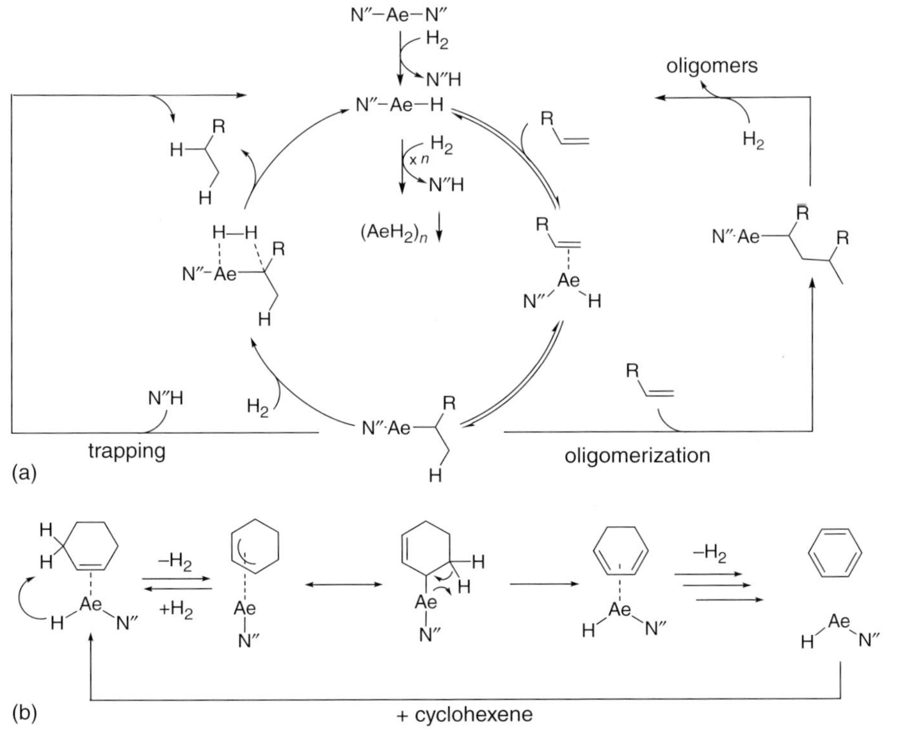

Figure 6.11 (a) Proposed mechanism for the catalytic hydrogenation of olefins with AeN''_2. (b) Proposed mechanism for double bond isomerization and dehydrogenation.

thus preventing polymerization. However, increasing the concentration of free amine by deliberately adding excess $HN(SiMe_3)_2$ slowed down the conversion significantly. A higher $HN(SiMe_3)_2$ concentration shifts the proposed equilibrium to the catalytically inactive species: $AeN''_2 + H_2 \rightleftarrows HAeN''_2 + N''H$. The optimal concentration of free amine N″H is therefore a compromise between prevention of oligomerization and catalyst activity.

Especially the most reactive BaN''_2 catalyst shows a broad scope and was found to hydrogenate a range of styrenic substrates (1,1-diphenylethylene, 1,2-diphenylethylene, α-methyl styrene, and *p*-methoxy styrene) in less than one hour (120 °C, 6 bar). The C=C double bonds of these styrenic substrates are activated by the adjacent phenyl group. As mentioned previously, Okuda's cationic Ca hydride complexes were found to hydrogenate isolated C=C double bonds, such as that in 1-hexene, but this conversion remains challenging. Generally, higher catalyst loadings and longer reaction times are needed. Apart from that a side reaction such as the isomerization of 1-hexene to 2-hexene was found. Since the latter 1,2-substituted alkene could not be further reduced, the yields for hydrogenation are never quantitative.

It was found that especially the most active Ae amide catalyst, BaN''_2, is also able to hydrogenate isolated alkenes showing a surprisingly outstanding performance. For example, trimethylsilyl ethylene was completely reduced within 30 minutes. While the C=C bond in $Me_3SiCH{=}CH_2$ is not conjugated it is to some extent activated since the intermediate carbanion after hydride attack, $Me_3Si(Me)CH^-$, is stabilized by negative hyperconjugation. The isolated double bonds in norbornadiene could also be reduced, giving a mixture of norbornene, nortricyclene, and norbornane. Although the C=C bond in norbornadiene is partially activated by through-space homoconjugation, the observation of norbornane suggests that also truly unactivated C=C bonds may be reduced. Indeed, the BaN''_2 catalyst also hydrogenates the more challenging substrate 1-hexene. The poor conversion of 54% is due to partial isomerization, which, we propose, proceeds through a classical mechanism: deprotonation of 1-hexene gives the 1-*n*Pr-allyl anion, which is protonated by either H_2 or $HN(SiMe_3)_2$ to give 2-hexene. The latter 1,2-disubstituted alkene is inert toward further hydrogenation. This is in line with the observation of Okuda and coworkers that cyclohexene could not be hydrogenated using cationic Ca catalyst **9** [33]. The superb activity of BaN''_2 in hydrogenation catalysis is underscored by the fact that both cyclohexadiene and cyclohexene could be fully converted. However, besides cyclohexane also benzene has been observed. The formation of benzene is formally a dehydrogenation reaction that can be explained by successive deprotonation and ß-hydride elimination reactions (Figure 6.11(b)).

The previously mentioned myrcene hydrogenation by $(DMAT)_2Ca{\cdot}(THF)_2$ (**5**) (see Figure 6.5) suggested that only conjugated double bonds can be reduced. It was found, however, that the same catalyst could quantitatively convert 1-hexene into hexane without alkene isomerization (Table 6.2). This experiment clearly shows that there is a metal-dependent balance between substrate deprotonation, leading to isomerization, and hydride addition, giving reduction. The more important conclusion is that hydrogenation of isolated alkenes is not restricted to cationic Ca catalysts. The ease of C=C double bond reduction seems to be determined more by the alkene substitution pattern. As in transition metal catalyzed alkene hydrogenation, an increasing number of alkene substituents lowers the reactivity of the double bond. This explains why the tri-substituted isolated C=C bond in myrcene is inert toward Ca-catalyzed hydrogenation but 1-hexene can be successfully converted. The simple Ba amide complex BaN''_2 is the first group 2 metal catalyst that can also convert 1,2-substituted alkenes such as norbornene or cyclohexene.

6.3 Hydrogenation of C=N Double Bonds

The catalytic hydrogenation of C=N double bonds with molecular hydrogen is relatively unattended and remains a challenge. Harder and coworkers recently demonstrated that aldimines can be hydrogenated under mild conditions (80 °C, 1–6 bar H_2) using very simple, easily available Ae amide catalysts (AeN''_2; $N'' = N(SiMe_3)_2$), a discovery that was rather unexpected [34].

Looking at a hypothetical reaction mechanism for the Ae-catalyzed hydrogenation of imines (Figure 6.12), which is related to the catalytic cycle for olefin hydrogenation (Figure 6.5), the difficulties become immediately apparent. The last step of the catalytic cycle (σ-bond metathesis) is formally a deprotonation of H_2. While this deprotonation step was shown to be possible in olefin hydrogenation, it seemed questionable whether it would work in imine hydrogenation. In the latter catalytic cycle H_2 has to be deprotonated by an amide intermediate, which is a much weaker base than the Ca benzyl or alkyl intermediates formed during alkene hydrogenation. Considering the large difference in the pK_a values of H_2 (p$K_a \approx 49$ [37]) and amines (p$K_a \approx 35$) or N″H (pK_a = 25.8 [36]), efficient and rapid deprotonation of H_2 by the much weaker amide bases should be unlikely. In contrast to expectation, a range of catalysts AeN''_2 (Ae = Mg, Ca, Sr, Ba) showed catalytic activity in the reduction

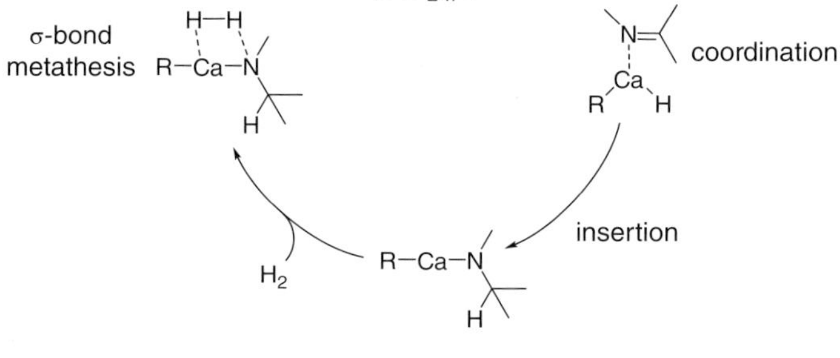

Figure 6.12 Hypothetical catalytic cycle for imine hydrogenation with a Ca catalyst.

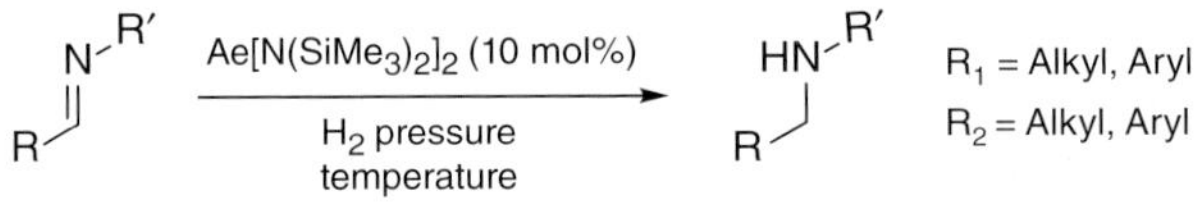

Figure 6.13 Catalytic hydrogenation of aldimines with alkaline earth metal amide catalysts AeN″$_2$.

Table 6.3 Alkaline earth metal mediated hydrogenations of Ph(H)C = N*t*Bu.

Catalyst (mol%)	H_2 (bar)	Temperature (°C)	Time (h)[a]
MgN″$_2$ (10)[b]	6	80	19.0
CaN″$_2$ (10)[b]	6	80	3.0
CaN″$_2$ (10)[b]	6	120	0.5
CaN″$_2$ (10)[b]	12	80	2.25
SrN″$_2$ (10)[b]	6	80	1.25
BaN″$_2$·(THF)$_2$ (10)[b]	6	80	0.75
(DMAT)$_2$Ca·(THF)$_2$ (10)[c]	6	80	< 0.25

a) Reaction time for >99% conversion.
b) N″ = N(SiMe$_3$)$_2$.
c) DMAT = 2-dimethylamino-α-trimethylsilyl-benzyl.

of aldimines to amines (Figure 6.13). The catalyst CaN″$_2$ resulted in, after three hours at 80 °C and 6 bar H_2 pressure, complete hydrogenation of the imine Ph(H)C = N*t*Bu, used in this study as the benchmark substrate (Table 6.3). Doubling the pressure slightly increased the conversion speed but influences of pressure were generally found to be small. Catalytic conversion, however, is strongly dependent on the temperature. The strongly polar solvent THF was found to be detrimental to conversion. This is likely due to metal THF ligation, which blocks coordination and the subsequent reduction of the imine. Also the influence of the amine N″H, liberated during catalyst initiation, was studied. Increasing its concentration led to slower conversion. Catalyst initiation can be considered an acid–base equilibrium and therefore any increase in the N″H concentration shifts the equilibrium to the side of the homoleptic amide complex AeN″$_2$, thus decelerating catalysis. Liberation of the inhibitor N″H during catalyst initiation can be avoided by using the catalyst (DMAT)$_2$Ca·(THF)$_2$ (**5**), which would give DMAT-H after catalyst initiation. The low acidity of DMAT-H makes the initiation reaction a unidirectional process, and consequently very fast and quantitative conversion is observed. Functional groups such as Cl- or MeO-substituents in *para*-position of the phenyl group in Ph(H)C = N*t*Bu are tolerated. Although various combinations of alkyl or aryl substituents at C or N are allowed, the substrate scope is currently limited to aldimines. Ketimines, a much more challenging group of substrates, could not be hydrogenated to date. The AeN″$_2$ catalysts become more active descending down the group of Ae metals: Mg < Ca < Sr < Ba. The latter Ba amide catalyst, BaN″$_2$·(THF)$_2$, showed a superior performance despite the presence of inhibiting THF ligands.

The catalyst loadings are generally high and could only be lowered to 5 mol%. This major drawback is explained by the formation of larger Ae hydride clusters. In the first step, the precatalyst AeN''_2 reacts with H_2 to HAeN″ and N″H. Although this step is quite endothermic, aggregation and ligand exchange of HAeN″ may give larger clusters of the form $Ae_x(H)_y(N'')_{2-y}$, a process that is exothermic. The formation of larger aggregates is therefore the driving force for the generation of catalytically active metal hydride species. Aggregation to larger species, however, also lowers the concentration of the active catalyst present in solution, which may explain the need for higher catalyst loadings. The existence of larger metal hydride clusters could be observed by Diffusion-Ordered-Spectroscopy (DOSY), an NMR method in which measurement of diffusion speed provides information on particle size. Clusters with molecular weight up to 7500 could be detected. Further evidence for the presence of bigger aggregates could be obtained from the recent discovery of well-defined Ca, Sr, and Ba hydride clusters by Harder and coworkers (Figure 6.14) [27, 28].

Theoretical studies on the mechanism of imine hydrogenation with the CaN''_2 as the catalyst have been reported [34]. Three conceptually different reaction mechanisms were systematically evaluated (Figure 6.15). Mechanism A represents the classical metal hydride route, which presupposes the formation of a metal hydride species. Mechanism B involves a bifunctional catalyst, which, similar to the Shvo or Noyori catalysts [38], contains a hydridic $H^{\delta-}$ and a protic $H^{\delta+}$ for dual imine attack. This bifunctional catalyst is the first intermediate in the formation of HCaN″ and is probably short-lived. Mechanism C avoids the formation of a high-energy metal hydride intermediate. This mechanism, which is similar to that proposed by Berkessel et al. for ketone hydrogenation with a K*Ot*Bu catalyst [39], involves a six-membered ring transition state. Attempts to optimize the six-membered ring transition states proposed for B and C gave in all cases four-membered ring structures. Either a transition state for hydride-imine attack (B′) or for Ca hydride formation (C′) was found. This shows that the metal hydride mechanism A is the only true mechanism. The energies in the catalytic

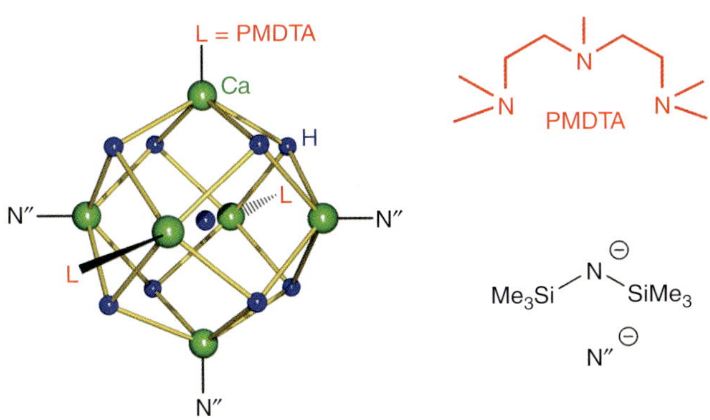

Figure 6.14 A complex Ca hydride cluster isolated from the reaction of CaN''_2 and $PhSiH_3$ in the presence of PMDTA.

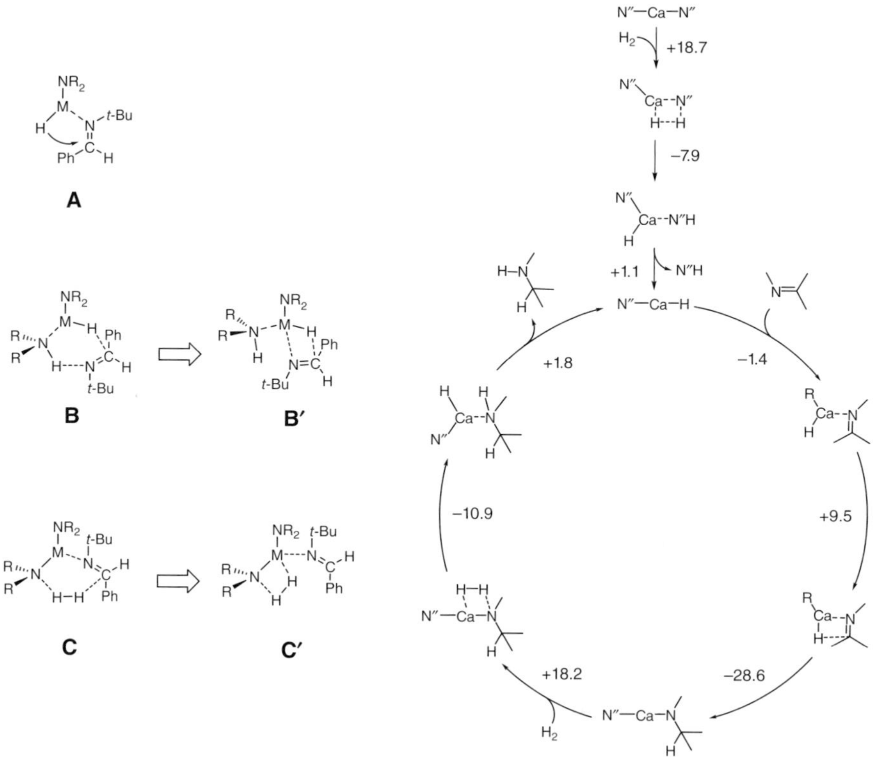

Figure 6.15 Calculated pathways for imine hydrogenation (**A**–**C**) and the proposed mechanism for the catalytic hydrogenation of imines with CaN″$_2$ and H_2; energies given as ΔG (80 °C, 6 bar) in kcal/mol at the level M06/6-311++G(d,p)//M06/6-31++G(d,p).

cycle (Figure 6.15) clearly demonstrate that the deprotonation of H_2 is the critical step (the activation energy for this step is c. 18 kcal/mol).

This method could not be extended to group 1 metal catalysts. The amide catalysts LiN″, NaN″, and KN″ (10 mol%) were tested in the catalytic hydrogenation of Ph(H)C = N*t*Bu (80 °C, 6 bar) but showed only low conversion (up to 29% after 24 hours for KN″). This shows that the higher Lewis acidity of the 2+ alkaline earth cations is essential for an efficient catalysis.

Although the metal amide catalysts CaN″$_2$ are very simple, easily accessible metal complexes, later work by the Harder group showed that imine hydrogenation can also be accomplished using catalytic quantities of $LiAlH_4$, an even simpler commercially available bulk commodity [40]. Under very harsh working conditions, $LiAlH_4$ was previously shown to be an active catalyst for alkyne reduction [12, 13]. Under much milder conditions it was applied as a catalyst for the reduction of alkenes and aldehydes/ketones by the much more reactive hydridoboranes [41]. Imine reduction by the bulk commodity H_2 using the simple catalyst $LiAlH_4$ (down to 2.5 mol%) under mild conditions (1 bar H_2 and 85 °C) was, however, unprecedented. Selected results are summarized in Table 6.4. An increase in pressure accelerates the conversion implying that the reaction

Table 6.4 $LiAlH_4$ catalyzed hydrogenations of Ph(H)C = N*t*Bu.

Catalyst (mol%)	H_2 (bar)	Temperature (°C)	Time (h)[a]
$LiAlH_4$ (10)	1	85	2
$LiAlH_4$ (5)	1	85	15
$LiAlH_4$ (2.5)	1	85	66
$LiAlH_4$ (5)	5	85	6
$LiAlH_4$ (5)	1	65	6 [39%]
$LiAlH_4$ (5)	1	100	6 [51%]

a) Reaction time for >99% conversion or, in case no full conversion was reached, time for conversion given in square brackets.

with H_2 is rate limiting. The reaction needs a higher temperature of 85 °C, but at 100 °C, however, hardly any conversion is found. This can be attributed to thermal decomposition of $LiAlH_4$ to Li_3AlH_6, Al^{0}, and H_2, which lowers the catalyst loading. The reaction cannot be catalyzed by a mixture of LiH and AlH_3. Alternative reducing agents such as RedAl ($LiAlH_2(OCH_2CH_2OMe)_2$) and $NaAlH_4$ decelerated the hydrogenation noticeably whereas $NaBH_4$ is inactive. This indicates a mechanism in which heterobimetallic cooperativity is important. The preliminary mechanism (Figure 6.16) is based on NMR observations, a crystal structure of a potential intermediate, preliminary density functional theory (DFT) calculations, and analogies to the cycle shown in Figure 6.15. Although the activities and scope of $LiAlH_4$-mediated imine hydrogenation are inferior to the previously discussed group 2 metal amide catalysis, its simplicity has far-reaching consequences. Classical imine-to-amine reduction uses stoichiometric quantities of $LiAlH_4$ and ether solvents. The final hydrolysis step is sometimes uncontrolled and risky (especially at a larger scale) and workup can be tedious due to formation of considerable quantities of Li and Al salts as a waste product. In contrast, catalytic imine reduction using H_2 and small amounts of $LiAlH_4$ is a solvent-free 100% atom-economical process that does not require tedious workup and produces much less waste. Since the proposed catalyst consists of different metal cations and different anions, there is plenty of room for catalyst optimization. It is likely that the activity, substrate scope, and perhaps even chemo- and stereoselectivities will be optimized further.

6.4 Hydrogenation of C=O Double Bonds

Interest in transition metal free hydrogenation catalysis was documented already in the 1960s when Walling and Bollyky chose KO*t*Bu as a strong base to hydrogenate benzophenone in *tert*-butanol (Figure 6.17) [42]. The development of the catalytic hydrogenation of C=O bonds with simple main group metal bases was based on two observations. First, it was found that bases catalyze H/D exchange between D_2 and H_2O. This implies the unusual chemical equilibrium shown in

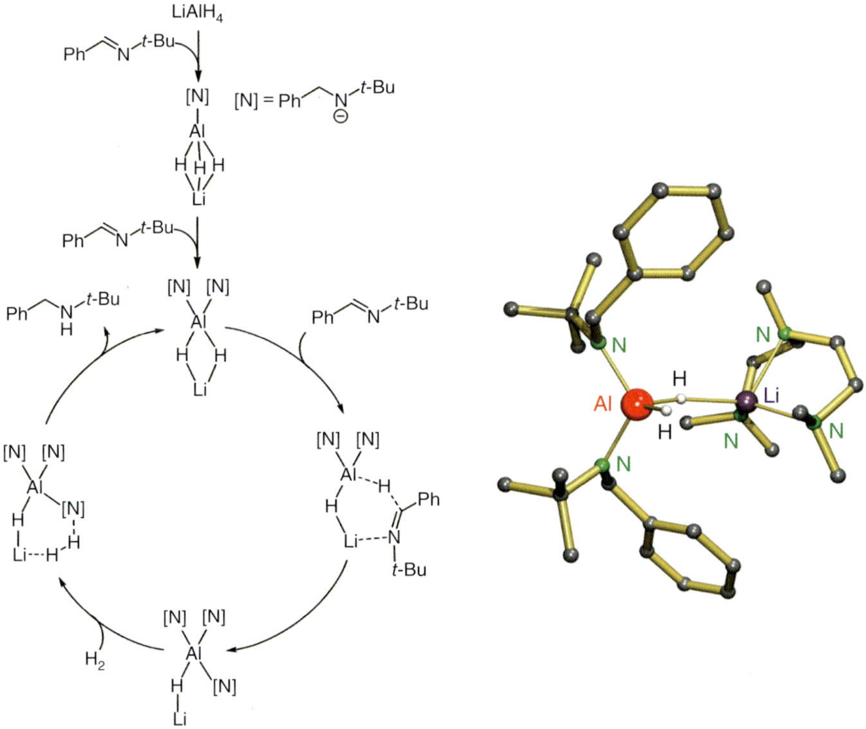

Figure 6.16 Proposed mechanism for the catalytic hydrogenation of imines with $LiAlH_4$ and H_2. The crystal structure of a possible catalyst is shown: $[N]_2AlH_2Li{\cdot}PMDTA$.

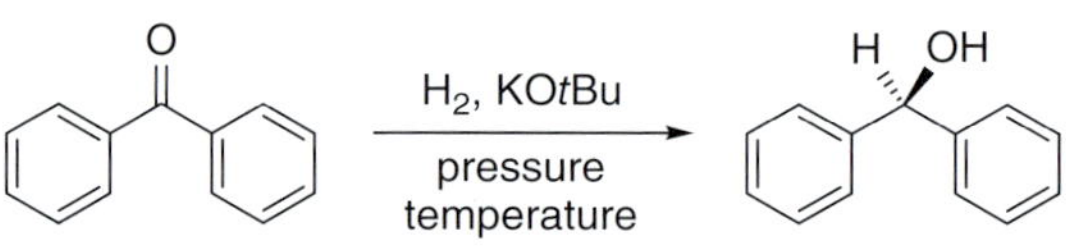

Figure 6.17 Catalytic hydrogenation of benzophenone with KO*t*Bu.

Eq. (6.1), which most unfavorably lies strongly to the side of D_2 [43]. Second, organic molecules can be dehydrogenated by strong bases such as in the Varrentrapp reaction [44]. This shows that under suitable conditions (e.g. high H_2 pressure) base-catalyzed reverse hydrogenation catalysis might be feasible and this led Walling et al. to propose an alkoxide-mediated ketone hydrogenation along the lines of Eqs. (6.3)–(6.4). Deprotonation of H_2 by an alkoxide would be highly unusual but a small percentage of active hydride may be sufficient to catalyze hydrogenation.

$$D_2 + OH^- \rightleftharpoons D^- + HOD \tag{6.1}$$

$$RO^- + H_2 \rightleftharpoons ROH + H^- \tag{6.2}$$

$$H^- + R_2C{=}O \rightleftharpoons R_2CH{-}O^- \tag{6.3}$$

$$R_2CHO^- + ROH \rightleftharpoons R_2CHOH + RO^- \tag{6.4}$$

It was indeed found that with the very simple alkali metal catalyst KO*t*Bu good conversion of benzophenone to benzhydrol can be reached, provided that harsh reaction conditions are applied (150–200 °C, 100 bar H_2). In general, reactions times are long and the catalyst loadings are high (Table 6.5). With a catalyst loading of 20 mol%, the reduction of benzophenone is completed within 25 hours (135 bar, 210 °C). Faster conversion at lower temperature can be achieved with a threefold excess of KO*t*Bu. Among the different solvents tested in this hydrogenation, diethylene glycol dimethyl ether (diglyme) was found to be most effective. The reason for the acceleration of the hydrogenation in diglyme could be the better solvation of the cation, making *t*-BuO^- a stronger base. The substrate scope of this method is limited to stable, non-enolizable ketones. In order to improve the method and reach broader applicability, Berkessel et al. investigated the mechanism of this catalytic hydrogenation of ketones in detail [39] and could expand the scope to ketones such as PhC(O)*t*Bu, $(t\text{-Bu})_2CO$, and 2,2,5,5-tetramethylcyclopentanone. The latter aliphatic substrate showed a lower conversion than aromatic substrates. The fast conversion of Ph ring containing ketones may be attributed to a K^+–Ph π interaction but may also be due to activation of the C=O bond by conjugation with the Ph ring. In contrast to the original procedure, Berkessel was not able to obtain any conversion in diglyme as solvent. A decrease in the reaction temperature also gave decreased reaction rates, as observed earlier. Kinetic investigations show that the reduction is first order in substrate, base, and H_2. Different alkaline metals have been tested, showing an increasing activity along the line $Li \ll Na \ll K \approx Rb < Cs$, which follows the covalence of the metal–oxygen bond. Experiments with deuterium revealed no effect on the kinetic course of the hydrogenation, thus excluding a kinetic isotope effect. Using D_2 for the hydrogenation of Ph_2CO gave only 60% deuteration incorporation in the final product benzhydrol. This indicates a base-catalyzed H/D exchange between D_2 and the solvent *t* BuOH, which could be confirmed in the absence of substrate. The experimental observations led to the conclusion that a mechanism related to Noyori's Ru-catalyzed asymmetric hydrogenation of ketones [45] could be assumed. This mechanism is based on a six-membered intermediate in which H_2 is activated by bridging between the alkoxide and the ketone C moieties (Figure 6.18). This transition state circumvents the highly endothermic formation of KH from KO*t*Bu and H_2. Most recently, however, calculations have shown that this process is not feasible.

Table 6.5 KO*t*Bu catalyzed hydrogenations of benzophenone with H_2 in *t*-BuOH.

Catalyst (mol%)	H_2 (bar)	Temperature (°C)	Time (h)[a]	Conversion (%)
KO*t*Bu (20)	135	210	25	98
KO*t*Bu (20)	102	153	50.5	47
KO*t*Bu (300)	96	150	14.5	98
KO*t*Bu (20)	100	170	18	52[a]
KO*t*Bu (300)	78	130	5	98[a]

a) Diglyme as solvent.

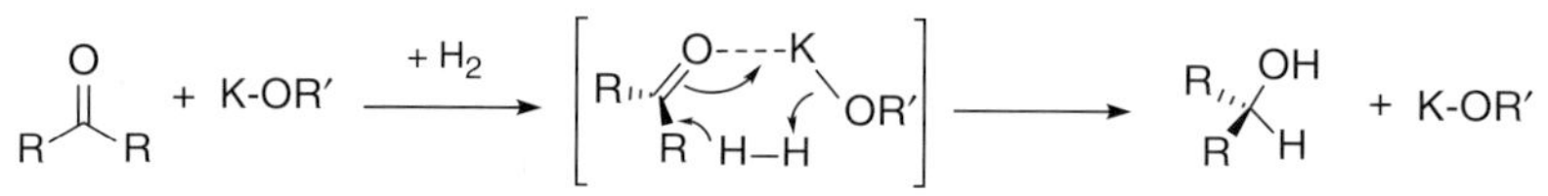

Figure 6.18 Base-catalyzed hydrogenation of ketones.

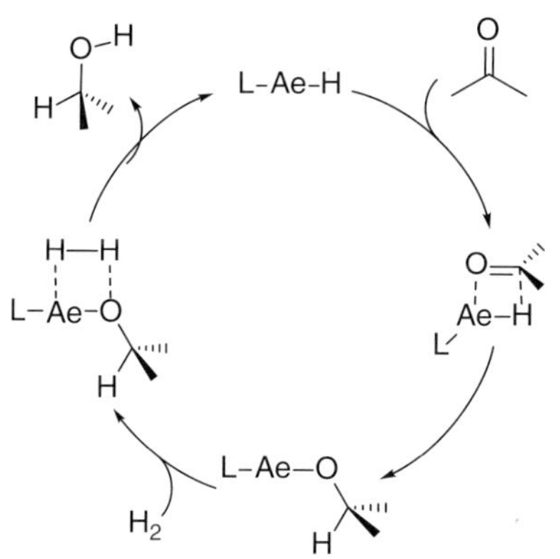

Figure 6.19 Possible mechanism for the catalytic hydrogenation of ketones with alkaline earth metals.

Instead, a two-step mechanism is proposed. In the first stage of the reaction KH is formed, which in the second step reacts with the ketone [46]. This indicates that, in contrast to chemical intuition, indeed a highly reactive short-living K hydride species can be formed by deprotonation of H_2 with the very weak base KO*t*Bu. A mechanism for the potential, hitherto unobserved, hydrogenation of ketones with Ae metal catalysts could therefore be similar to that of imine hydrogenation (Figure 6.19).

6.5 Summary and Perspectives

Although catalytic hydrogenation of unsaturated bonds is likely one of the most investigated processes, s-block metal catalyzed conversions are still a big challenge and hitherto underdeveloped. Compared to transition metal catalyzed hydrogenation, the scope and the activity of early main group metal complexes toward H—H bond activation are still limited.

The use of Ca, Sr, or even K catalysts in alkene hydrogenation broke the dogma that transition metals are needed for hydrogenation catalysis. The initial results were far from being competitive with commercially available transition metal hydrogenation catalysts. The scope seemed to be limited to activated alkenes. Conjugated double bonds, such as those in styrene or cyclohexadiene, are much easier targets for hydrogenation. This is due to the higher stability of the benzyl or allyl intermediates along the catalytic pathway. Other drawbacks are the low functional group tolerance and the observation of alkene oligomerization as an undesired side reaction (especially for alkenes sensitive to polymerization, e.g. styrene). Synthesis and isolation of cationic Ca hydride complexes gave impetus to the field. These cationic catalysts generally showed a somewhat higher reactivity but catalysis can be complicated by the low solubility of these species. The dicationic dimeric complex $[(Me_4TACD)_2Ca_2H_2]^{2+}$ (**9**) showed very high activity and allowed for hydrogenation of unactivated (nonconjugated) alkenes such as 1-hexene; however, internal (doubly substituted) C=C bonds as in cyclohexene were unaffected. It was postulated that the cationic nature of **9** makes the metals more Lewis acidic and therefore more efficient in activating the C=C bond for hydrogenation. Later observations showed that also neutral Ca catalysts such as $(DMAT)_2Ca{\cdot}(THF)_2$ (**5**) are able to hydrogenate isolated double bonds, e.g. 1-hexene. As part of a general trend to use simpler catalysts, the readily available group 2 metal amides (AeN''_2, $N'' = N(SiMe_3)_2$) have been investigated in alkene hydrogenation. Using metal amide complexes as precatalysts in alkene hydrogenation has a major advantage. During the initiation step the free amine is formed as a side product: $AeN''_2 + H_2 \rightarrow HAeN'' + N''H$. The latter amine is relatively acidic and rapidly traps any Ae alkyl intermediate, thus suppressing alkene oligomerization as an undesired side reaction. Apart from their simplicity and facile synthetic access, these catalysts have another advantage. The heaviest barium amide complex BaN''_2 showed an unsurpassed reactivity. Preliminary catalytic tests demonstrated that it not only converts isolated alkenes such as 1-hexene but also doubly substituted alkenes such as cyclohexene or norbornene. Therefore, unactivated double bonds may be more difficult targets in hydrogenation catalysis but also neutral Ae metal complexes are able to transform these functionalities. It seems that the number of substituents at the double bond influences the reactivity. As in transition metal catalyzed alkene hydrogenation, the reactivity of alkenes decreases with the number of substituents. While activated double bonds with two substituents can be hydrogenated easily (*cf*. DPE or cyclohexadiene), doubly substituted unactivated double bonds are a challenge that so far has only been met by the Ba catalyst BaN''_2. The hydrogenation of cyclohexadiene led to the formation of benzene. The observation of this *de*hydrogenation product suggested that the catalytic cycle for hydrogenation is fully reversible, which would pave the way for transfer hydrogenation catalysis. Indeed, during the writing of this chapter the first successful alkene transfer hydrogenation using 1,4-cyclohexadiene as a proton source and simple AeN''_2 catalysts was reported [47]. DFT calculations support a mechanism that proceeds through an intermediate with an unstabilized Meisenheimer anion ($C_6H_7^-$) that rapidly transfers a hydride to the metal (Figure 6.20). This reactivity provides an alternative pathway to formation of the metal hydride catalyst. Catalytic activity for BaN''_2 is generally substantially higher than for CaN''_2 and apart from activated alkenes (styrene, diphenylethylene) also the isolated double bond in 1-hexene could be cleanly reduced without isomerization.

Simple catalysts such as AeN''_2 were also shown to be active in imine hydrogenation with an increasing reactivity down the group $Mg < Ca < Sr < Ba$. The method is limited to aldimines but in some cases even the commercially available bulk metal hydride reagent $LiAlH_4$ could be used in catalytic quantities down to 2.5 mol% and at relatively mild reaction conditions (80 °C and 1 bar H_2) [40]. The latter heterobimetallic catalyst offers great potential for variation and catalyst optimization by choosing the best combination of s- and p-block metals.

Ketone hydrogenation is a much less developed field. This is due to the very difficult formation of metal hydride species by the reaction of metal alkoxides (an

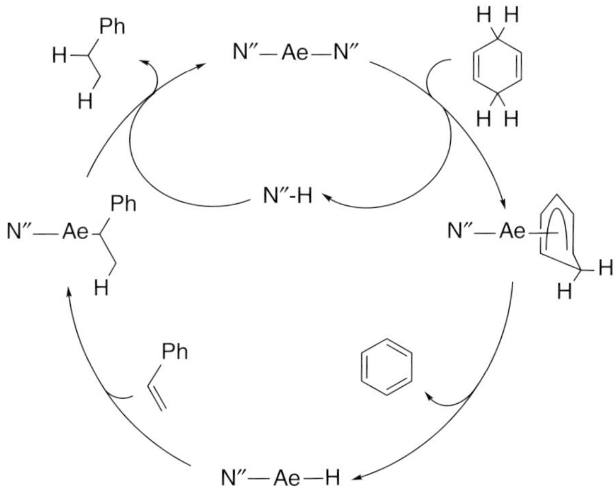

Figure 6.20 Alkene transfer hydrogenation with AeN''_2 catalysts exemplified by the reduction of styrene using 1,4-cyclohexadiene as a proton source.

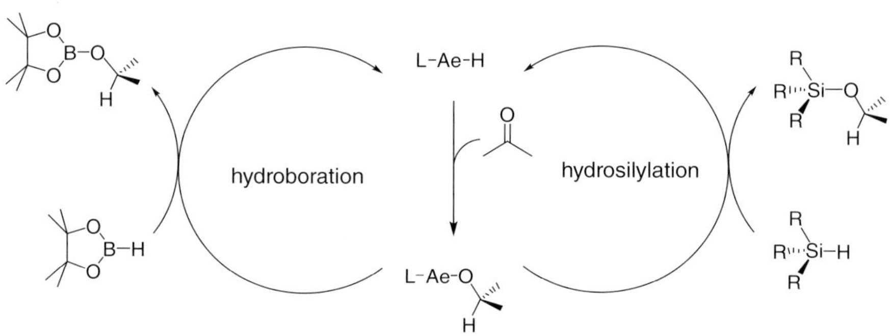

Figure 6.21 Reduction of ketones by catalytic hydroboration and hydrosilylation.

intermediate in the catalytic cycle) with molecular H_2. Consequently, harsh reaction conditions are hitherto always a prerequisite. For the reduction of ketones it is much easier to use the more polar silanes (R_3SiH) as a reducing reagent (Figure 6.21) as has been demonstrated by Harder et al. [48]. Apart from hydrosilylation, also hydroboration using polar hydridoboranes (HBpin) can be easily accomplished as has been reported by Hill and coworkers (Figure 6.21) [49]. The driving force in these catalytic cycles is the very high oxophilicity of B or Si.

All main breakthroughs in early main group metal catalyzed hydrogenation catalysis have been reported in the twenty-first century, which shows that this is still a very young field with full potential for future investigations. It would be an exciting goal to include electron transfer or redox processes in hydrogenation catalysis. Main group metals are not known for a wide variety of oxidation states but the recent isolation of Mg(I) complexes (Figure 6.22) by Jones and coworkers [50] may provide opportunities for such redox-based catalysis. Unlike low valent Ni^0 complexes, compounds of the type L–Mg–Mg–L so far did not show any redox reactivity with molecular H_2 [51]. Heterobimetallic complexes with polarized metal–metal bonds may be the key to the heterolytic cleavage of H_2 but the design of a catalytic cycle with an oxidative H—H bond activation step presupposes also a catalytic metal reduction step, which likely is the largest challenge. For that matter, introduction of redox non-innocent ligands may be a more fruitful approach to such chemistry.

Although the limitations of s-block metal catalyzed hydrogenation hitherto pose an open question, the increasing number of young researchers entering the field promises a rapid development. It is likely that the near future will see important breakthroughs and improvements in substrate scope and functional group tolerance.

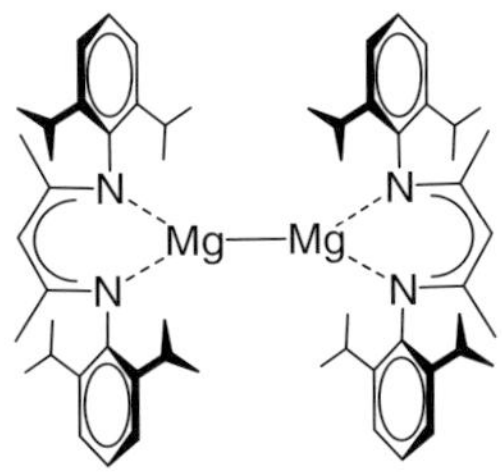

Figure 6.22 Mg(I) complex.

References

1 D. Steinborn, *Fundamentals of Organometallic Catalysis*, vol. 1. Weinheim, Germany, Wiley-VCH, 2012, 49–50.

2 M.P. von Wilde, *Chem. Ber.* 1874, 7, 352–357.

3 (a) P. Sabatier, *La Catalyse en chimie organique*. Paris, Béranger, 1913. https,//gallica.bnf.fr/ark,/12148/bpt6k9615479v/f15.image.texteImage; (b) P. Sabatier, M.J. Nye, *Chem. World* 2004, 1, 46–49.

4 S. Nishimura, *Handbook of Heterogeneous Catalytic Hydrogenation for Organic Synthesis*. Weinheim, Germany, Wiley, 2001.

5 (a) G. Wilkinson, *J. Am. Soc. (A)* 1966, 1711–1732; (b) G. Wilkinson, *Bull. Soc. Chim. Fr.* 1968, 12, 5055–5058.

6 (a) R.R. Schrock, J.A. Osborn, *J. Am. Chem. Soc.* 1976, 98, 2134–2143; (b) R.R. Schrock, J.A. Osborn, *J. Am. Chem. Soc.* 1976, 98, 2143–2147; (c) R.R. Schrock, J.A. Osborn, *J. Am. Chem. Soc.* 1976, 98, 4450–4455.

7 R.H. Crabtree, *Acc. Chem. Res.* 1979, 12, 331–337.

8 (a) C. Zirngibl, R. Hedderich, R.K. Thauer, *FEBS Lett.* 1990, 261, 112–116; (b) R.K. Thauer, A.R. Klein, G.C. Hartmann, *Chem. Rev.* 1996, 96, 3031–3042; (c) A. Berkessel, *Curr. Opin. Chem. Biol.* 2001, 5, 486–490; (d) S. Shima, R.K. Thauer, *Chem. Rec.* 2007, 7, 37–46.

9 (a) E.J. Lyon, S. Shima, G. Burrman, S. Chowdhuri, A. Batschauer, K. Steinbach, R.K. Thauer, *Eur. J. Biochem.* 2004, 271, 195–204; (b) M. Korbas, S. Vogt, W. Meyer-Klaucke, E. Bill, E.J. Lyon, R.K. Thauer, S. Shima, *J. Biol. Chem.* 2006, 281, 30804–30813.

10 (a) S. Shima, O. Pilak, S. Vogt, M. Schick, M.S. Stagni, W. Meyer-Klaucke, E. Warkentin, R.K. Thauer, U. Ermler, *Science* 2008, 321, 572–575; (b) T. Hiromoto, K. Ataka, O. Pilak, S. Vogt, M.S. Stagni, W. Meyer-Klaucke, E. Warkentin, R.K. Thauer, S. Shima, U. Ermler, *FEBS Lett.* 2009, 583, 585–590.

11 (a) J.A. Wright, P.J. Turrell, C.J. Pickett, *Organometallics* 2010, 29, 6146–6156; (b) K.F. Kalz, A. Brinkmeier, S. Dechert, R.A. Mata, F. Meyer, *J. Am. Chem. Soc.* 2014, 136, 16626–16634.

12 L.H. Slaugh, *Tetrahedron* 1966, 22, 1741–1746.

13 L.H. Slaugh, *J. Org. Chem.* 1967, 32, 108–113.

14 G. Jeske, H. Lauke, H. Mauermann, H. Schumann, T.J. Marks, *J. Am. Chem. Soc.* 1985, 197, 8111–8118.

15 C.M. Haar, C.L. Stern, T.J. Marks, *Organometallics* 1996, 15, 1765–1784.

16 Y. Obora, T. Ohta, C.L. Stern, T.J. Marks, *J. Am. Chem. Soc.* 1997, 119, 3745–3755.

17 C.A. Willoughby, S.L. Buchwald, *J. Am. Chem. Soc.* 1994, 116, 11703–11714.

18 P.A. Chase, G.C. Welch, T. Jurca, D.W. Stephan, *Angew. Chem. Int. Ed.* 2007, 46, 8050–8053.

19 Z.M. Heiden, D.W. Stephan, *Chem. Commun.* 2011, 47, 5729–5731.

20 L. Greb, P. Oña-Burgos, B. Schirmer, S. Grimme, D.W. Stephan, J. Paradies, *Angew. Chem. Int. Ed.* 2012, 51, 10164–10168.

21 (a) D.W. Stephan, *Chem. Commun.* 2010, 436, 8526–8533; (b) L.J. Hounjet, D.W. Stephan, *Org. Process Res. Dev.* 2014, 18, 385–391; (c) D.W. Stephan, *Acc. Chem. Res.* 2015, 48, 306–316; (d) D.W. Stephan, G. Erker, *Angew. Chem. Int. Ed.* 2015, 54, 6400–6441; (e) D.W. Stephan, *Science* 2016, 354, 1248–1256.

22 Y. Wang, W. Chen, Z. Lu, Z.H. Li, H. Wang, *Angew. Chem. Int. Ed.* 2013, 52, 7496–7499.

23 (a) P. Jochmann, D.W. Stephan, *Angew. Chem. Int. Ed.* 2013, 52, 9831–9835; *Chem. Eur. J.* 2014, 20, 8370–8378.

24 J.A. Hatnean, J.W. Thomson, P.A. Chase, D.W. Stephan, *Chem. Commun.* 2014, 50, 301–303.

25 J. Spielmann, F. Buch, S. Harder, *Angew. Chem. Int. Ed.* 2008, 47, 9434–9438.

26 S. Harder, J. Brettar, *Angew. Chem. Int. Ed.* 2006, 45, 3474–3478.

27 B. Maitland, M. Wiesinger, J. Langer, G. Ballmann, J. Pahl, H. Elsen, C. Färber, S. Harder, *Angew. Chem. Int. Ed.* 2017, 56, 11880–11884.

28 M. Wiesinger, B. Maitland, C. Färber, G. Ballmann, C. Fischer, H. Elsen, S. Harder, *Angew. Chem. Int. Ed.* 2017, 56, 16654–16659.

29 (a) A. Weeber, S. Harder, H.-H. Brintzinger, K. Knoll, *Organometallics* 2000, 19, 1325–1332; (b) F. Feil, K. Knoll, S. Harder, *Angew. Chem. Int. Ed.* 2001, 40, 4261–4264.

30 A.S.S. Wilson, C. Dinoi, M.S. Hill, M.F. Mahon, L. Maron, *Angew. Chem. Int Ed.* 2018, 57, 15500–15504.

31 P. Jochmann, J.P. Davin, T.P. Spaniol, L. Maron, J. Okuda, *Angew. Chem. Int. Ed.* 2012, 51, 4452–4455.

32 V. Leich, T.P. Spaniol, L. Maron, J. Okuda, *Angew. Chem. Int. Ed.* 2016, 55, 4794–4797.

33 D. Schuhknecht, C. Lhotzky, T.P. Spaniol, L. Maron, J. Okuda, *Angew. Chem. Int. Ed.* 2017, 56, 12367–12371.

34 H. Bauer, M. Alonso, C. Färber, H. Elsen, J. Pahl, A. Causero, G. Ballmann, F. De Proft, S. Harder, *Nat. Catal.* 2018, 1, 40–47.

35 H. Bauer, M. Alonso, C. Fischer, B. Rösch, H. Elsen, S. Harder, *Angew. Chem. Int. Ed.* 2018, 57, 15177–15182.

36 R.R. Fraser, T.S. Mansour, S. Savard, *J. Org. Chem.* 1985, 50, 3232–3234.

37 K. Abdur-Raschid, T.P. Fong, B. Greaves, D.G. Gusev, J.G. Hinman, S.E. Landau, A.J. Lough, R.H. Morris, *J. Am. Chem. Soc.* 2000, 122, 9155–9171.

38 J.S.M. Samec, J.-E. Bäckvall, *Chem. Eur. J.* 2002, 8, 2955–2961.

39 A. Berkessel, T.J.S. Schubert, T.N. Müller, *J. Am. Chem. Soc.* 2002, 124, 8693–8698.

40 H. Elsen, C. Färber, G. Ballmann, S. Harder, *Angew. Chem. Int. Ed.* 2018, 57, 7156–7160.

41 (a) V.A. Pollard, S.A. Orr, R. McLellan, A.R. Kennedy, E. Hevia, R.E. Mulvey, *Chem. Commun.* 2018, 54, 1233–1236; (b) A. Bismuto, S.P. Thomas, M.J. Cowley, *ACS Catal.* 2018, 8, 2001–2005.

42 (a) C. Walling, L. Bollyky, *J. Am. Chem. Soc.* 1961, 83, 2968–2969; (b) C. Walling, L. Bollyky, *J. Am. Chem. Soc.* 1964, 86, 3750–3752.

43 (a) K. Wirtz, K.F. Bonhoeffer, *Z. Physik. Chem.* 1936, 177A, 1–6; (b) W.K. Wilmarth, J.C. Dayton, J.M. Fluornoy, *J. Am. Chem. Soc.* 1953, 75, 4549–4553; (c) S.L. Miller, D. Rittenberg, *J. Am. Chem. Soc.* 1958, 80, 64–65.

44 F. Varrentrapp, *Liebig's Ann.* 1840, 35, 196.

45 (a) R. Noyori, M. Yamakawa, S. Hashiguchi, *J. Org. Chem.* 2001, 66, 7931–7944; (b) R. Noyori, T. Ohkuma, *Angew. Chem. Int. Ed.* 2001, 40, 40–73; (c) T. Ohkuma, M. Koizumi, H. Ikehira, T. Yokozawa, R. Noyori, *Org. Lett.* 2000, 2, 659–662; (d) M. Yamakawa, H. Ito, R. Noyori, *J. Am. Chem. Soc.* 2000, 122, 1466–1478.

46 P.A. Dub, N.J. Henson, R.L. Martin, J.C. Gordon, *J. Am. Chem. Soc.* 2014, 136, 3505–3521.

47 H. Bauer, K. Thum, M. Alonso, C. Fischer, S. Harder, *Angew. Chem. Int. Ed.* 2019, 58, early view.

48 J. Spielmann, S. Harder, *Eur. J. Inorg. Chem.* 2008, 1480–1486.

49 M. Arrowsmith, T.J. Hadlington, M.S. Hill, G. Kociok-Köhn, *Chem. Commun.* 2012, 48, 4567–4569.

50 S.P. Green, C. Jones, A. Stasch, *Science* 2007, 318, 1754–1757.

51 S.J. Bonyhady, C. Jones, S. Nembenna, A. Stasch, A.J. Edwards, G.J. McIntyre, *Chem. Eur. J.* 2010, 16, 938–955.

7

Frustrated Lewis Pair-Catalyzed Reductions Using Molecular Hydrogen

Jan Paradies and Sebastian Tussing

Paderborn University, Department of Chemistry, Warburger Strasse 100, 33098 Paderborn, Germany

7.1 Introduction

The term "frustrated Lewis pair" (FLP) has been introduced in 2006 by Douglas W. Stephan with the observation that a borane-based strong Lewis acid and a phosphine-derived Lewis base did not form the corresponding classical Lewis adduct. Stephan reported in his seminal work the synthesis of a rigid tetrafluorophenylene-bridged phosphonium fluoroborate, which was subsequently transformed into the corresponding borohydride by the reaction with chlorodimethylsilane (Figure 7.1) [1].

By heating the phosphonium borate above 100 °C, the molecule liberated molecular hydrogen (H_2), forming the corresponding phosphinoborane. The rigidity of the linker prevented the intramolecular formation of the Lewis adduct (*quenching*) while the steric bulk of the substituents of phosphorus and boron prohibited the *intermolecular* adduct formation leaving the individual reactivity intact. Remarkably, this molecule readily activated molecular H_2 at room temperature (rt) and regenerated the zwitterionic phosphonium borate. It became quickly obvious that the spatial restriction of the Lewis pair combined with high Lewis acidity and Lewis basicity is responsible for the observed reactivity with H_2.

Naturally, this reactivity did not remain long unrecognized and more examples were elaborated. The second example of an FLP for H_2 activation was introduced by Erker. The flexible ethylidene-bridged phosphinoborane also accomplished the H_2 activation at room temperature and is still one of the most active FLP catalysts to date [2]. More examples of *intermolecular* "frustrated" Lewis pairs in concert with catalytic applications were reported [3], which laid the foundation for this field of research.

Ever since the unquenched reactivity of a Lewis acid in the presence of a Lewis base was described as "frustrated" Lewis pair, this animated organic, inorganic, physical, and theoretical chemists to investigate new applications [4–9]. Furthermore, the deep impact of this activation method by main group element compounds has been the subject of reviews [4, 5, 7, 8, 10–20], books [21, 22], and book chapters [21–25].

Homogeneous Hydrogenation with Non-Precious Catalysts, First Edition. Edited by Johannes F. Teichert.

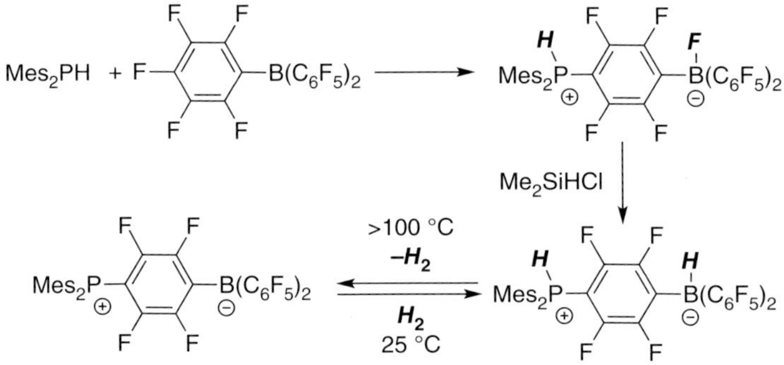

Figure 7.1 Stephan's intramolecular frustrated Lewis pair.

This chapter is intended to give an overview of FLP reactivity with respect to dihydrogen and the associated catalytic hydrogenations. Also, aluminum- [26–34], ruthenium- [35–37], zirconium- [38–43], or even carbon-based [44–46] Lewis acids are active in the H_2 activation; however, their synthetic and catalytic scope is not yet fully explored. Therefore, these examples will not be included in this chapter. Furthermore, the reaction of FLPs with small molecules, e.g. oxides of carbon, sulfur, or nitrogen, will not be discussed and the reader is referenced to excellent reviews [10, 47].

7.2 Mechanistic Considerations

The phrase "frustrated Lewis pair" illustrates the inherent attraction of a Lewis basic and a Lewis acidic site, resulting in the formation of a Lewis adduct, which, however, is prevented by steric and/or electronic prerequisites. Comparable unquenched reactivity has already been observed by H.C. Brown in 1942 in his study of "Steric Strains as a Factor in the Relative Stability of Some Coordination Compounds of Boron" [48]. He found that steric factors are primarily responsible for the presence or absence of Lewis adducts. An illustrative prototype of this effect is seen in the reaction of trimethylborane with either pyridine (py) or 2,6-lutidine (Figure 7.2).

Trimethylborane readily forms the Lewis adduct with pyridine, whereas the methyl groups in 2,6-lutidine prevent the formation of the corresponding

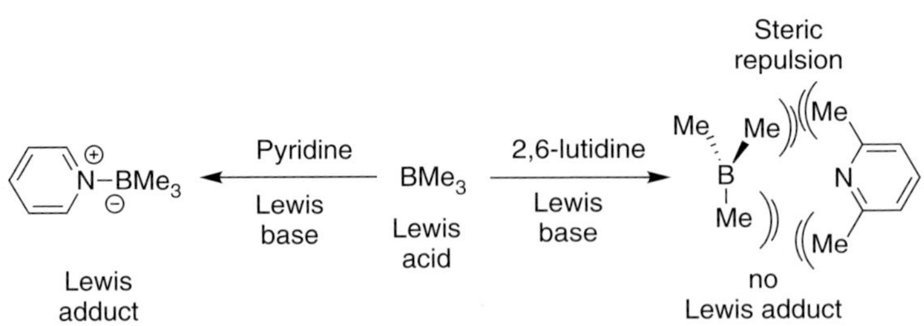

Figure 7.2 Reactivity of trimethylborane with pyridine-derived Lewis bases.

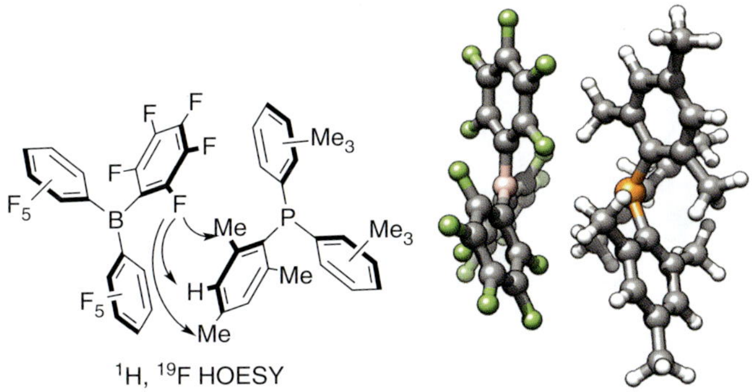

Figure 7.3 NMR structure of $PMes_3/B(C_6F_5)_3$ determined by ^1H, ^{19}F HOESY (molecular structure representation generated with Chimera). (all F atoms show HOESY NMR interactions; only one F-aryl interaction is depicted for clarity) [57].

Lewis adduct, thus leaving the reactive centers available for chemical reactions. Consequently, although entropically less favored it is reasonable to assume that an FLP features non-covalent interactions through Coulomb and dispersion forces, which ultimately leads to a suitable orbital setup for H_2 activation (compare Figure 4a). Such aggregation was first described as "Encounter Complex" [2, 49–55], whose highly fluctuating structure makes it almost impossible to be studied experimentally. So far, the solid-state structure of an *intermolecular* FLP has not been examined but the first evidence in solution has been obtained. The association of trimesityl phosphine with a tris(pentafluorophenyl)borane was investigated by nuclear magnetic resonance (NMR) spectroscopy (Figure 7.3) [56]. The corresponding ^1H, ^{19}F HOESY (heteronuclear overhauser effect spectroscopy), and diffusion NMR studies support the interaction between the phosphine and the borane. The equilibrium constant was determined to be $K = 0.3–0.7\,M^{-1}$ which is in agreement with slightly endergonic encounter complex formation (ΔG^0 (298K) = $+0.4 \pm 0.2$ kcal/mol).

Quantum chemical data support that the formation of the encounter complex is responsible for the pre-polarization of the hydrogen molecule [53, 58] within the electric field (EF) prior to its orbital interactions with the Lewis pair's heteroatoms [59, 60]. The orbital interactions consist of the donation of electron density from the Lewis base into the σ $(H_2)^*$ orbital of the pre-polarized H_2, which leads to destabilization of the H—H bond according to an electron transfer (ET) process. A clear cut between the pure ET and EF mechanism for active FLPs is not possible, and it is more reasonable to consider that both processes occur in the course of the H_2 activation, but at different reaction coordinates [60]. Pre-polarization through the EF mechanism is certainly required to stabilize the charge separation and to lower the $\sigma(H_2)^*$ for efficient ET from the corresponding Lewis base and charge stabilization by the Lewis acid (compare Figure 7.4a).

This optimal geometry can be readily realized, for example, by Erker's *intramolecular* FLP $Mes_2P–CH_2–CH_2–B(C_6F_5)_2$ [2, 62], which is probably the reason for its unmatched reactivity in H_2 activation.

Only recently, electrophilic phosphonium cations (EPCs) were found to activate C—F bonds [63–65] and molecular hydrogen [61]. EPCs are isoelectronic

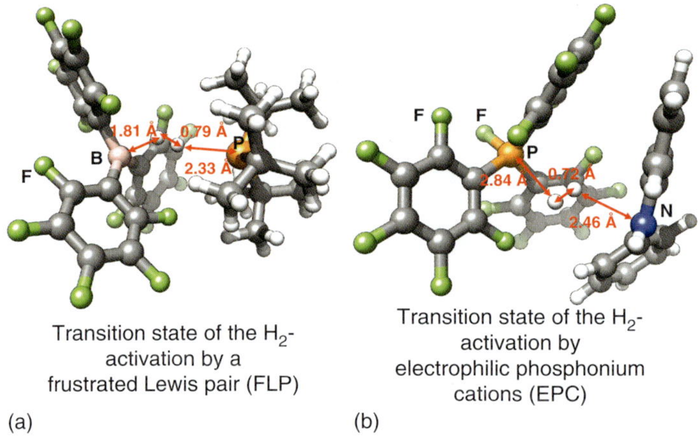

Figure 7.4 DFT-calculated transition states of the heterolytic splitting of H_2 by an (a) FLP consisting of $B(C_6F_5)_3$ and P*t*Bu_3 and (b) an electrophilic phosphonium cation (EPC); [61] (molecular structure representation created with Chimera).

to boranes, resulting in comparable reactivity [66, 67] and similar mode of H_2 activation (compare Figure 7.4a,b) [61]. Again, the Lewis acidity is strongly increased by the fluorine atoms, leading to the stabilization of the phosphorus-(V) species, which in turn functions as hydride source. In this light it is not surprising that also carbenium ions [68] or silylium ions [69] are active in the H_2 splitting in combination with a suitable base. However, suitable downstream processes of such derived FLPs are to date unavailable, rendering these interesting activation methods incompatible with catalytic hydrogenations.

Also, Lewis bases other than Mes_3P (**1**) or *t*-Bu_3P (**3**) should be active in H_2 activation according to the general picture of the H_2 activation by FLPs described above. Indeed, a large number of Lewis bases for such purpose have been identified ranging from bisphosphines [70, 71], phosphinimines [72], imines [73–76], amines [77, 78], diethyl ether [79] over carbenes [80–86] to carbanions [87]. The most commonly used FLPs for H_2 activation consist of an amine or phosphine as Lewis base and $B(C_6F_5)_3$ (**2**) as Lewis acid.

The most striking aspect of FLP-catalyzed hydrogenations is the formation of the hydride-donor directly from the heterolytic splitting of molecular hydrogen. Borohydrides are versatile reduction agents and an arsenal of methodologies for the reduction of several functional groups have been elaborated [88, 89]. The advent of FLP-catalysis started with the reduction of imines with H_2 [90]. Usually, the reduction of imines requires their activation by a Lewis acid or alternatively by a Brønsted acid. The onium borohydride resulting from the FLP-mediated H_2 activation undergoes proton transfer to the imine liberating the Lewis base with concomitant hydride transfer generating the amine (Figure 7.5a).

Quickly, it was realized that the corresponding substrate, here the imine, should also function as viable Lewis base in the H_2 activation (Figure 7.5b). Accordingly, most FLP-catalyzed reductions of aldimines, ketimines, or nitrogen-containing heterocycles do not require the addition of phosphines or other Lewis bases since the substrate fulfills their part in the heterolytic

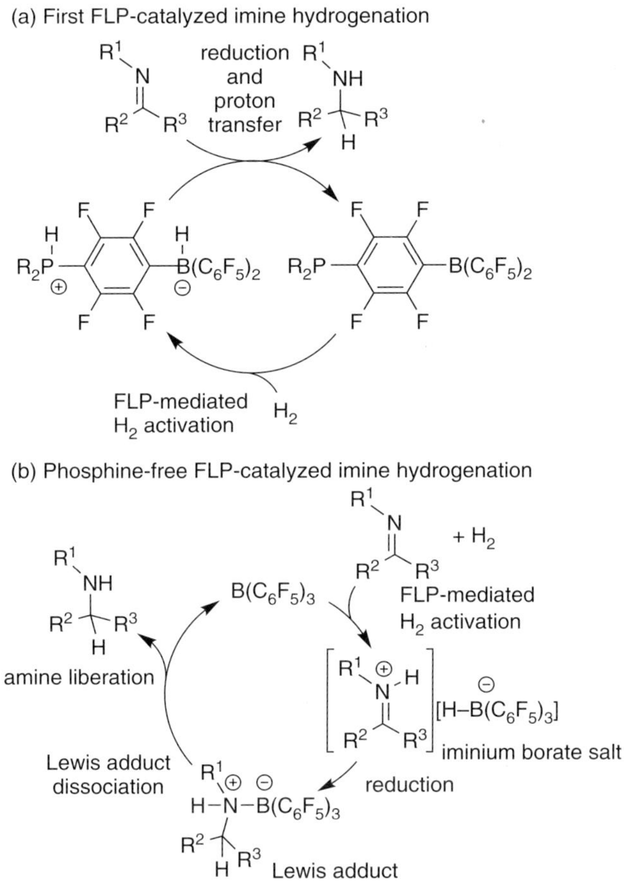

Figure 7.5 (a) Catalytic cycle for the FLP-catalyzed hydrogenation of imines [90]; (b) catalytic cycle for the phosphine-free FLP-catalyzed hydrogenation of nitrogen-containing substrates [91].

splitting of H_2 [91]. With this in mind, one must come to the conclusion that autoinduced catalytic processes should be operative. Indeed, this important feature was first mentioned as a potential process [92] and has been supported by quantum mechanical investigations [52]. However, detailed kinetic experiments of the imine reduction were only recently available and disclosed important information for FLP catalysis (*vide infra*) [74, 75].

7.3 Influence of the Lewis Acid and Lewis Base on Hydrogenation Reactivity

One of the most important features of *intermolecular* FLPs is that the Lewis acid and Lewis base component can be varied independently. This directly guides us to the question of which combination is the best catalyst for the specific purpose. Although a straightforward answer is highly desirable, it cannot be provided. Clearly, from the abovementioned factors a very strong Lewis acid and very strong electron-donating Lewis base will provide a very active FLP for the H_2 activation. However, the splitting of molecular hydrogen is an important but not the only part in a catalytic hydrogenation. Suitable downstream processes are required, which in turn strongly depend on the ability to accommodate a hydride of certain nucleophilicity and a proton of specific pK_a. Consequently, the balance between Lewis acidity and Lewis basicity must be found to ensure efficient catalytic turnover.

As discussed in section 7.2, the Lewis base is responsible for retrieving the proton from the heterolytic H_2 activation. However, another significant role of the Lewis base is to transfer the proton to the substrate. Hence, the pK_a of the corresponding onium species is quite relevant for the reaction rate and/or for the suppression of undesired reaction pathways. A general prerequisite is that the Lewis base does not form or only reversibly forms a Lewis adduct with the Lewis acid. This requirement can be met by introduction of steric bulk and/or electronic modifications (Figure 7.6).

It has been the general consensus that only strong electron-donating Lewis bases are capable of activating H_2 in combination with, e.g. $B(C_6F_5)_3$. This notion is supported by the experimental observation and characterization of the H_2 activation products by NMR spectroscopy or X-ray crystallography.

In contrast, it was found that also the very weak electron-donating fluorinated phosphines (compare Figure 7.6 and Figure 7.7) activate H_2 in the presence of $B(C_6F_5)_3$ [93, 94]. At room temperature the H_2 activation is highly reversible and only the coexisting phosphine and borane were observed in solution (Figure 7.7).

At low temperatures (−20 to −80 °C) the corresponding phosphonium borohydrides resulting from the H_2 activation were characterized by ^1H, ^{31}P, ^{11}B, and ^{19}F NMR spectroscopy. The correlation of the pK_a of the conjugate Brønsted acid arising from the heterolytic H_2 activation with the detection temperature of H_2 splitting shows a linear trend (Figure 7.7a), giving rise to the first structure–reactivity relationship in FLP-catalyzed hydrogenations [93].

As a consequence, the onium borohydride is formed transiently under hydrogenation conditions (25–120 °C), which is sufficient to perform catalytic reactions [79, 93–95]. However, the nature of the Lewis base must be adjusted to the catalytic reaction of interest. So far, a general prediction of the best Lewis base for a specific reaction cannot be made but the inspection of pK_a differences in the substrate and product seem to be a suitable measure.

7.3.1 Choice of Lewis Acid

The easier the selection of suitable Lewis base from a library the more difficult it is to synthesize new Lewis acids for FLP catalysis. The Lewis acidity determines the strength of the $\sigma(H_2)$ orbital interaction. For borane-based FLPs the commercially available perfluorated borane $B(C_6F_5)_3$ is most commonly utilized. Its Lewis acidity can be determined by the Gutmann–Beckett [96–99] or Childs and coworkers [100] methods and was arbitrarily set to 100%. Other Lewis acids can

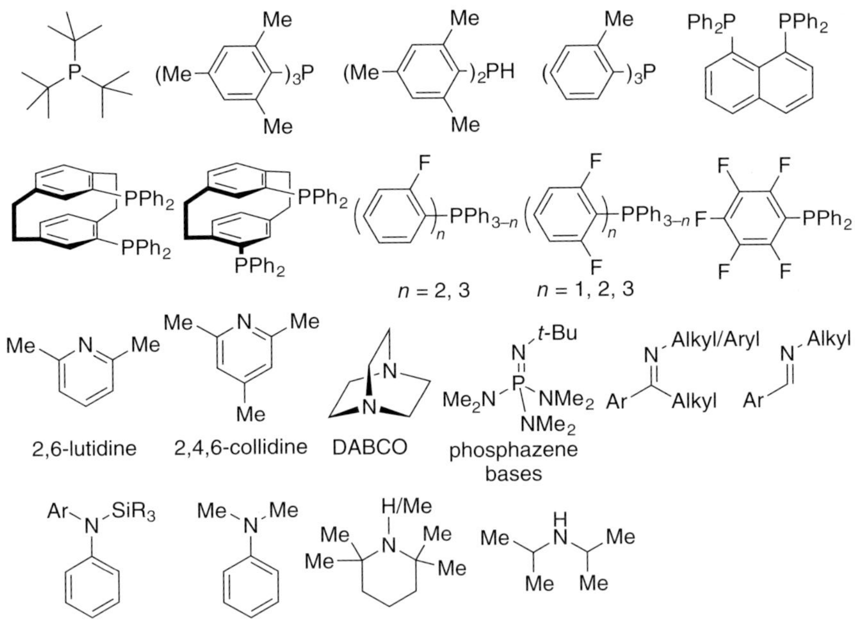

Figure 7.6 Lewis bases for the FLP-mediated H_2 activation.

therefore be ranked in stronger or weaker Lewis acidic boranes. A selection of Lewis acids that have been used in H_2 activation are depicted in Figure 7.8.

In selecting a Lewis acid for the specific reaction one needs to take into account that the increased Lewis acidity will result in reduced nucleophilicity of the generated hydride although it seems appealing to use strong Lewis acids to enhance the H_2 activation step.

The slightly reduced Lewis acidity of 95% of **B** compared to **A** (100%) might suggest comparable reactivity. Surprisingly, this minimal difference leads to very different chemistry as seen in the H_2 activation using *o*-Tol_3P as Lewis base (Figure 7.9) [101].

While the $B(C_6F_5)_3$ system splits hydrogen at room temperature irreversibly (no H_2 liberation was observed) the slightly less Lewis acidic borane $B(C_6F_4H)_3$ was not only able to activate H_2 in the presence of *o*-Tol_3P but also released H_2 under reduced pressure at ambient temperature.

7.4 Balance Between Lewis Acidity and Lewis Basicity

Experience showed that the most appealing feature of an *intermolecular* FLP is also the most challenging one. As Lewis acid and Lewis base can be varied independently it is crucial to balance acidity and basicity for the optimization of the desired reaction. To garner a better understanding of the different requirements of the individual FLP components it is helpful to investigate a simple reaction system in greater detail.

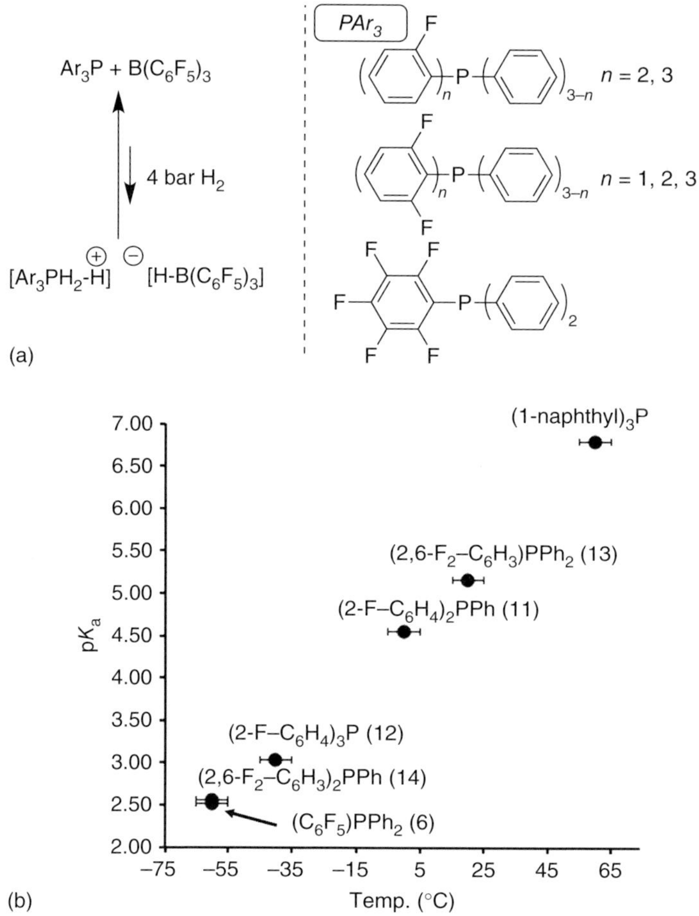

Figure 7.7 (a) Highly reversible H_2 activation using fluorinated triphenylphosphine derivatives; (b) Temperature dependency of H_2 activation detection of various phosphines vs. the pK_a of the conjugate acid [93, 94].

As highlighted before, the hydrogenation of *N*-*t*-butylimines is one of the milestones in FLP catalysis, yet simple enough for a sophisticated study on the influence of Lewis acid and Lewis base [90–92]. In this case, the reaction system consists of a catalytic Lewis acid and of an imine, which acts equally as Lewis base and as substrate (Figure 7.10). Activation and heterolytic splitting of molecular hydrogen leads to an iminium borohydride salt, which collapses to the corresponding *N*-*t*-butylamine (simple catalytic cycle). The amine is also able to activate H_2 in combination with the borane, leading to a second catalytic pathway (autoinduced catalytic cycle) [52, 74, 75]. Proton transfer to an imine generates the same iminium borohydride salt and results in the formation of additional amine. This leads to an autoinduced catalytic reaction mechanism where the reaction product acts as catalyst.

A drift toward the autoinduced cycle should be observed although the overall concentration of Lewis base remains unchanged during the reaction ($c_{t,\text{imine}} + c_{t,\text{amine}} = c_{0,\text{imine}} + c_{0,\text{amine}}$) and should be strongly dependent on the nature of the Lewis acid and of the substrate. Indeed, experimental data shows that reactions catalyzed by the very strong Lewis acid $B(C_6F_5)_3$ follow first order kinetics, and weaker Lewis acids, e.g. $B(2,4,6\text{-}F_3\text{–}C_6H_2)_3$ or $B(2,6\text{-}F_2\text{–}C_6H_3)_3$, follow the mechanism of autoinduced catalysis. Despite the basicity difference of the imine and of the amine (ΔpK_a), the strength of the Lewis acid influences which catalytic pathway dominates the overall reaction. The reaction system consists of two competing reaction cycles. The equilibria of hydrogen activation are slow and are considered as rate-determining steps. However, the hydrogen activation should be favored by the more basic amine, leading to autoinduced

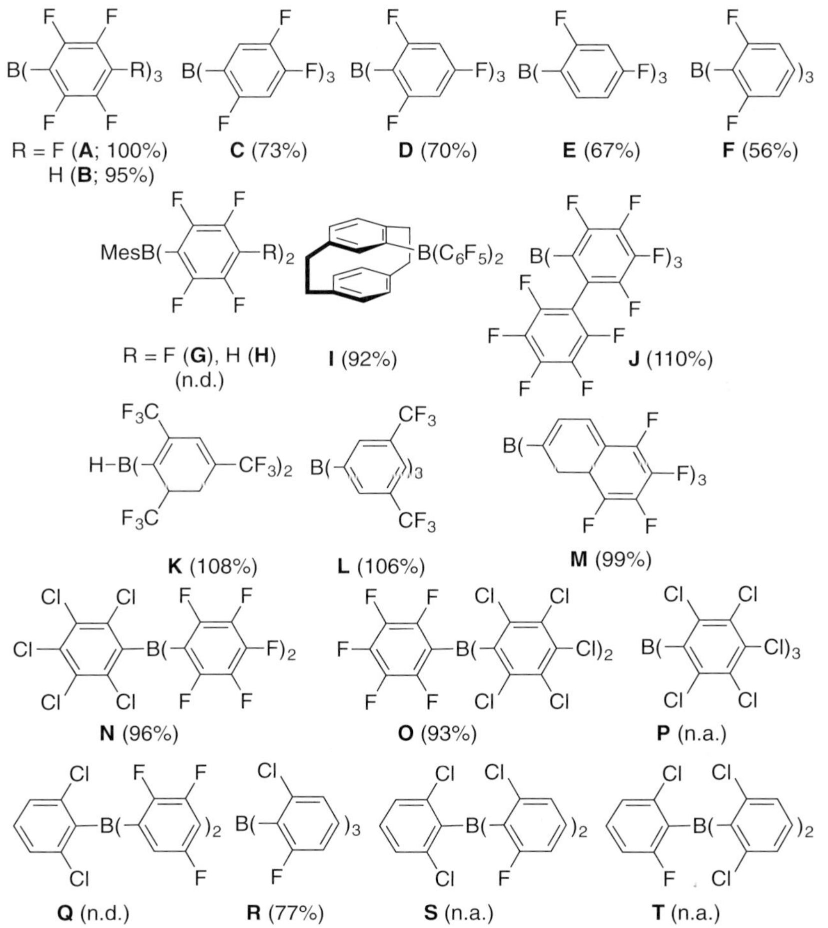

Figure 7.8 Lewis acids **A–T** for FLP-mediated H_2 activation (percentage in parentheses corresponds to the Lewis acidity with reference to $B(C_6F_5)_3$ (100%, Gutmann–Beckett method)). **A** [96]; **B** [101]; **C** [102]; **D** [102]; **E** [102]; **F** [102, 103]; **G** [104]; **H** [104]; **I** [105]; **J** [106]; **K** [107]; **L** [108]; **M** [109]; **N** [110]; **O** [110]; **P** [110]; **Q** [111]; **R** [112]; **S** [112]; **T** [112].

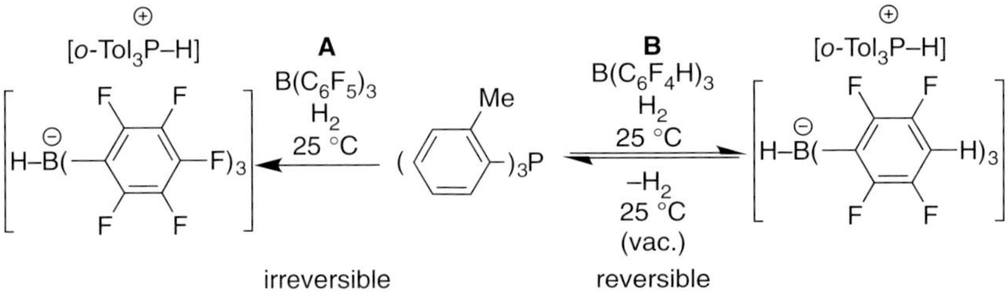

Figure 7.9 Comparison of $B(C_6F_5)_3$ and $B(C_6F_4H)_3$ in the H_2 activation [101].

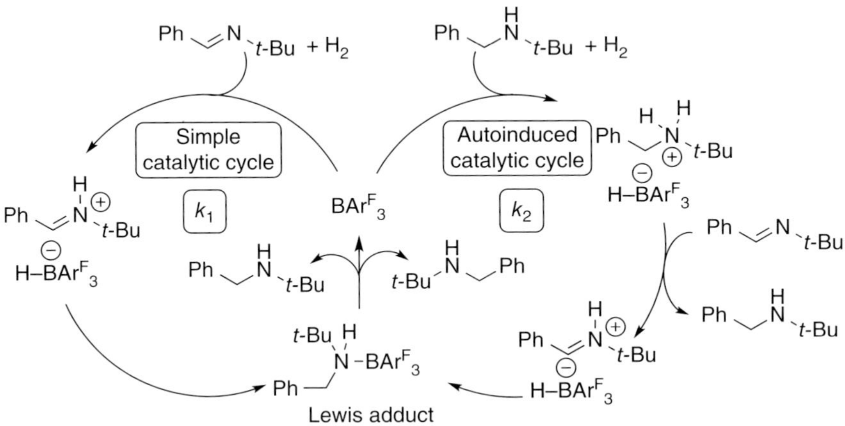

Figure 7.10 Reaction mechanism of the autoinduced catalytic hydrogenation of *N*-*t*-butylimines with boranes. Left cycle: H_2 activation with imine as Lewis base. Right cycle: H_2 activation with amine as Lewis base [74, 75].

catalysis. The key is the amount of available "free borane," which is affected by the position of equilibria. Only if the equilibrium of hydrogen activation with the imine (simple cycle) favors free borane over borohydride, the autoinduced catalytic cycle will dominate the overall reaction and an autoinduced kinetic behavior will be observed. This is the case when weaker Lewis acids are used [74, 75].

To provide a more general picture of FLP reactivity in hydrogenations the electronic influence of substituents on the imine in order to elaborate a structure–reactivity relationship was investigated: The associated rate constants k_1 and k_2 for the standard and autoinduced catalytic cycle respectively (compare Figure 7.10) were determined by kinetic experiments, which in turn enabled Hammett analysis. This gives the direct connection between catalytic reactivity and structural information in FLP catalysis. The basicity of the imines and the amines is quantified as pK_a of the conjugate Brønsted acids. The difference in basicity of imine and amine (ΔpK_a) is of special interest, since this parameter will dramatically influence the kinetic profile of the reaction.

Changing substituents on the imine's phenyl (Ph) ring changes the electron density in the conjugated imine bond and, therefore, at the nitrogen atom. This leads to a linear correlation between Hammett substituent constant σ and pK_a of

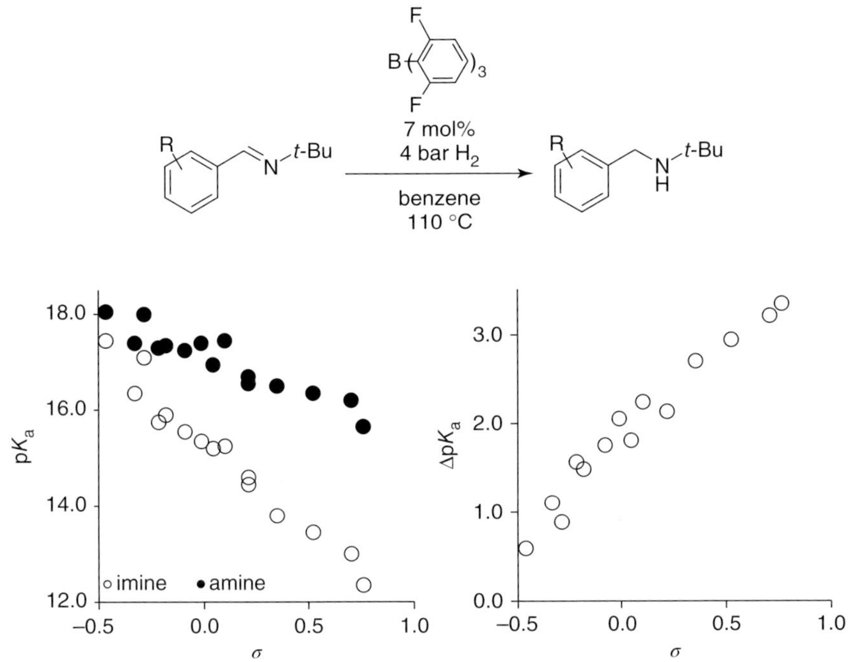

Figure 7.11 Top: Catalytic hydrogenation of an imine library with R = 4-OPh, 4-O*t*Bu, 4-OMe, 4-*t*Bu, 4-Me, 3-Me, 4-H, 4-F, 3-OMe, 4-Cl, 4-Br, 3-Cl, 4-CF_3, 4-NO_2, 4-SO_2Me (with increasing σ). Left: pK_a values of the corresponding Brønsted acids for imines and amines vs. Hammett substitution parameter. Right: ΔpK_a as difference in basicity between imines and amines [75].

the iminium ion spanning six pK_a units. Since the benzyl (Bn) group has solely an inductive effect on the N atom, the dependence is much smaller (Figure 7.11, left). The combination is a linear correlation between Hammett substituent constant σ and ΔpK_a, where both can be utilized as structural parameter in the structure–reactivity relationship (Figure 7.11, right). Both ΔpK_a and σ result in linear correlation with the logarithmic rate constants for the simple reaction (k_1) and the autoinduced reaction (k_2) (Figure 7.12 left, middle). By knowing this correlation, it is now possible to calculate rate constants and even predict the reactivity from structural data such as (calculated) pK_a values (Figure 7.12). The structure–reactivity relationship shows two important features: First, the greater the difference in the basicities of the imine and the amine (ΔpK_a), the more present is the autoinduced catalytic pathway (k_2/k_1 is large, Figure 7.12, right), since the amine is more basic than the imine. This effect does not necessarily imply that reactions that feature autoinduced reaction profiles are faster reactions. Second, the smaller the absolute pK_a values (less basic imines and amines), the more pronounced is the autoinduced mechanism. This observation accounts for the competing reaction cycles. For a less basic imine the position of the H_2 activation equilibrium favors borane over borohydride, boosting the autoinduced cycle.

From this systematic study it is possible to draw conclusions for the autoinduced hydrogenation and to generalize these findings for other

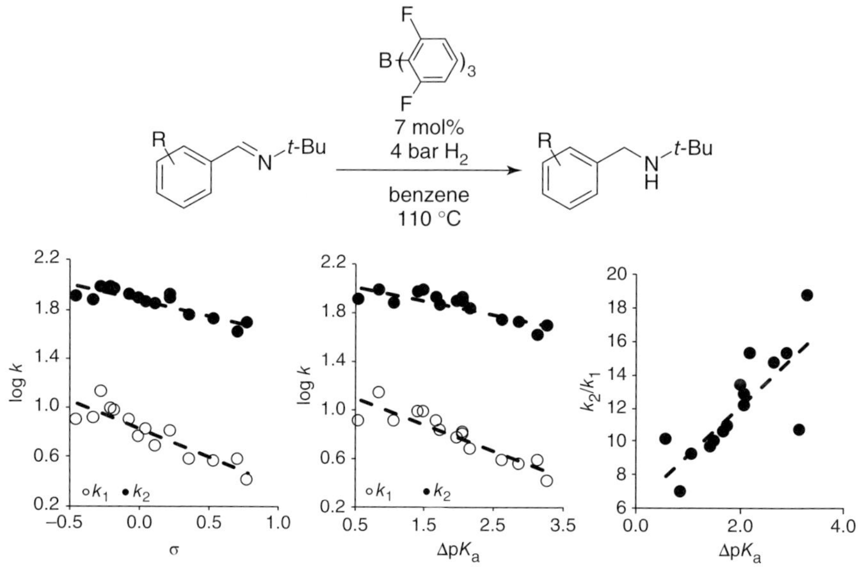

Figure 7.12 Structure–reactivity relationship of autoinduced imine hydrogenation with rate constants for simple catalytic cycle (k_1) and autoinduced catalytic cycle (k_2) with reference to Hammett σ parameter (left) or difference in pK_a of imines and amines (middle). Influence of autoinduced cycle shown as k_2/k_1 (right) [75].

FLP-catalyzed reactions. There are two important measures that influence the autoinduced reaction: On the one hand, the difference in the basicities of the starting material and the product (expressed as pK_a of the corresponding Brønsted acids) plays a crucial role. On the other hand, careful modulation of the overall difference between Lewis acidity of the borane and Lewis basicity of the individual base (in the example above influencing the position of H_2 activation equilibrium) introduces great possibilities, especially when the following reaction steps must be considered. As pointed out before, using a strong Lewis acid and a strong Lewis base in H_2 activation leads to a complete formation of onium-borohydrides but generates weak hydride nucleophiles and weak Brønsted acids. In fact, the combination of a strong Lewis acid and a weak Lewis base has the same capability to activate dihydrogen as the combination of a weak Lewis acid with a strong Lewis base. To promote the following reaction steps choosing a minimal difference between Lewis acidity and Lewis basicity for H_2 activation is beneficial. Thus, nucleophilicity of borohydride or Brønsted acidity can be maximized. Balancing these four variables is far from straightforward and shall be demonstrated in the following examples.

7.4.1 Hydrogenation of Olefins

In realizing a successful hydrogenation reaction, the H_2 activation is a "simple" task and is solely dependent on the choice of a Lewis acid and a Lewis base of suitable strength and steric restraint. More challenging is the reaction of H_2 activation products with the desired double bond.

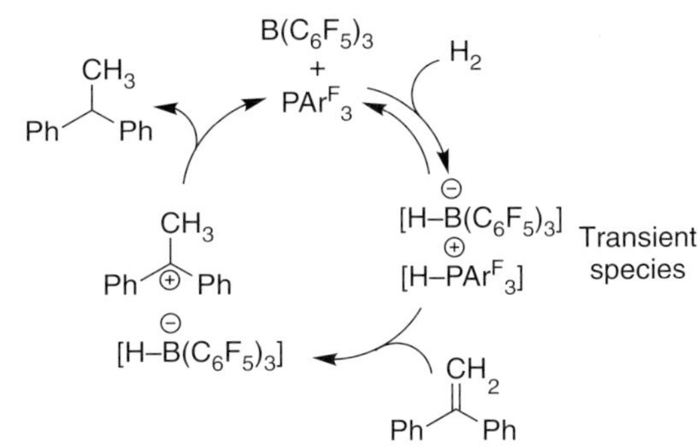

Figure 7.13 Catalytic hydrogenation of 1,1-diphenylethene with catalytic $B(C_6F_5)_3$ and weakly basic $P(2\text{-F–}C_6H_4)_3$ ($pK_{a,MeCN}(HP^+) = 3.03$).

In case of electron-rich double bonds as in unsubstituted olefins, styrene derivatives [93, 94], or silyl enol ethers [70], protonation of the double bond is the initial step followed by nucleophilic attack of the borohydride (Figure 7.13). Therefore, depending on the nature of the substrate molecule, a Brønsted acid of specific strength is needed. The generated carbocation reacts readily with weak hydride nucleophiles justifying the use of a strong Lewis acid. In this case, the strategy is to pick a Lewis base weak enough to yield a Brønsted acid in the desired pK_a range and pair it with a Lewis acid just strong enough to ensure a minimal amount of H_2 activation (transient H_2 activation). So, the FLP of choice consists of a strong Lewis acid and a weak Lewis base.

Things change when nitroolefins [103], acrylates, or malonates [102, 103] are targeted (Figure 7.14). Similar to common carbonyl reduction by borohydrides, the strength of the hydride nucleophile is of utmost importance and demands for a weak boron Lewis acid. This reduced Lewis acidity requires the counterbalancing by the Lewis base to achieve H_2 activation. This calls for an FLP consisting of a weak Lewis acid and a strong Lewis base.

7.4.2 Dehydrogenative Coupling

The dehydrocoupling of Si–H and N–H fragments provides an environmentally benign access to silyl-protected amines [33]. These ubiquitous structural motifs are usually obtained by the reaction of halosilanes with deprotonated amines,

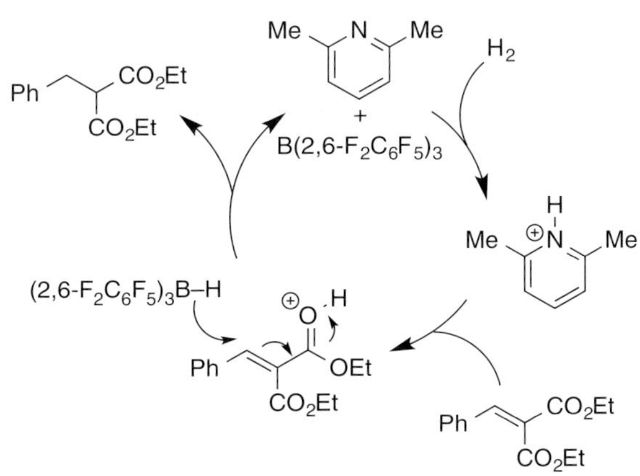

Figure 7.14 Catalytic hydrogenation of a diethylmalonate derivative with catalytic amounts of $B(2,6\text{-}F_2\text{–}C_6H_3)_3$ and 2,6-lutidine ($pK_{a,MeCN}(HN+) = 14.13$) [113].

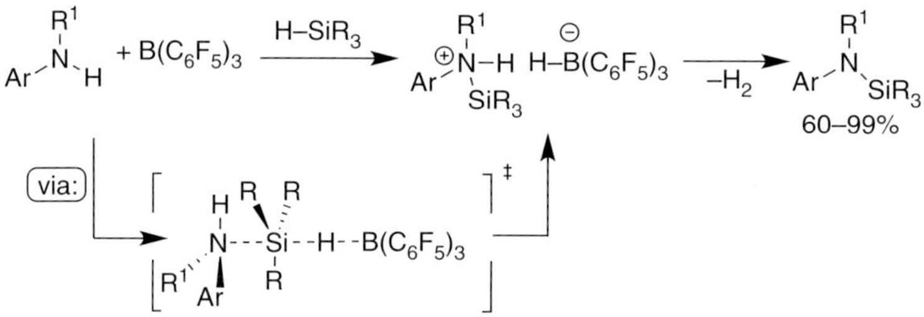

Figure 7.15 Acceptorless dehydrocoupling of aryl amines with hydrosilanes. $pK_{a,MeCN}(NH^+) = 10.62$.

whose generation often requires strong bases. The reaction of primary and secondary amines with hydrosilanes can be readily catalyzed in the presence of 1 mol% $B(C_6F_5)_3$ with concomitant liberation of molecular hydrogen to yield the silylamines (Figure 7.15) [34].

This reaction is initiated by hydride coordination by the borane and transfer of the polarized silicon group to the NH nucleophile. The resulting ammonium species is prone to hydrogen release, if the equilibrium of H_2 activation favors borane over borohydride as observed in the hydrogenation of olefins [23, 35]. It should be noted that the steric hindrance of the Lewis base plays a minor part, due to the high reactivity of the borane and hydrosilane and the low basicity of aryl amines in general.

The activation of molecular hydrogen led to interesting reactivity and the concept behind is well understood. The backward reaction – the recombination of proton and hydride from an onium-borohydride species – is far less studied, although the same principles can be applied. An impressive example is the dehydrogenative coupling of diarylamines, anilines, and indoles with silanes using catalytic amounts of $B(C_6F_5)_3$ [114]. This reaction is initiated by hydride abstraction of borane yielding a silicon electrophile, which is readily attacked by an NH nucleophile. The resulting ammonium species will be capable of releasing dihydrogen, if the equilibrium of H_2 activation favors borane over borohydride. This is true for a certain pK_a range of the ammonium and limits the substrate scope to arylamines. It should be noted that the steric hindrance of the Lewis base plays a minor part, due to the high reactivity of borane and hydrosilane and the low basicity of arylamines in general. The process can only be improved in a small scope by reducing Lewis acidity and/or Lewis basicity. While the first will reduce the electrophilicity of the borane and slow down hydride abstraction, the latter will narrow the substrate scope. Nevertheless, dehydrogenative coupling works for a broad range of arylamines and paves the way for the acceptorless dehydrogenation.

7.4.3 Acceptorless Dehydrogenation

From a thermodynamic point of view, a hydrogenation catalyst should also be able to perform the reverse reaction. Transfer hydrogenations are comparably

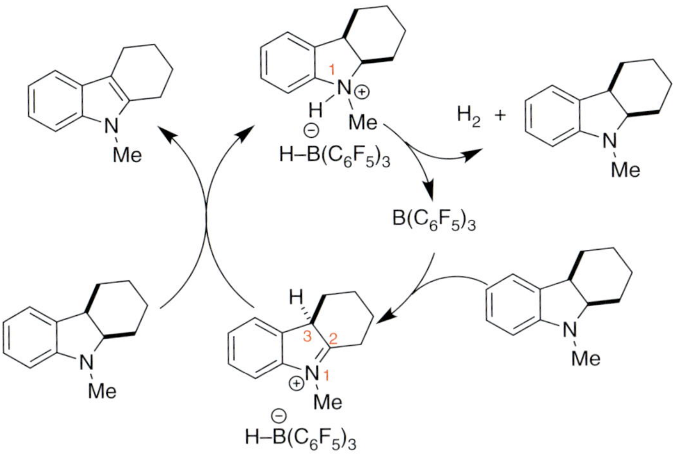

Figure 7.16 Acceptorless dehydrogenation of *N*-methylhexahydrocarbazole with catalytic amount of $B(C_6F_5)_3$.

abundant, whereas the related acceptorless dehydrogenations are considered as highly challenging, even for transition metal catalysts.

While the abstraction of a hydride atom from activated CH bond, e.g. in alkyl amines, is feasible by strong Lewis acids, the release of molecular hydrogen is, in fact, the rate-limiting step. Depending on the substrate, the hydride abstraction demands for a strong Lewis acid. On that account, the release of molecular hydrogen should be facilitated by a suitable Lewis base.

In the example in Figure 7.16, [115] hydride abstraction leads to a strongly Brønsted acidic 3-*H*-indolium, which should facilitate the release of molecular hydrogen. Unfortunately, the proton transfer to the ever-present starting material is kinetically more favored than direct deprotonation. The resulting 1-*H*-indolinium is a much weaker Brønsted acid, which results in the slower liberation of molecular hydrogen. Influencing the position of H_2 activation equilibrium to promote dehydrogenation would require a weaker Lewis acid (stronger hydride donor) or weaker Lewis base (stronger Brønsted acid). Since already two Brønsted acids of different strength are present and in dependence on the substrate, the only way to solve this predicament is by tuning the Lewis acidity.

This problem is solved by introducing a slightly less acidic borane in addition to $B(C_6F_5)_3$. The less acidic $B(2,4,6\text{-}F_3{-}C_6H_2)_3$ does not interfere with the catalytic reaction mechanism but forms an equilibrium with the ammonium borohydride through hydride exchange (Figure 7.17). Even though the position of equilibrium does not favor $[HB(2,4,6\text{-}F_3{-}C_6H_2)_3]$, the reaction rate of the overall dehydrogenation increases by more than twofold, illustrating an ideal example of cooperative borane catalysis.

7.4.4 Intramolecular Frustrated Lewis Pairs

Intramolecular FLPs were the first examples showing reversible metal-free H_2 activation since steric constraints and suitable orbital orientation within the FLP

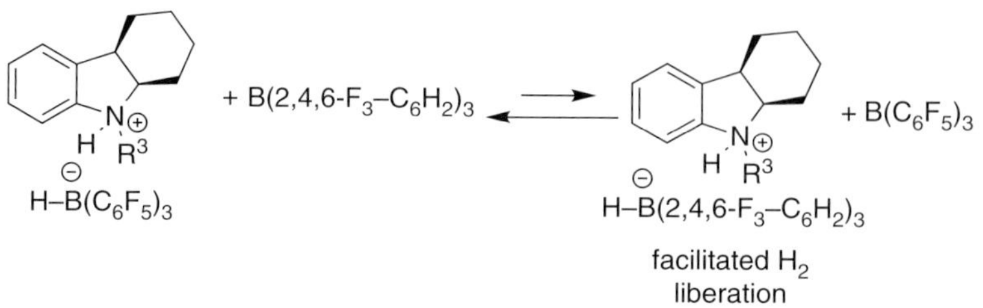

Figure 7.17 Equilibrium of H_2 activated *N*-methylhexahydrocarbazole with $B(C_6F_5)_3$ and the less Lewis acidic B(2,4,6-F_3–$C_6H_2)_3$.

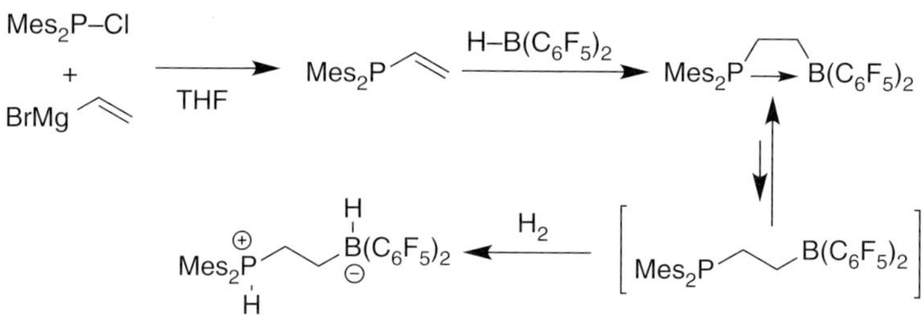

Figure 7.18 Synthesis of ethylene-bridged phosphino borane.

were considered crucial. The synthesis of *intramolecular* FLPs is by far more challenging compared to the synthesis of *intermolecular* ones. *Intramolecular* FLPs require the installation of the Lewis acid and Lewis base in one molecular entity. Several approaches to meet this challenge have been reported. The most straightforward and most elegant strategy takes advantage of the high reactivity of Piers' borane ($HB(C_6F_5)_2$) [116–118] toward olefins in the sense of a hydroboration reaction. The reaction of vinyl-bis(mesityl)phosphine with $HB(C_6F_5)_2$ furnished the ethylene-bridged phosphino borane in good yield (Figure 7.18) [2, 62, 119].

The phosphinoborane forms a four membered P/B-heterocycle [53] by a dative bond from the phosphorus to the boron atom. This intramolecular Lewis adduct is in equilibrium with its "open" form, which in turn is active in the heterolytic H_2 splitting, leading to the phosphonium borohydride salt.

Alternatively, other heterovinyl compounds, e.g. cyclohexenyl phosphines or simple enamines, can be utilized in the hydroboration to generate the corresponding *intramolecular* FLP systems. Accordingly, the reaction of $HB(C_6F_5)_2$ with these heterosubstituted olefins furnishes novel P/B [120] and N/B systems [121] in good yields (Figure 7.19).

A second approach toward the synthesis of intramolecular FLPs uses the halogen–lithium exchange of aryl bromides and subsequent treatment with bis(pentafluorophenyl)boron chloride ($(C_6F_5)_2BCl$) to introduce the Lewis acidic center. Utilizing this methodology, a selection of intramolecular FLPs were successfully accessed in good yields (Figure 7.20).

Accordingly, this approach can also be utilized for the synthesis of aniline-derived aminoboranes in excellent yields (Figure 7.21), which was later used in the metal-free *syn*-selective hydrogenation of alkynes [77, 123, 124].

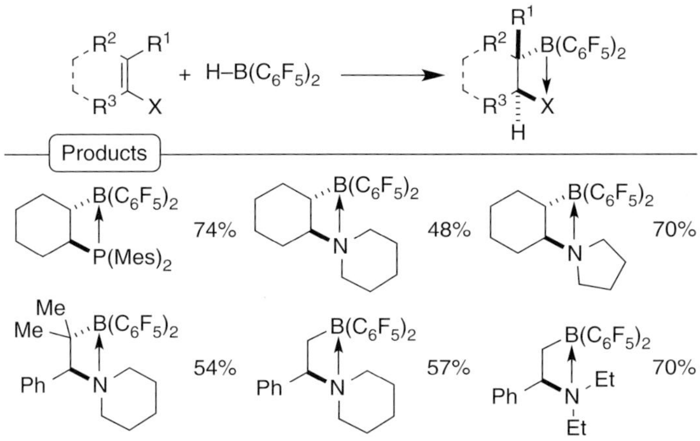

Figure 7.19 Synthesis of vicinal Lewis pairs by hydroboration.

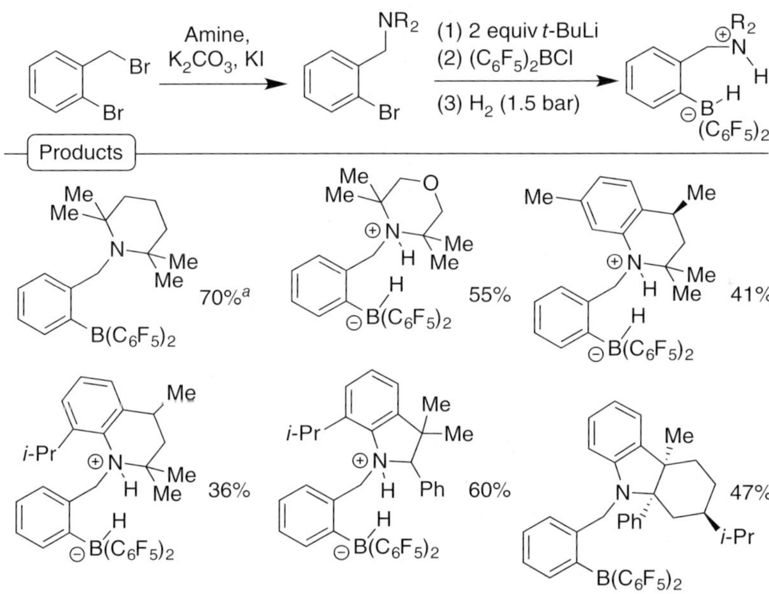

Figure 7.20 Synthesis of intramolecular FLPs bearing aryl boranes (yields correspond to the final step in the synthesis; [a]product isolated after second step as aminoborane).

In a similar manner, a chiral 1,1′-binaphthyl derivative was synthesized by the stepwise alkylation of 2′-iodo-[1,1′-binaphthalen]-2-amine with isopropyl iodide and methyl iodide followed by final metalation and reaction with the perfluorated chloroborane (Figure 7.22).

Compared to the other aminoboranes, the binaphthyl derivative displayed remarkable enantioselectivity and reactivity in the asymmetric hydrogenation of imines and enamines (see below) [125]. Apart from vicinal intramolecular FLPs also geminal FLPs have been reported. These intramolecular FLPs can be readily

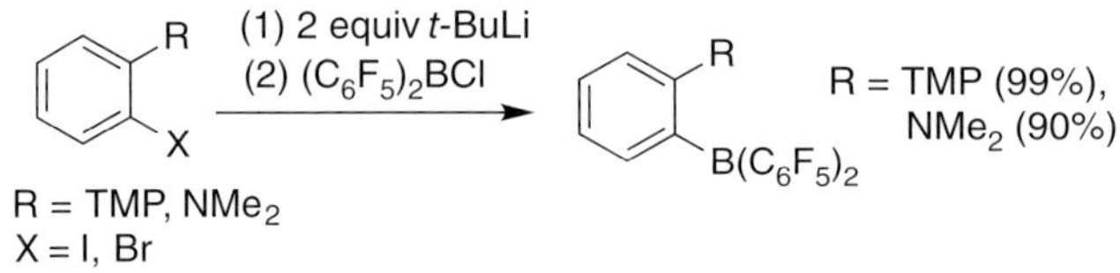

Figure 7.21 Synthesis of aniline-derived amino boranes (TMP = 2,2,6,6-tetramethylpiperidyl).

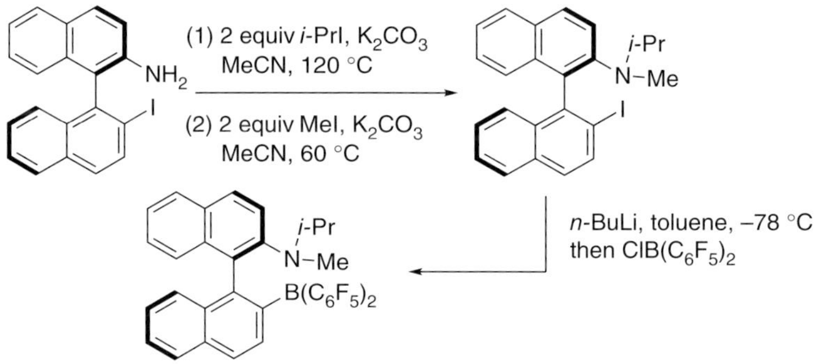

Figure 7.22 Synthesis of a chiral intramolecular frustrated Lewis pair.

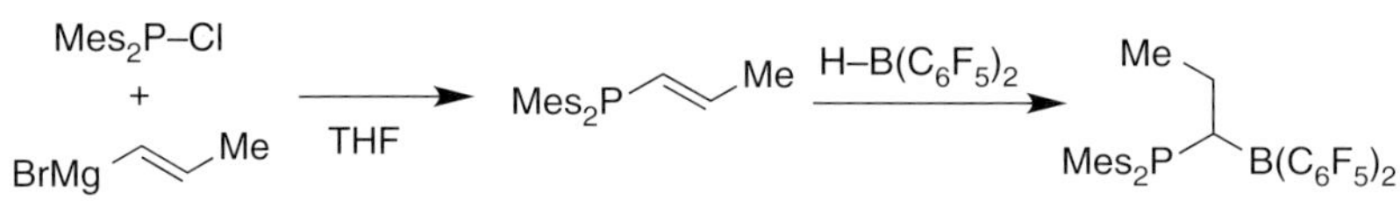

Figure 7.23 Synthesis of geminal P/B frustrated Lewis pairs.

accessed by the reaction of bis(mesityl)propenyl phosphine with $HB(C_6F_5)_2$ (Figure 7.23) [126, 127].

Crucial for the successful synthesis of such FLPs is the implementation of an electron-donating group at the olefin with the aim to achieve the formation of the Markovnikov product. Even though these systems did not prove to be active FLPs so far, they show unique addition chemistry to alkenes, alkynes, carbonyls, and azides [128–132].

7.4.5 Air-Stable FLPs

The increase of steric bulk in the Lewis acid can improve the air and moisture stability of boranes as one very well knows from mesityl borane fragments, for example, and their application in molecular electronics.

In accord with this the two sterically encumbered boranes **G** and **H** revealed increased tolerance toward laboratory atmosphere and, consequently, also toward functional groups (Figure 7.24) [104, 133, 134].

After being exposed for 30 minutes to laboratory atmosphere the sterically encumbered mesityl-substituted borane only slightly loses its hydrogenation activity whereas the perfluorated borane is completely deactivated after exposure to laboratory atmosphere.

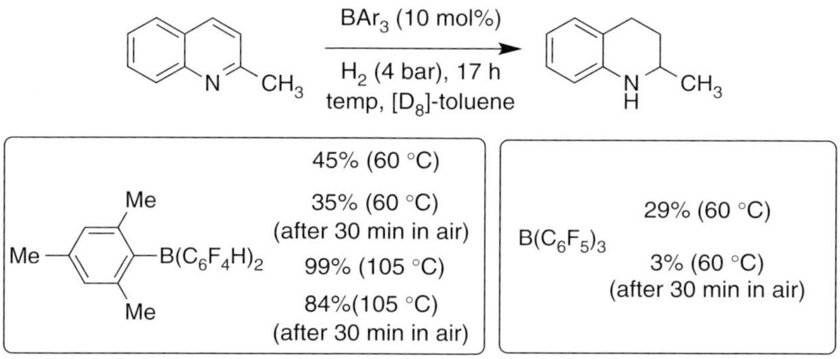

Figure 7.24 Comparative study of $B(C_6F_5)_3$ and $MesB(C_6F_4H)_2$ in the hydrogenation of 2-methyl quinolone.

The reduction of the Lewis acidity can also result in significantly increased tolerance toward functional groups. Weaker Lewis acids might still show adduct formation with substrates; however, as long as this process is reversible significant amounts of free Lewis acid are available for H_2 activation. Accordingly, triarylboranes featuring reduced fluorine content are active in the hydrogenation of substrates bearing strongly Lewis basic sites, e.g. malonates [102, 135], sulfoxides [102], or nitro groups [102, 103]. Notably, the corresponding tetrahydrofurane (THF)-adduct of $B(2,6\text{-}F_2{-}C_6H_3)_3$ could even be applied as catalyst in the hydrogenation of thiophenyl- or furyl-substituted nitroolefins [103]. $B(C_6F_5)_3$ is irreversibly coordinated by such functional groups (furyl, nitro, small esters, etc.) and catalytic hydrogenations are out of reach.

Only recently the combination of both increased steric bulk and electronic modification led to significant breakthroughs in FLP-catalyzed hydrogenations. The formal substitution of fluorine atoms by chloride atoms within the triaryl borane system has major impact on the reactivity and stability of the borane. This influence can be divided into electronic and steric reasons. On the one hand, chlorine atoms on aromatic rings have an increased electron-withdrawing effect compared to fluoride atoms, which is attributed to the reduced back-donating ability of the chloride atom. On the other hand, chlorine atoms are significantly larger than fluoride atoms, which in turn leads to enhanced shielding of the boron center, particularly when placed in the *ortho* position. The balance between these two effects leads to the development of moisture-tolerant borane-based FLPs for hydrogenations. The long-standing interest of the Ashley group laid the foundations to these significant developments [110, 136, 137]. One of the first catalytically active examples was the bis(pentafluorophenyl)-pentachlorophenyl borane $(C_6F_5)_2B(C_6Cl_5)$ in the series of chlorinated triaryl boranes (Figure 7.25).

The synthesis of this series of boranes was achieved by the addition of C_6Cl_5 species of zinc or lithium with BCl_3 followed by C_6F_5 transfer from the corresponding copper reagent [110].

It was shown that the $(C_6F_5)_2B(C_6Cl_5)$ in combination with THF generated a highly active, moisture-tolerant FLP catalyst for a variety of hydrogenations (Figure 7.26).

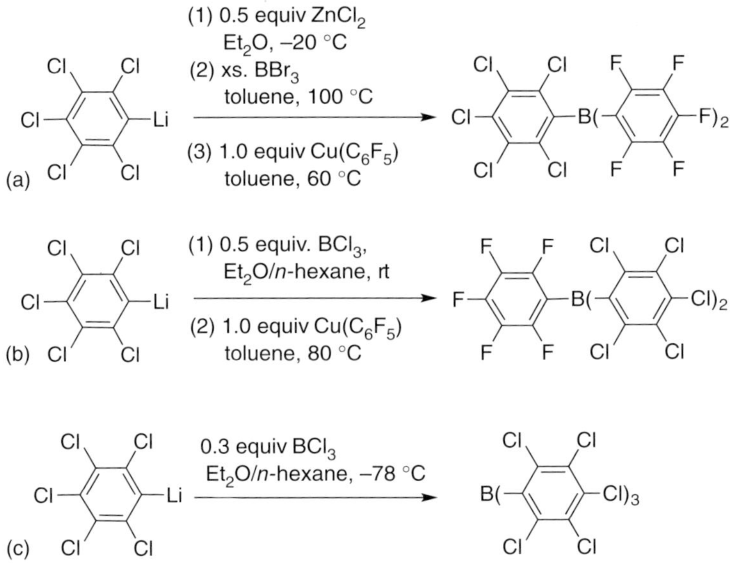

Figure 7.25 Selective synthesis of chlorinated triaryl boranes.

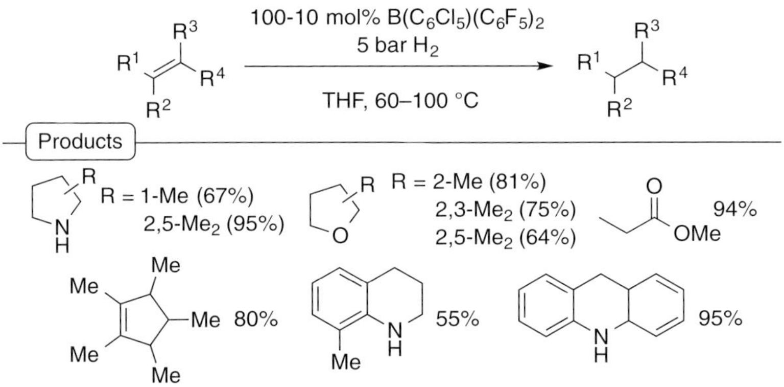

Figure 7.26 FLP-catalyzed hydrogenation using the moisture-tolerant $(C_6F_5)_2B(C_6Cl_5)$/THF system.

Further electronic modification of this system led to new reactivity. The Soos group was earlier interested in increasing the functional group tolerance of FLPs in hydrogenations (compare Figure 7.24) and investigated the corresponding chlorinated systems (Figure 7.27) [111].

A series of heteroleptic boranes on the basis of one chlorinated and two fluorinated aryls was synthesized by the addition of the Grignard reagent of the fluorinated compound to the corresponding potassium trifluoroborate salt. Interestingly, all four boranes from Figure 7.27 were active in the H_2 activation and were active catalysts in the hydrogenation of aldehydes and ketones (Figure 7.27). Later on, this synthesis was applied for the synthesis of triaryl boranes bearing chloro and fluoro substituents (Figure 7.27b and c) [112].

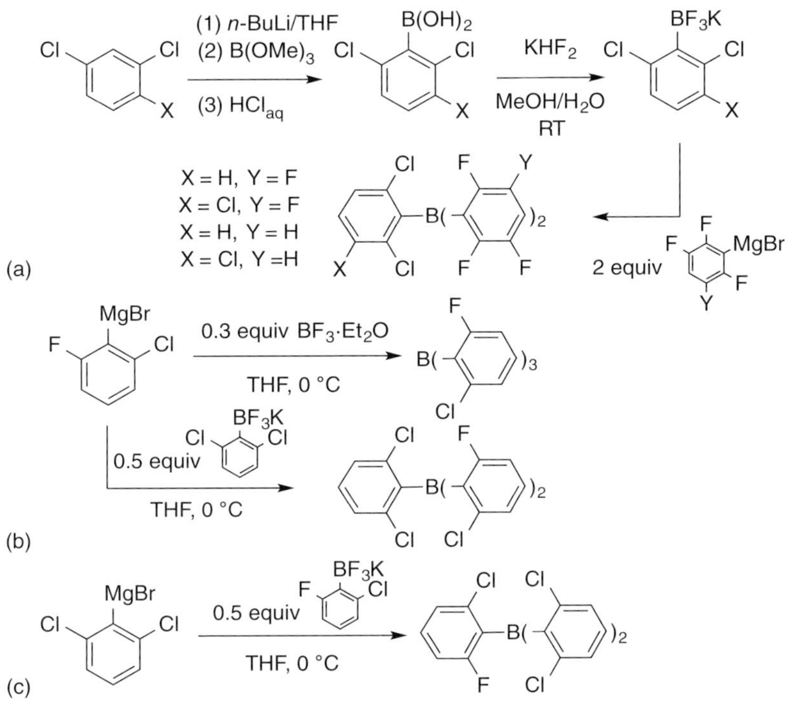

Figure 7.27 Electronically modified moisture-tolerant triaryl boranes.

7.5 Application of Frustrated Lewis Pairs in Hydrogenations

7.5.1 Hydrogenation of Aldimines and Ketimines

The hydrogenation of imines was the first application, and remained for a long time the prototype for FLP-catalyzed hydrogenation and was first reported using the original intramolecular phosphonium borate salt as catalyst (Figure 7.28) [90].

Imines bearing bulky substituents were quickly hydrogenated to the corresponding amines in excellent yields. However, imines bearing more electron-withdrawing moieties, e.g. sulfonyl groups (SO_2Ph), required higher temperature and prolonged reaction times, which may be attributed to the reversible adduct formation with the Lewis acid. Also $B(C_6F_5)_3$-activated nitriles and other substrates were susceptible to hydrogenation by the aryl-linked phosphonium borate FLP system [62, 90, 119].

However, the limited access to the aryl-bridged phosphonium borate urged the scientists to investigate *intermolecular* FLPs as hydrogenation catalysts due to the comparably easier access. As an interesting feature in the imine hydrogenation, the addition of a Lewis base is not always required (compare Figures 7.5, 7.10–7.12, section 7.3). Also, the substrate can act as Lewis base for the splitting of molecular hydrogen in combination with a suitable borane (Figure 7.29) [91].

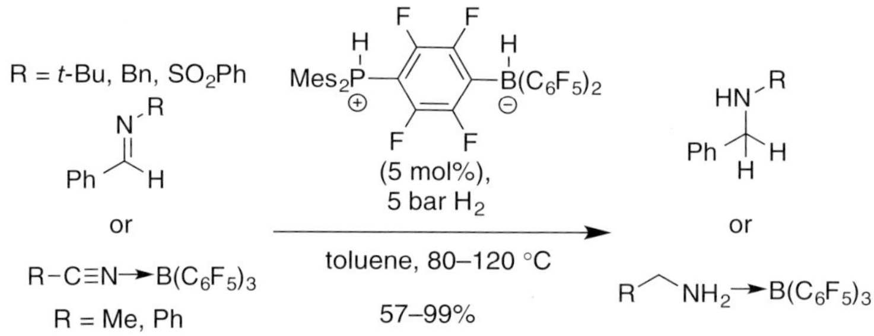

Figure 7.28 Catalytic reduction of imines using Stephan's frustrated Lewis pair.

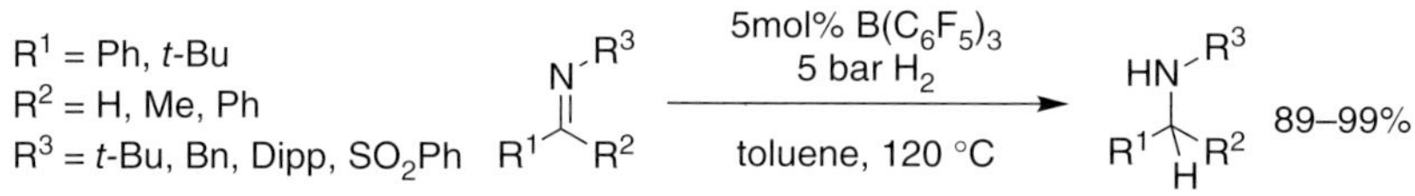

Figure 7.29 Intermolecular FLPs for the hydrogenation of imines (Dipp = 2,6-diisopropylphenyl).

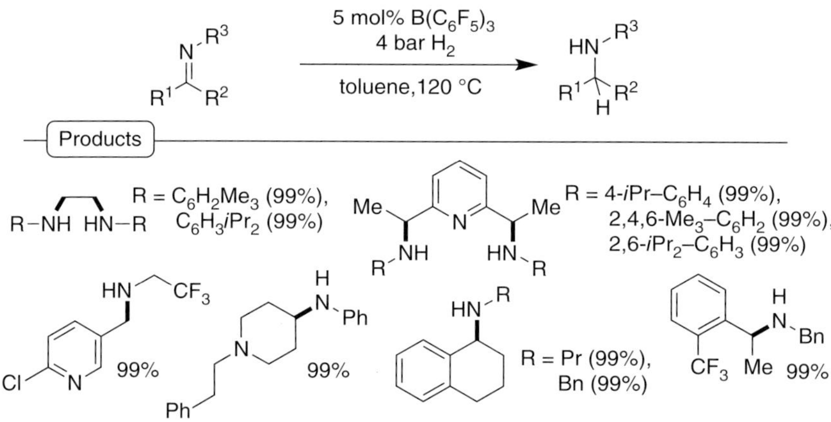

Figure 7.30 Functionalized substrates in the FLP-catalyzed hydrogenation.

Quickly, more challenging substrates were approached in the FLP-catalyzed hydrogenation using the phosphine-free conditions (Figure 7.30) [73].

Symmetric as well as unsymmetrically substituted ketimines were converted into the corresponding amines in high yield leaving other functionalities, e.g. 2-chloropyridines or CF_3-groups, which are susceptible to reduction under transition metal catalysis, untouched.

Generally, the reduction of prochiral ketimines furnishes chiral amines that are highly desirable compounds for organic synthesis and medicinal applications. Consequently, the asymmetric hydrogenation of such substrates using FLP chemistry is highly desirable. One approach is the diastereoselective hydrogenation of enantiopure substrates (Figure 7.31) [73].

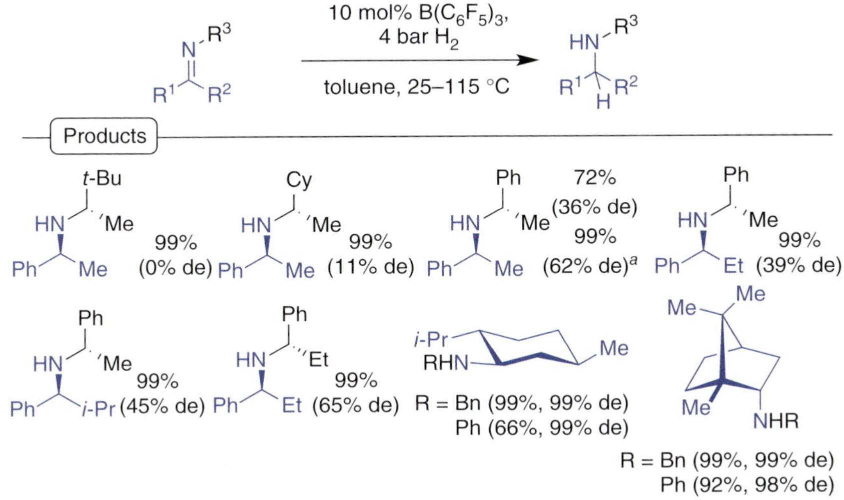

Figure 7.31 Diastereoselective hydrogenation of ketimines by FLP ([a]reaction performed with 115 bar H_2 pressure).

Figure 7.32 First asymmetric FLP-catalyzed hydrogenation.

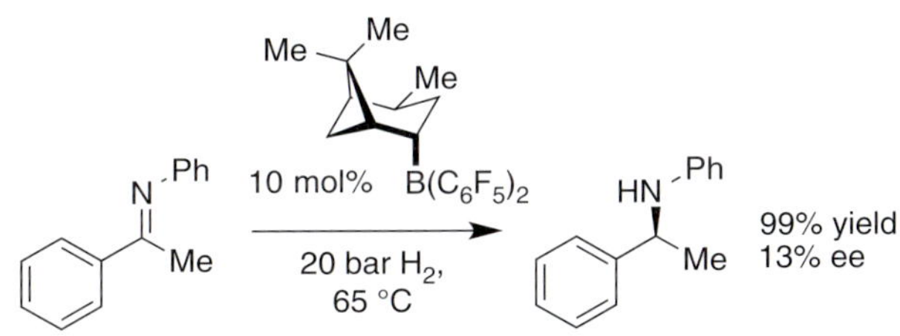

Low diastereoselectivities (11–65% de) were obtained for substrates bearing the chiral information in the *N*-alkyl-imino side chain. For substrates featuring the chiral information in the imine part, excellent diastereoselectivities (98–99% de) were obtained in combination with high yields.

The first highly desirable asymmetric hydrogenation using an FLP and molecular hydrogen was realized by Jürgen Klankermayer. His first chiral borane was derived from α-pinene and the hydrogenation product obtained in quantitative yields, however, with only disappointingly low enantiomeric excess of 13% (Figure 7.32) [92].

Without any doubt, this study clearly demonstrated that the asymmetric FLP-catalyzed hydrogenation is possible but it remained a challenge to obtain higher selectivity values. This was mastered by the same group by application of camphor-derived borane structures (Figure 7.33) [92, 138–140].

Modification of the parent camphor-derived borane to the related intramolecular chiral FLP also provided an active catalyst for the asymmetric hydrogenation of methyl aryl ketimines (Figure 7.34) [140].

Although displaying slightly diminished stereoselectivity of the hydrogenation compared to the parent system, the intramolecular FLP has a significant advantage: it can be easily recovered from the reaction mixture by simple precipitation

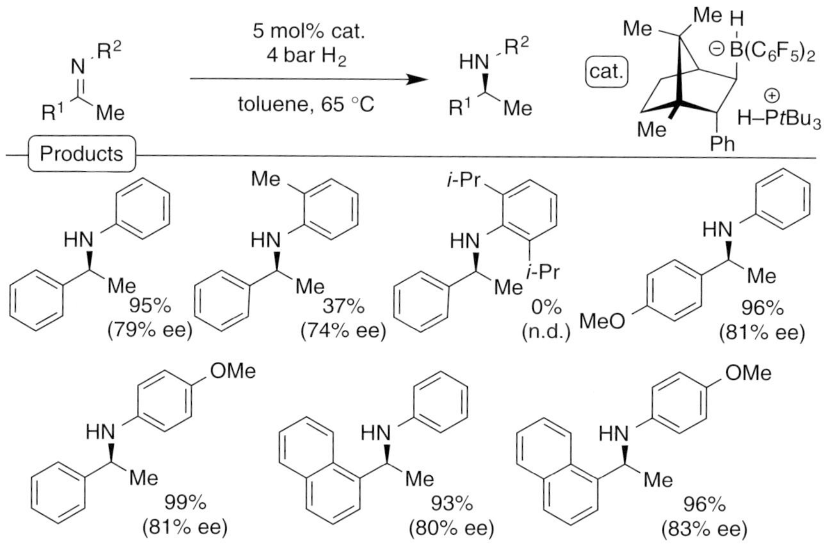

Figure 7.33 Asymmetric hydrogenation of ketimines.

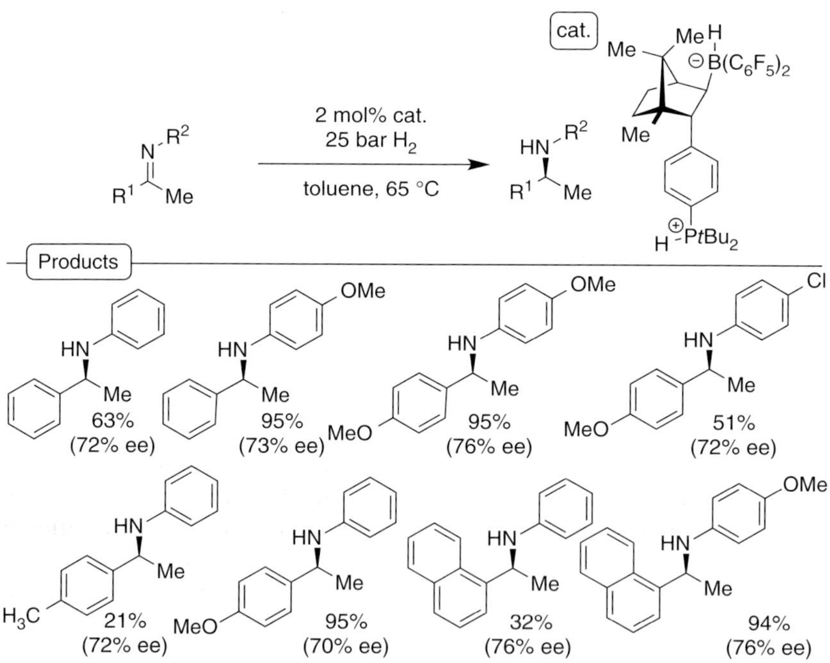

Figure 7.34 Asymmetric hydrogenation by an easily recyclable chiral intramolecular frustrated Lewis pair.

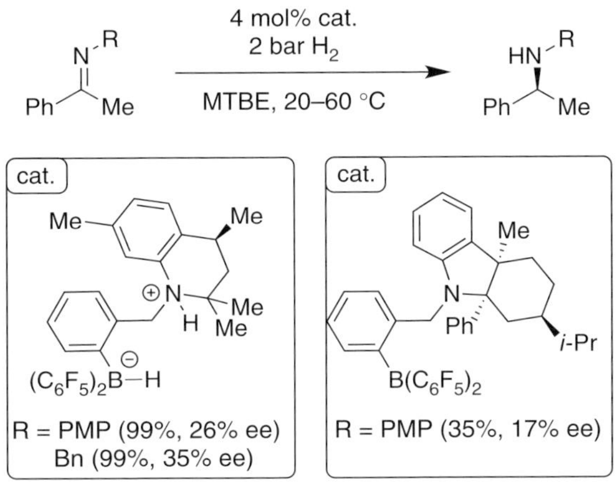

Figure 7.35 Intramolecular frustrated Lewis pairs in the asymmetric hydrogenation of ketimines.

with pentane in air. The solid catalyst was recycled five times without loss of enantioselectivity; only a slight decrease in activity was noticed (>99% to 70% conversion).

The synthesis of chiral *intramolecular* FLPs is highly challenging and substantial work has been devoted to this topic. The Lewis pairs derived from readily accessible chiral indolines or quinolines was applied in the asymmetric hydrogenation of selected ketimines (Figure 7.35) [122].

Despite straightforward derivatization of the *ansa*-ammonium borates to modulate the steric properties, only low enantioinduction in the hydrogenation of imines was observed. The intramolecular FLP based on the 1,1′-bisnaphthyl-derived aminoborane displayed higher enantioinduction in the asymmetric hydrogenation of imines and enamines (Figure 7.36) [141].

The enantioselective hydrogenation of ketimine-derived substrates displayed high dependency on the substitution pattern; nevertheless, moderate to high enantioselectivities were obtained (32% ee up to 83% ee). However, enamines proved to be more suitable substrates for the FLP-catalyzed hydrogenation, so that quantitative yields and moderate to excellent enantioselectivities were achieved under mild conditions applying 2 bar dihydrogen pressure.

A related 1,1′-binaphthyl system was developed using the *in situ* hydroboration of a bisolefin with Piers' borane. The 3,3′-position was efficiently modified to obtain a large library of chiral bis-boranes [142]. Indeed, the implementation of a very bulky tetra-*tert*-butyl terphenyl group ensured high to excellent enantioselectivity (74–89% ee) and high yields (63–99%) at mild conditions (Figure 7.37).

An innovative approach was selected using Piers' borane and the chiral *tert*-butylsulfinimide for the *in situ* assembly of a chiral catalyst. The authors suggest that the active species is an adduct of the Lewis acid and Lewis base, which initiates a concerted H^+/H^- transfer to the imine (Figure 7.38) [143].

The chiral products were obtained in high yields and high enantioselectivity of 84–95% ee using ammonia-borane as stoichiometric hydrogen source for transfer hydrogenation. The application of molecular hydrogen is also possible but the stereoinduction drops significantly. However, this readily available catalyst

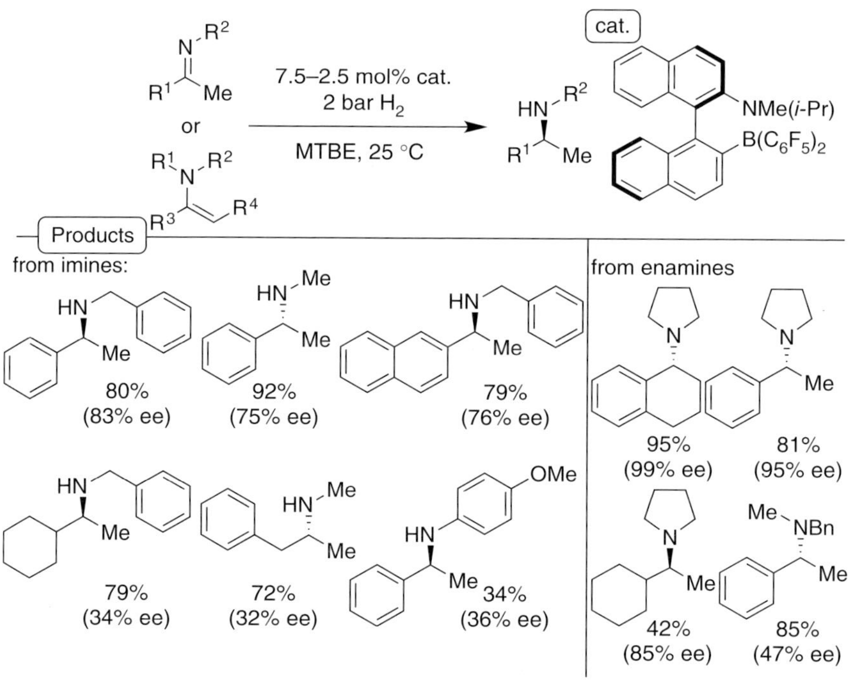

Figure 7.36 Asymmetric hydrogenation by the ansa-aminoborane.

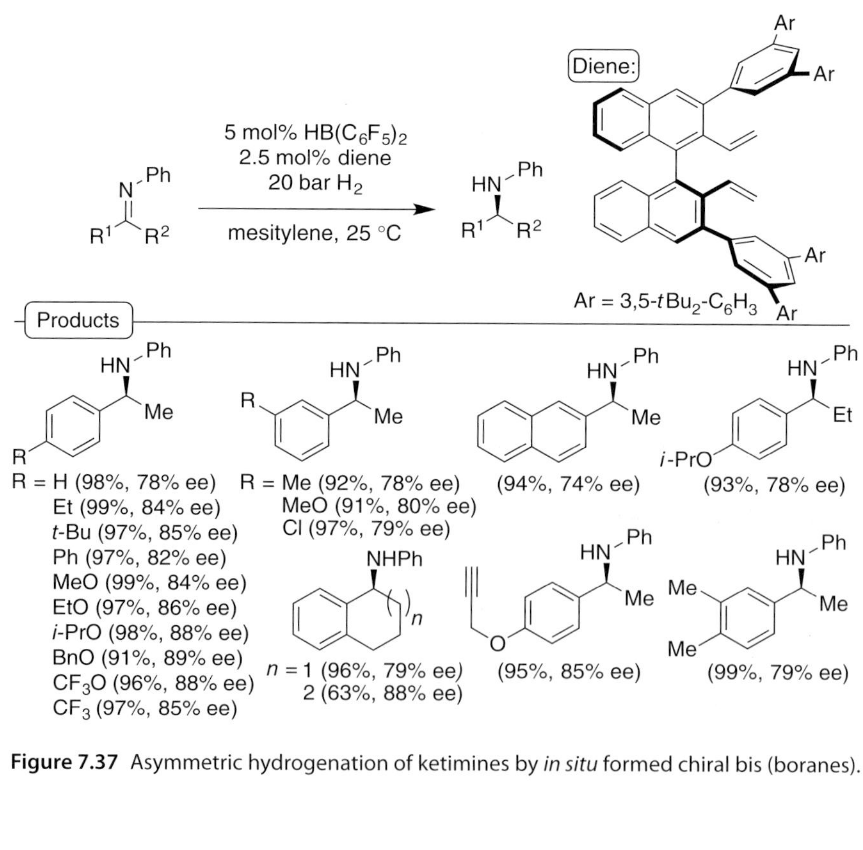

Figure 7.37 Asymmetric hydrogenation of ketimines by *in situ* formed chiral bis (boranes).

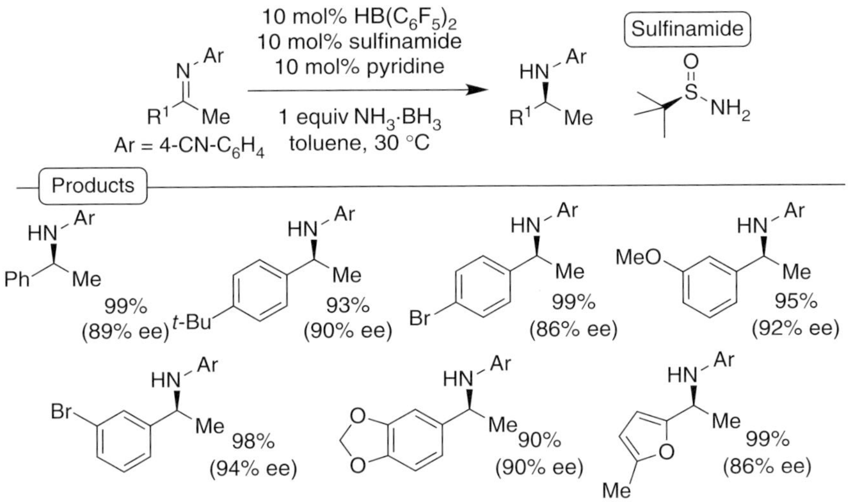

Figure 7.38 *In situ* formation of a chiral sulfinimide/borane catalyst for the enantioselective hydrogenation of imines.

system was also applicable to the asymmetric transfer hydrogenation of quinazolines (*vide infra* [144]). Only recently, the availability of moisture-resistant boranes sparked the rapid expansion of the imine hydrogenation methodology. Reductive amination is one of the key technologies for the synthesis of secondary and tertiary amines producing only water as the stoichiometric by-product. In 2017 the Soós group employed their water-tolerant borane to the reductive amination of a series of aldehydes (Figure 7.39) [112].

A large selection of aldehydes and primary or secondary amines were investigated and the corresponding products were obtained in good to excellent yields applying 80 °C and 20 bar H_2 pressure. This new development of FLP-catalyzed hydrogenation in one of the key applications, namely the reductive amination, clearly displays the large potential of FLP chemistry for organic synthesis as an enabling technology.

Besides aldimines or ketimines also the corresponding oxime ethers are susceptible to FLP-catalyzed hydrogenation using the standard borane $B(C_6F_5)_3$ as catalyst (Figure 7.40) [145]. Notably, the presence of a bulky substituent at the ether oxygen atom was crucial for the efficient conversion of the oximes into the corresponding hydroxylamines. The use of the triisopropylsilyl (TIPS)-protecting group ensured high yielding reductions in combination with a robust but easily removable silyl-protecting group to liberate the corresponding hydroxylamines.

7.5.2 Hydrogenation of Enamines and Silylenol Ethers

The hydrogenation of enamines was the second reported application of an FLP-catalyzed reaction. This was achieved with Erker's *intramolecular* frustrated ethylene-bridged phosphino borane (Figure 7.41) [2, 62, 146]. The hydrogenation is likely to proceed by the initial protonation of the enamine at

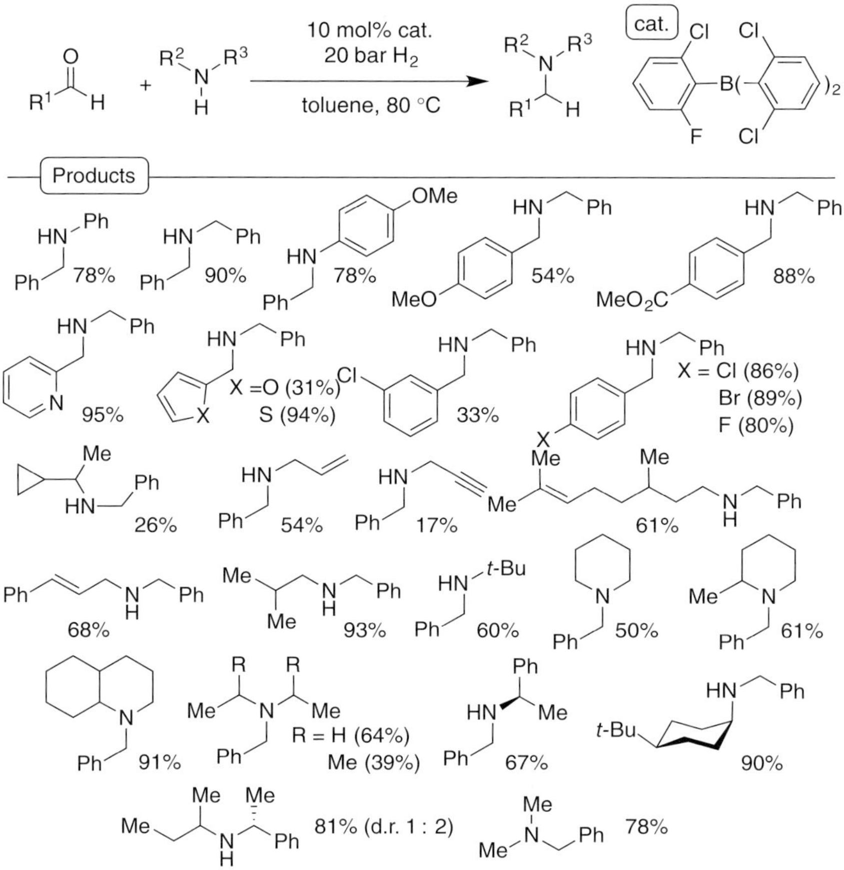

Figure 7.39 Reductive amination of aldehydes by a water-tolerant borane-derived FLP.

the carbon atom, leading to an iminium species, which then undergoes the final hydride transfer providing the saturated product.

This mechanistic view is strongly sustained by the stepwise protonation/reduction sequence of the [3] ferrocenophane-derived dienamine (Figure 7.42) [119]. The selective protonation of the dienamine with hydrochloric acid occurred at the 4-position resulting in an electrophilic iminium species. Subsequent reduction with $[H–B(C_6F_5)_3]^-$ furnished the partially and fully reduced [3] ferrocenophane derivative.

A very efficient access to vicinal B/N-based *intramolecular* FLPs was realized by the hydroboration of simple enamines [121]. The 1-phenylethylene-bridged ammonium borate turned out to be the most versatile catalyst for the hydrogenation of a small selection of enamines (Figure 7.43) [121].

Also, other *intramolecular* [77, 123] and *intermolecular* [104, 134] FLPs were evaluated as catalysts for the hydrogenation of 1-piperidylcyclohexene or 1-morpholinocyclohexene.

A related substance class to enamines are silyl enol ethers. Accordingly, these substrates undergo FLP-catalyzed hydrogenations through the proposed

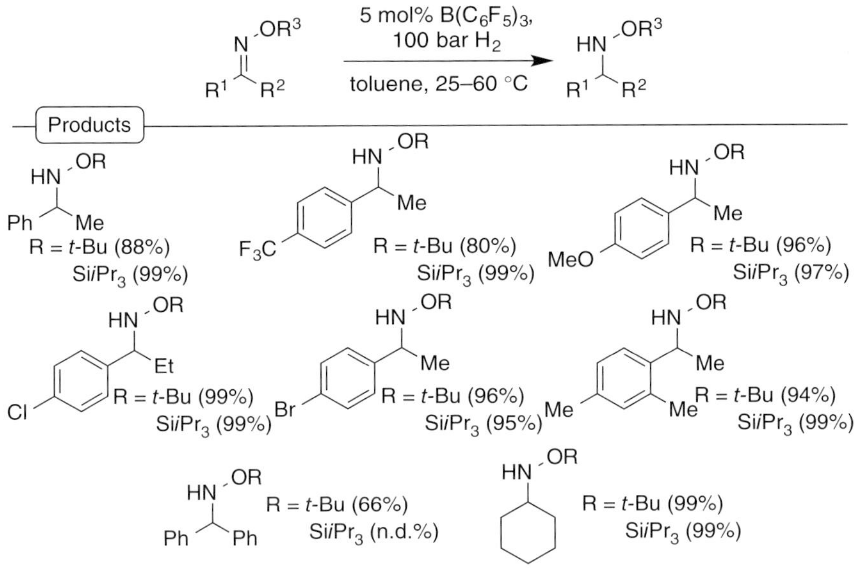

Figure 7.40 FLP-catalyzed hydrogenation of oxime ethers.

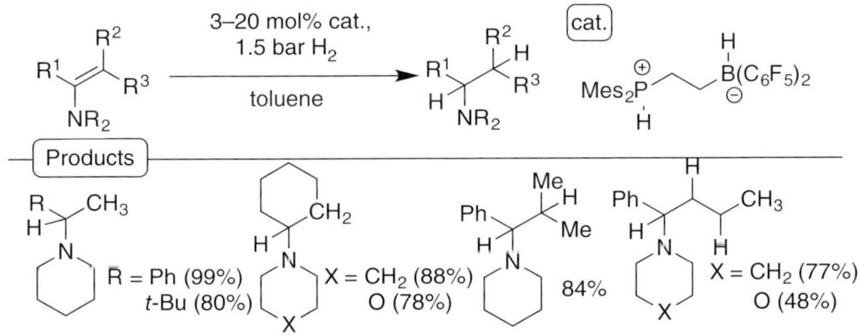

Figure 7.41 FLP-catalyzed hydrogenation of enamines.

proton/hydride transfer mechanism. Particularly bisphosphines in combination with $B(C_6F_5)_3$ have proved as suitable catalysts for the enol ether hydrogenation. 1,8-bis(diphenylphosphino)naphthalene activated H_2 in the presence of $B(C_6F_5)_3$ reversibly at 60 °C [70]. The reactivity was exploited for the catalytic hydrogenation of substituted silyl enol ethers directly furnishing silyl-protected secondary alcohols (Figure 7.44) [70].

This reactivity was later implemented in the first FLP-catalyzed domino reaction [147, 148] in which the required enol ether was generated *in situ* by the Lewis acid catalyzed 1,4-hydrosilylation [149] of an enone. The resulting enol ether was subsequently hydrogenated by a [2.2]paracyclophane bisphosphine derived FLP (Figure 7.45) [71, 105].

Also tetrasubstituted silyl enol ethers were reduced in excellent yields with high diastereoselectivity in favor of the *cis*-diastereomers.

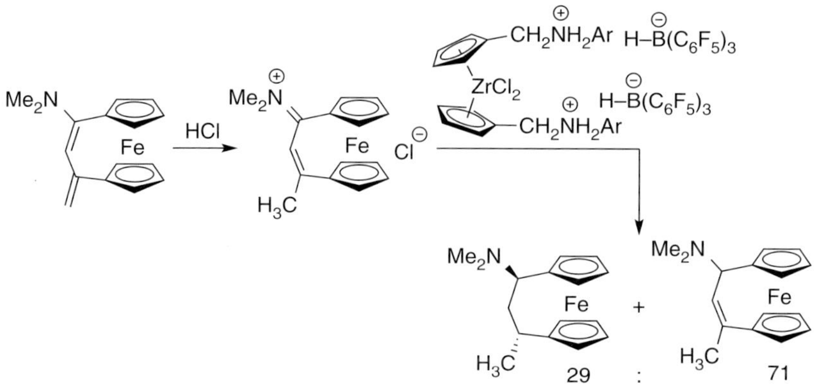

Figure 7.42 Mechanistic rationalization for the hydrogenation of enamines by proton/hydride transfer.

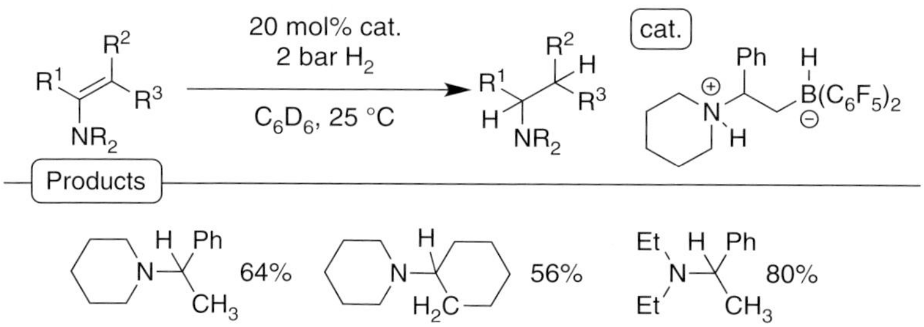

Figure 7.43 Hydrogenation with enamine-derived intramolecular FLPs.

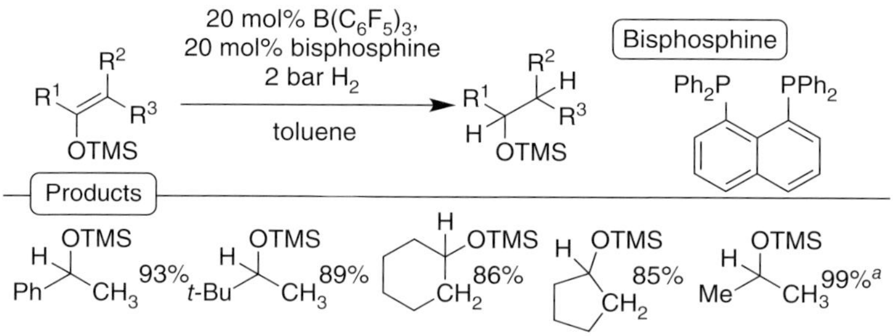

Figure 7.44 FLP-catalyzed hydrogenation of silylenol ethers ([a]performed with 60 bar H_2 pressure).

The asymmetric hydrogenation of trimethylsilyl (TMS)-protected enol ethers was achieved by the *in situ* assembly of the reactive bisalkenyl-derived bisborane (Figure 7.46) [150].

The secondary alcohols were obtained in most cases in excellent yields and in almost enantiopure form after deprotection with TBAF (tetrabutylammonium fluoride).

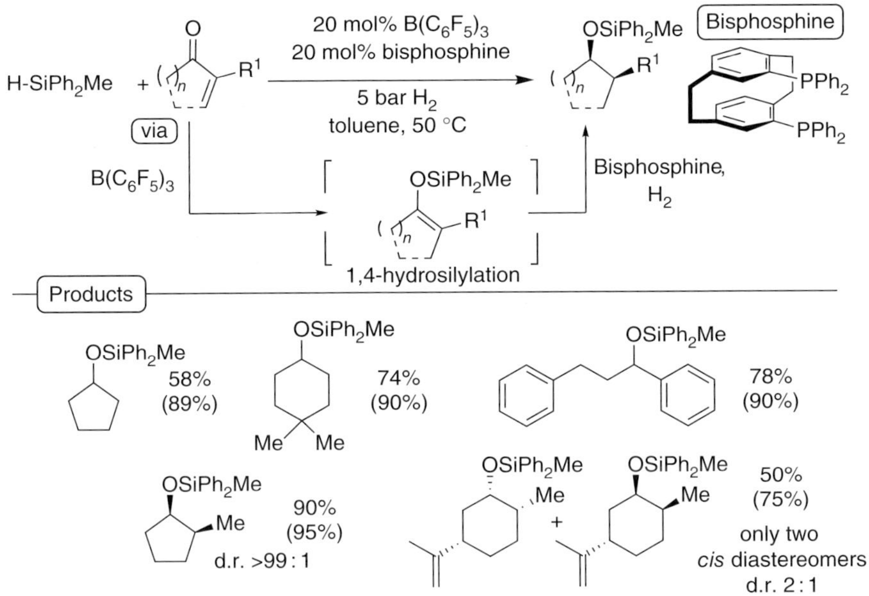

Figure 7.45 FLP-catalyzed domino hydrosilylation/hydrogenation of enones (values in *parentheses* correspond to yields determined by NMR spectroscopy).

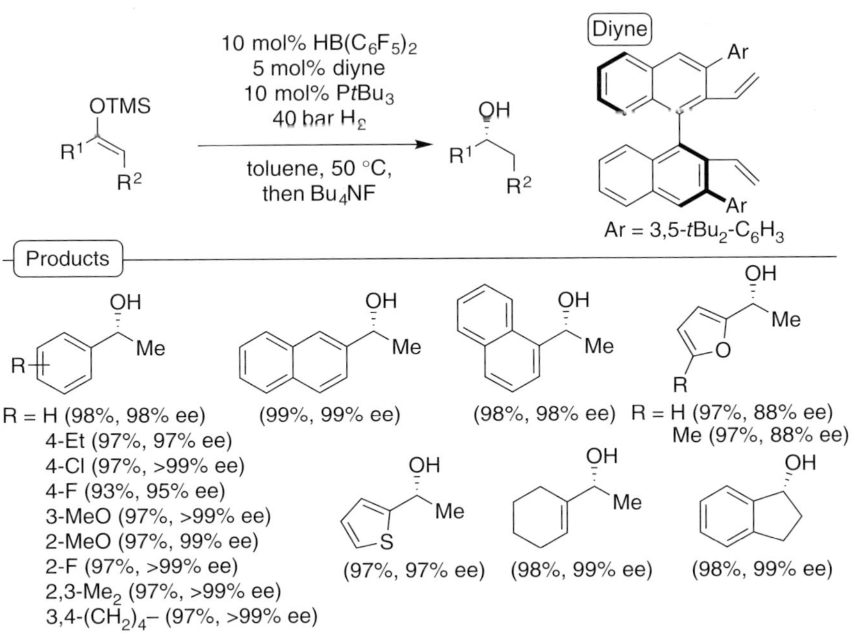

Figure 7.46 Enantioselective hydrogenation of silyl enol ethers.

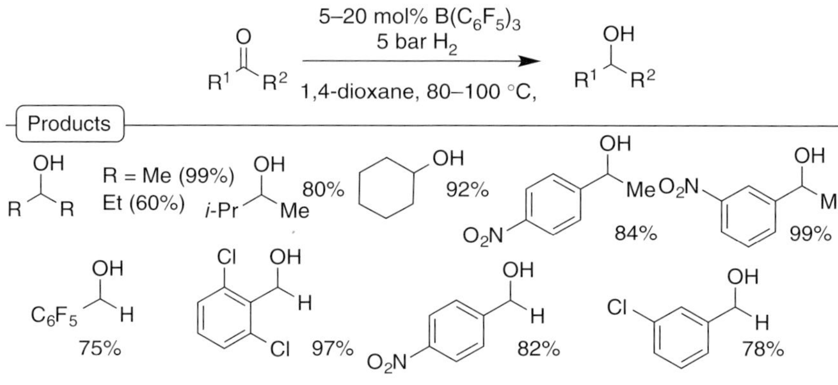

Figure 7.47 FLP-catalyzed hydrogenation of ketones and aldehydes.

7.5.3 Hydrogenation of Ketones

One of the biggest challenges in FLP-catalyzed hydrogenations remained the reduction of carbonyl compounds, e.g. ketones and aldehydes. This can be clearly attributed to the propensity of the reduction products to irreversibly bind to the corresponding Lewis acid, which may ultimately result in catalyst deactivation. Many attempts have been made to optimize the nature of the Lewis acid with foresight to reversible binding of alcohols. Astonishingly, it was found that it was rather crucial to optimize the Lewis base for this application. The application of ethereal solvents, e.g. Et_2O, i-Pr_2O, or 1,4-dioxane, in combination with the readily available Lewis acid $B(C_6F_5)_3$ provided the solution to this problem. Independently, two groups found that also ethers are viable Lewis bases for the activation of H_2 with $B(C_6F_5)_3$ (Figures 7.47 and 7.48) [137, 151].

The major difference between the two methods is found in the amount of applied H_2 pressure. In the presence of higher H_2 pressure the reaction time is significantly reduced to 12–14 hours. Additionally, the increased H_2 pressure allowed to conduct the reaction at lower temperatures, which minimized undesired side reactions such as condensation or elimination of the secondary alcohols. However, in both cases the primary and secondary alcohols were obtained in good to excellent yields.

In the search for water-tolerant boranes for FLP-catalyzed hydrogenations it was observed that partially chlorinated boranes are equally active in the hydrogenation of ketones and aldehydes (Figure 7.49) [111]. Again, cyclic ethers proved as the best solvent for the hydrogenation while the reduced catalyst inhibition by the water-tolerant borane allowed to perform the reaction at lower temperatures.

7.5.4 Reductive Deoxygenations

Deoxygenations are particularly challenging reactions even for elaborated transition metal catalysts and usually require drastic conditions above 150 °C. Oxygen-trapping reagents are generally used to convert oxygen functional groups into reactive intermediates, e.g. by treatment with hydrosilanes, phosgene derivatives [152], trific anhydrides [153], or oxalyl chloride [154]. The

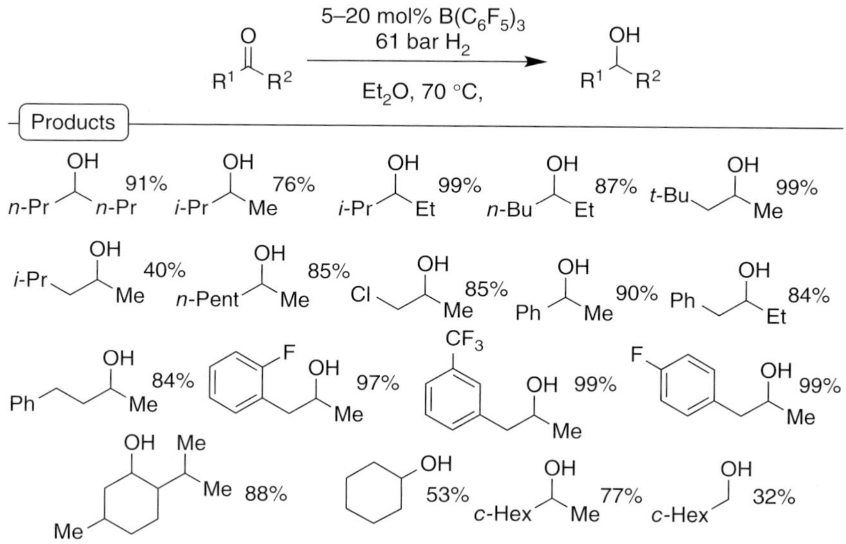

Figure 7.48 FLP-catalyzed hydrogenation of ketones.

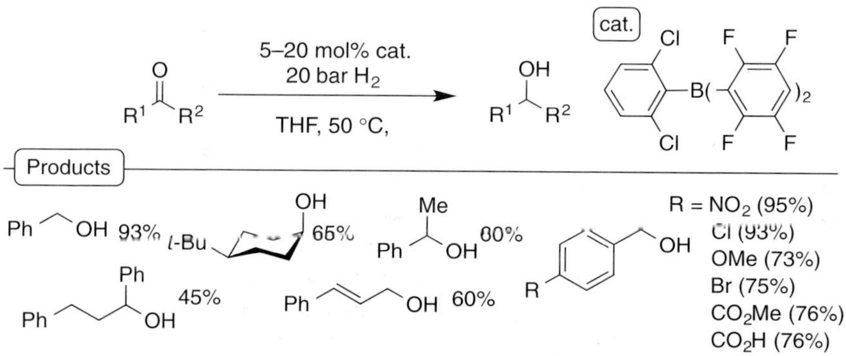

Figure 7.49 Hydrogenation of aldehydes and ketones by a moisture-stable borane.

reduction of phosphine oxides, e.g. triphenylphosphine oxide ($O{=}PPh_3$), is of great importance since it is produced on ton scale in the industrial application of the Wittig reaction for the synthesis of vitamins [155]. Typical procedures for the reduction of $O{=}PPh_3$ rely on the application of hydrosilanes at elevated temperatures [156–165] or on the conversion of the phosphine oxide to the corresponding dichlorophosphorane followed by reduction with elemental aluminum or silicon [166–168]. A reaction using H_2 as stoichiometric reductant is therefore highly desirable since only water or HCl is formed as by-product. With respect to an FLP-catalyzed hydrogenation, this reaction is particularly demanding. Firstly, the phosphine oxides, water, or HCl and also the produced phosphines act as strong donors for boranes and may result in the deactivation of the Lewis acid. Secondly, the stoichiometric amount of HCl may decompose the

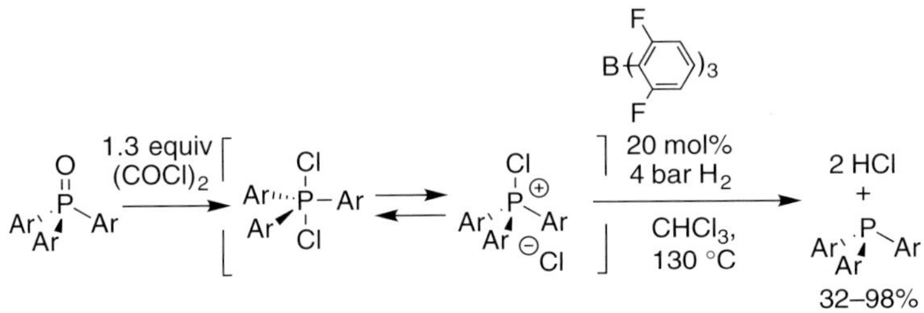

Figure 7.50 FLP-catalyzed reduction of phosphine oxides.

boranes by protodeborylation. All in all, these challenges can only be overcome by the careful selection of a suitable Lewis acid.

The conversion of phosphine oxides to the corresponding dichlorophosphoranes with oxalyl chloride $(COCl)_2$ under CO and CO_2 liberation is well known, and also the subsequent reduction with borohydride is reported [167–170]. Therefore, we decided to use a weak Lewis acid in order to generate borohydride donors *in situ* and to avoid the deactivation of the borane by the phosphine oxide, the produced hydrogen chloride, or by the phosphine. Indeed, the combination of $B(2,6\text{-}F_2{-}C_6H_3)_3$ and 2,6-lutidine (20 mol%) in the presence of 4 bar H_2 reduced the dichlorophosphorane Ph_3PCl_2 (generated from the reaction of $O{=}PPh_3$ with oxalyl chloride) to PPh_3 in 95% yield. However, control experiments showed that the additional Lewis base is not necessary and traces of the phosphine oxide in the reaction mixture served as an evenly potent Lewis base for the H_2 activation. This methodology is applicable for the reduction of phosphine oxides in the presence of 1.3 equiv of oxalyl chloride (Figure 7.50).

Very much to our surprise, the borane is not required if the reduction is conducted at 80 bar, whereas the reaction did not proceed at 4 bar in the absence of Lewis acid. Again, traces of phosphine oxide were required for efficient hydrogenation of the dichlorophosphorane so that we conclude that the corresponding oxide acts as Lewis base. From the EPC chemistry (*vide infra*) established by Stephan [63], it is known that phosphonium ions can act as Lewis acid and can, in fact, split H_2 according to the FLP mechanism [61, 172]. Accordingly, we suggest the activation of H_2 by the FLP consisting of tetravalent phosphonium chloride and phosphine oxide (Figure 7.51)[171].

The substrate scope of this reaction is comparable to the low-pressure variant but yield and reaction time are significantly improved including the reduction of bisphosphines, which is not achievable with the low-pressure variant.

Similarly, oxalyl chloride can be used for the *in situ* formation of chloroiminium chlorides [169, 173–175]. Such species are highly electrophilic and should be susceptible to FLP-catalyzed reduction. This process allows the reduction of amides by H_2 in the absence of transition metal complexes. Indeed, the reaction of the amides with 1.5 equiv of oxalyl chloride produced the chloroiminium chlorides *in situ*, which were reduced to the corresponding amines in the presence of 2 mol% $B(2,6\text{-}F_2{-}C_6H_3)_3$ and 80 bar at 40 to 70 °C (Figure 7.52) [176].

Again, control experiments showed that the addition of a supporting Lewis base, such as 2,6-lutidine, is not necessary. Initially, it was anticipated that either

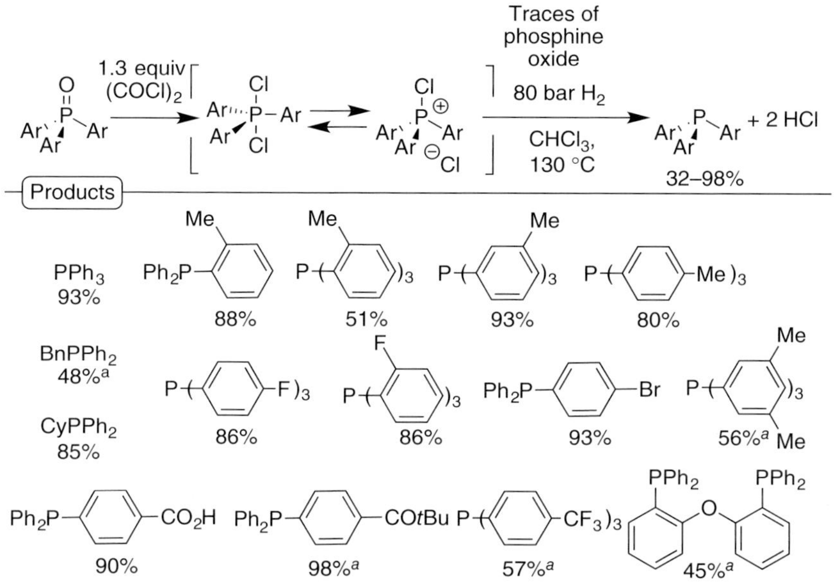

Figure 7.51 High-pressure reduction of phosphine oxides by oxalyl chloride and 80 bar H_2. [a]The addition of 20 mol% $B(2,6\text{-}F_2\text{-}C_6F_5)_3$/2,6-lutidine was required.

a reaction intermediate or the produced amine fills this part. However, reaction intermediates were not active as Lewis base in the H_2 activation and the product is furnished as hydrochloride salt. Spectacularly, we observed that the electronically modified borane and the chloride are able to activate H_2 as evidenced by the scrambling of a H_2/D_2 mixture to HD. This astonishing reactivity is not only limited to chloride but also bromide or iodide initiated the H_2/D_2 scrambling, however, higher temperatures were required. Quantum mechanical investigations strongly support this picture of halides being a capable Lewis base in the transient, endergonic process over an activation barrier of 22.1, 24.3, and 27.6 kcal/mol for chloride, bromide, and iodide respectively.

7.6 Hydrogenation of Heterocycles

Owing to the structural relationship of imines and enamines to N-heterocycles it is not surprising that such structures are also susceptible to FLP-catalyzed hydrogenations. However, some considerations must be taken into account since N-heterocycles can be fairly strong Lewis bases, which might lead to Lewis acid deactivation. Consequently, most of the investigated heterocycles feature substituents in *ortho* position to the heteroatom to guarantee efficient shielding of the heteroatomic site.

In analogy to the phosphine-free imine hydrogenation, the FLP-catalyzed hydrogenation of quinolone derivatives requires only catalytic amounts of the corresponding Lewis acid. In the presence of 5–10 mol% $B(C_6F_5)_3$ a variety of N-heterocycles were cleanly reduced to the tetrahydro derivatives in high yields (Figure 7.53) [177].

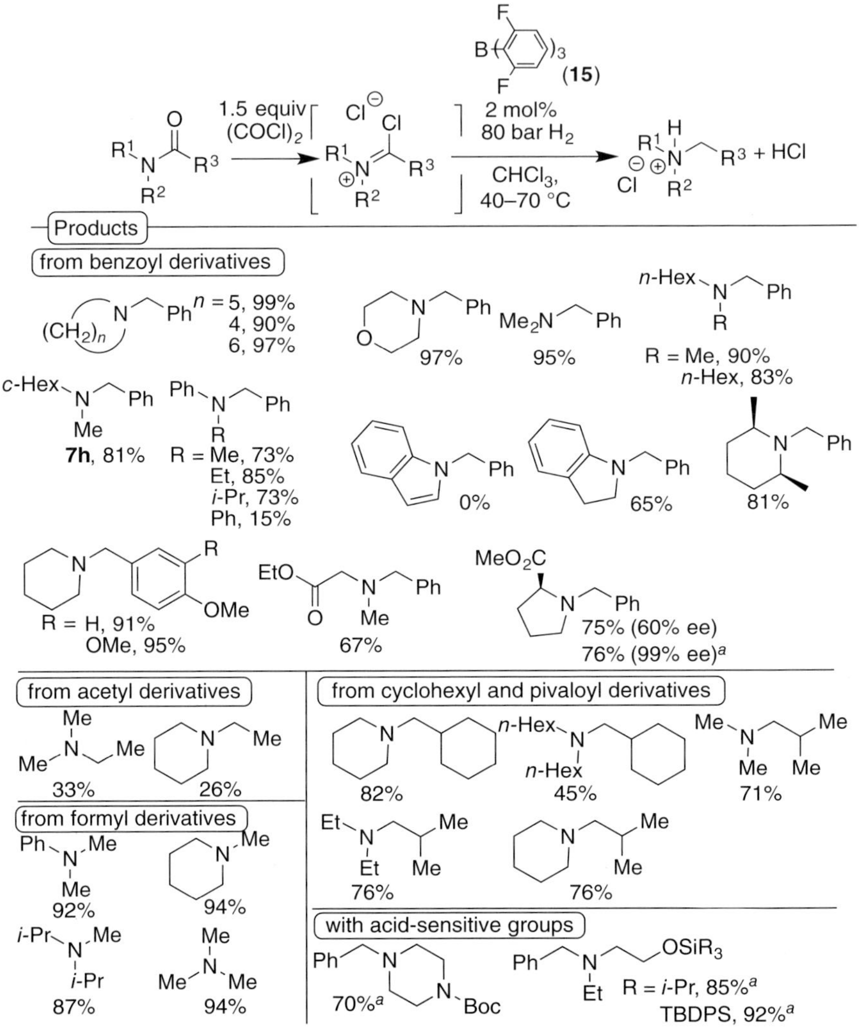

Figure 7.52 FLP-catalyzed hydrogenation of amides. [a]With 1 equiv 2,6-lutidine.

Similarly, the hydrogenation of a selection of quinoline derivatives was achieved by the use of a more sterically hindered boranes, e.g. $MesB(C_6F_4H)_2$, which also provided for increased functional group tolerance (Figure 7.54) [133, 134].

The corresponding tetrahydroquinolines were obtained in good to excellent yields and for some exquisite examples even the hydrogenation of a styryl functionality was observed. This methodology was exploited in the efficient three-step synthesis of the natural product cuspareine (Figure 7.55) [133]. The second step of the natural product synthesis is the FLP-catalyzed hydrogenation of the pyridine ring concurrently with the adjacent double bond to give access to the core structure of the natural product precursor in 91% yield. Final N-methylation provided

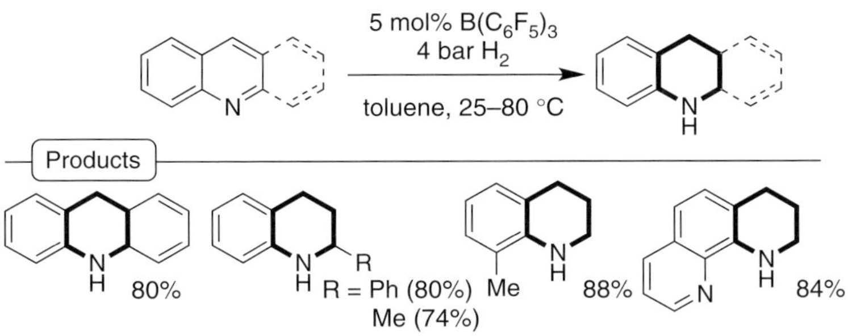

Figure 7.53 FLP-catalyzed hydrogenation of quinoline derivatives.

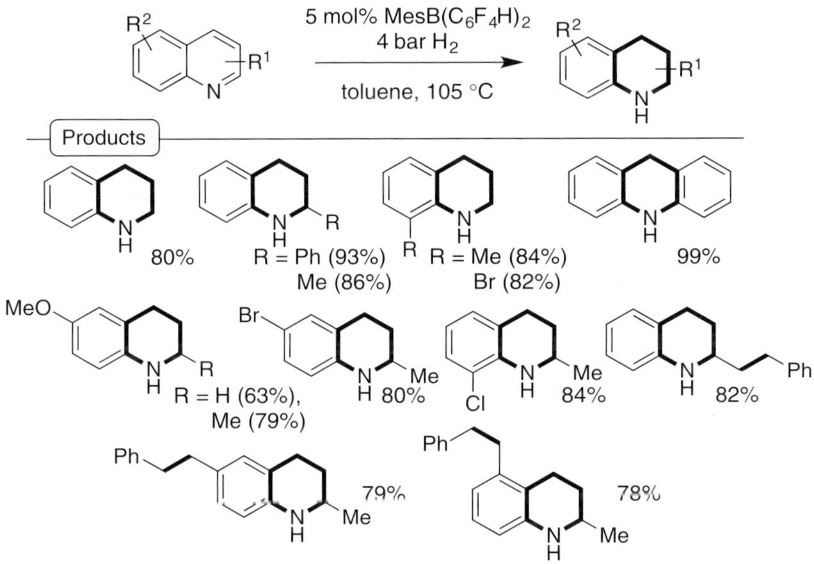

Figure 7.54 $MesB(C_6F_4H)_2$ catalyzed hydrogenation of quinolines.

rac-cuspareine in an overall yield of 52%. This short reaction sequence clearly demonstrates that FLP catalysis can emerge to a synthetically useful tool in the synthesis of valuable organic compounds.

The mechanism of the FLP-catalyzed quinoline hydrogenation was also investigated by quantum chemical and experimental methods. The analyses support a multistep mechanism comprising H_2 activation by the FLP consisting of quinolone and $B(C_6F_5)_3$ (Figure 7.56). The protonated quinoline resulting from the FLP-mediated H_2 activation is reduced by the hydride addition to C2. The generated 1,2-dihydroquinoline is a strong reducing agent and converts another molecule of quinoline into the 1,4-dihydroquinoline. Finally, the 1,4-dihydroquinoline is fully reduced to the corresponding tetrahydroquinoline by protonation and hydride addition.

The reduction of other N-heterocyclic compounds was first achieved with stoichiometric amounts of the Lewis acid at high temperatures (115 °C) [178]. Under

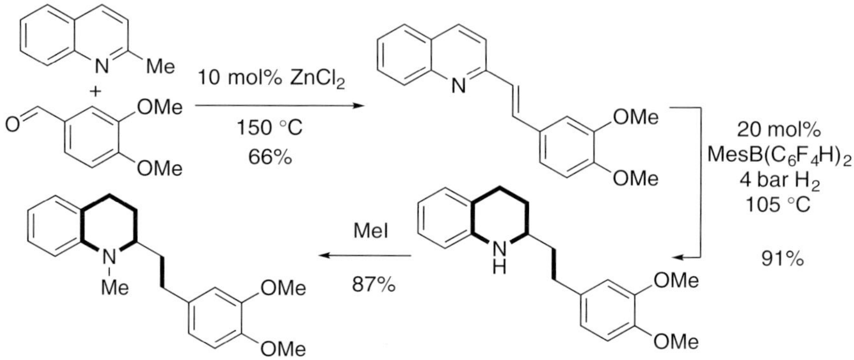

Figure 7.55 Synthesis of (*rac*)-cuspareine.

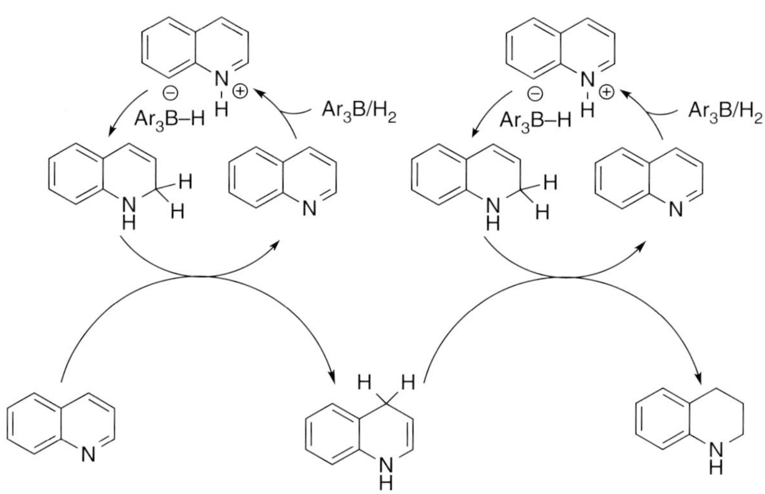

Figure 7.56 Mechanism of the FLP-catalyzed hydrogenation of quinoline.

these conditions not only the reduction of the nitrogen-containing heterocycles was accomplished but also the saturation of the annulated benzo-system in good to excellent yields (Figure 7.57).

A catalytic process showing a very broad substrate scope was elaborated using the *in situ* formation of a catalytically active borane. The tertiary borane C_6F_5–CH_2CH_2–$B(C_6F_5)_2$ was assembled by the hydroboration of CH_2=$CH(C_6F_5)$ with Piers' borane ($HB(C_6F_5)_2$) and used as Lewis acid in the diastereoselective FLP-catalyzed hydrogenation of mono- and disubstituted pyridines (Figure 7.58) [179].

More than 30 pyridine derivatives were diastereoselectively hydrogenated offering unparalleled access to *cis*-2,6-substituted piperidines in excellent yields. The potential of the methodology was demonstrated by the strikingly simple one-step synthesis of isosolenopsin A, an antibacterial and *anti*-HIV active small molecule (Figure 7.59) [179].

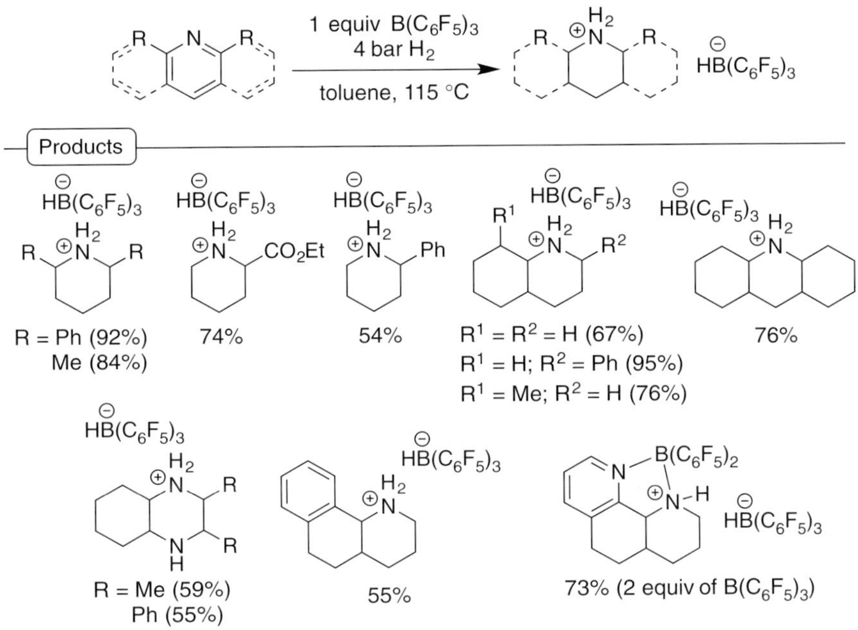

Figure 7.57 Stoichiometric reduction of pyridines, (benzo)quinolines, and quinoxalines.

Also, enantioselective FLP-catalyzed hydrogenations of heterocycles have been reported. 2,3-Disubstituted quinoxalines could be reduced under mild conditions (25 °C, 20 bar H_2) in good yields and high to excellent enantioselectivities (Figure 7.60) [180].

The hydrogenation proceeded with excellent diastereoselectivity and exclusively *cis*-tetrahydroquinoxalines were obtained. Slight modifications of the catalyst system to the 2-O*i*Pr-5-*t*Bu-C_6H_3-derivative enabled the highly stereoselective hydrogenation of the corresponding di- and trisubstituted quinolones (Figures 7.61 and 7.62) [181, 182]. However, the realization of the stereoselective hydrogenation of 2,3-disubstituted quinolines was more challenging compared to the corresponding 2,4-disubstituted quinolines (Figure 7.61) [181].

The hydrogenation of both substrate classes proceeded with high *cis*-selectivity (94 : 4 to >99 : 1) giving rise to efficient access to tetrahydroquinolines. The enantioselectivity of the reduction was generally higher for the 2,4-substituted quinolines (86–98% ee). The stereoselective hydrogenation of the corresponding 2,3-disubstituted quinolines was realized at reduced temperatures of 0 °C. Surprisingly, the enantioselectivity dropped from 87% ee or 89% ee to 45% ee or 56% ee respectively, when the 4-substituent was formally translocated to the 3-position. Also, the reduction of 2,3,4-trisusbstitued quinolones was investigated using the *i*-Pr-derivative of the bisborane (Figure 7.62) [182].

Also, asymmetric transfer hydrogenations of 3-arylquinoxalines using ammonia-borane as stoichiometric reductant were realized. The catalyst consisting of Piers' borane and the chiral *tert*-butyl-sulfinimide in 30 mol% catalyst loading at 30 °C converted the quinoxalines to the tetrahydroquinoxalines

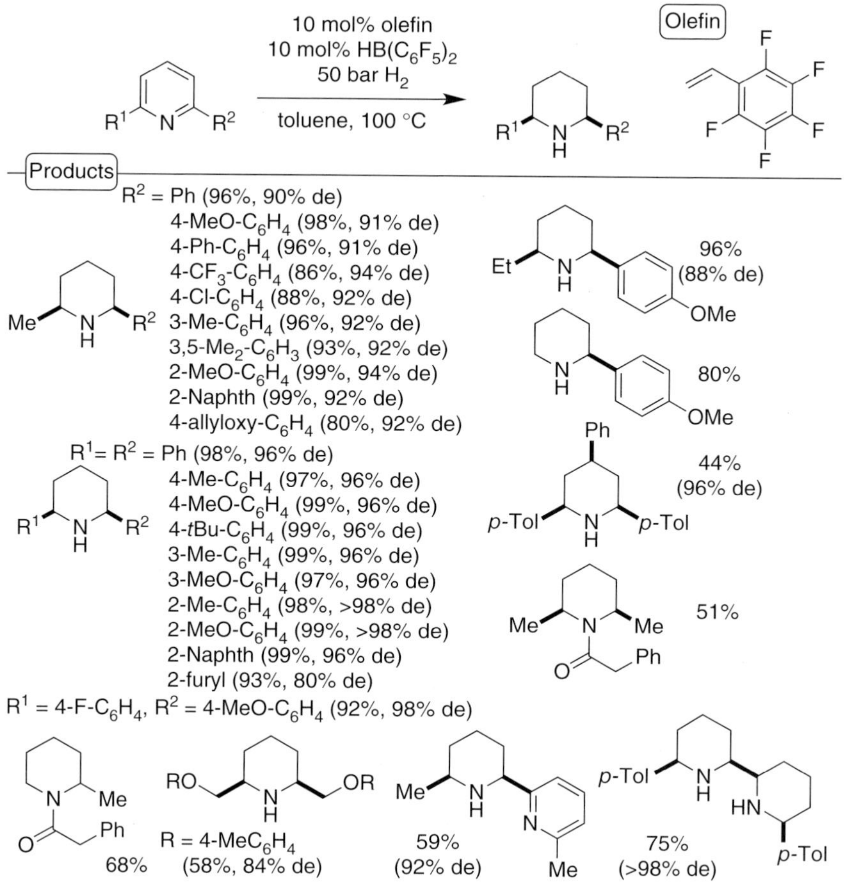

Figure 7.58 FLP-catalyzed hydrogenation of pyridines.

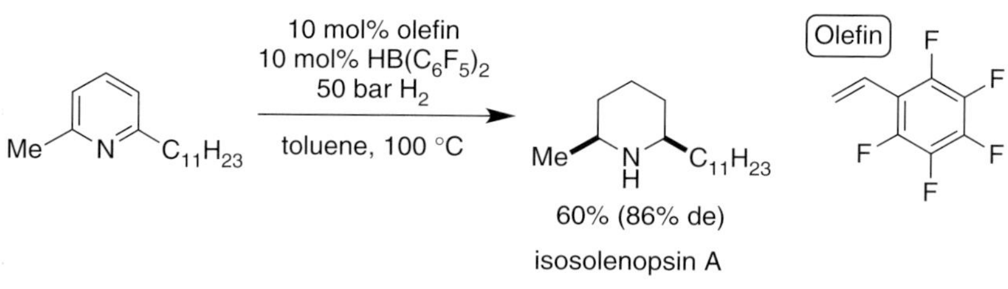

Figure 7.59 Synthesis of isosolenopsin A by FLP-catalyzed hydrogenation.

in high to excellent yield and high ee as a predominantly *cis*-diastereomer (Figure 7.63) [144].

Surprisingly, the corresponding 2,3-dialkylquinoxalines, which were unreactive in the asymmetric FLP-catalyzed hydrogenation, gave now the corresponding tetrahydro derivatives in high yield. Additionally, the diastereoselectivity was in favor to the *trans*-diastereomer as compared to the aryl derivatives, which favor the *cis* configuration.

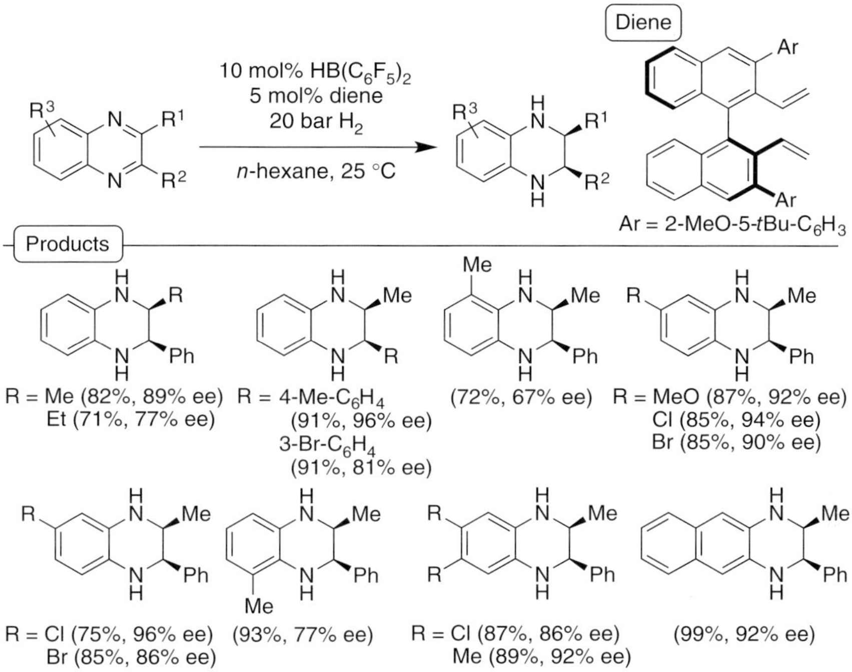

Figure 7.60 FLP-catalyzed enantioselective hydrogenation of quinoxalines.

7.7 Hydrogenation of Enones, Alkylidene Malonates, and Nitroolefins

The FLP-catalyzed hydrogenation of functionalized, especially oxygen-containing functional groups, is highly challenging due to the strong Lewis basic character combined with usually insufficient steric shielding. As a consequence, catalyst inhibition is one of the most problematic issues in FLP-mediated reductions of enones, malonates, or nitroolefins.

The hydrogenation of (*S*)-carvone was achieved using an electronically and sterically modified FLP consisting of $MesB(C_6F_5)_2$ in combination with DABCO (Figure 7.64) [104, 133]. Although the reaction time of 6 d strikes as quite long, the conditions are mild (4 bar H_2, 25 °C) and encompassed the first example of a catalytic reduction of an enone by an FLP.

Also, ynones can be reduced by FLP-catalyzed hydrogenations but the addition of the FLP to the triple bond is sometimes an undesired side reaction [128, 130, 183, 184].

The stoichiometric hydrogenation of ynones to the corresponding *cis*-enones proceeds in high yields with the phosphonium borate salts of the *intermolecular* FLPs $t\text{-}Bu_3P/B(C_6F_5)_3$ or *cis*-tBuCH=C(C_6F_5)B(C_6F_5)$_2$. However, in the presence of catalytic amounts of FLP (DABCO/*cis*-tBuCH=C(C_6F_5)B(C_6F_5)$_2$, 20 mol%) the substrate is transformed in 80% yield to a mixture of *trans*-enone, *cis*-enone, and ketone in a ratio of 50 : 1.5 : 1 (Figure 7.65).

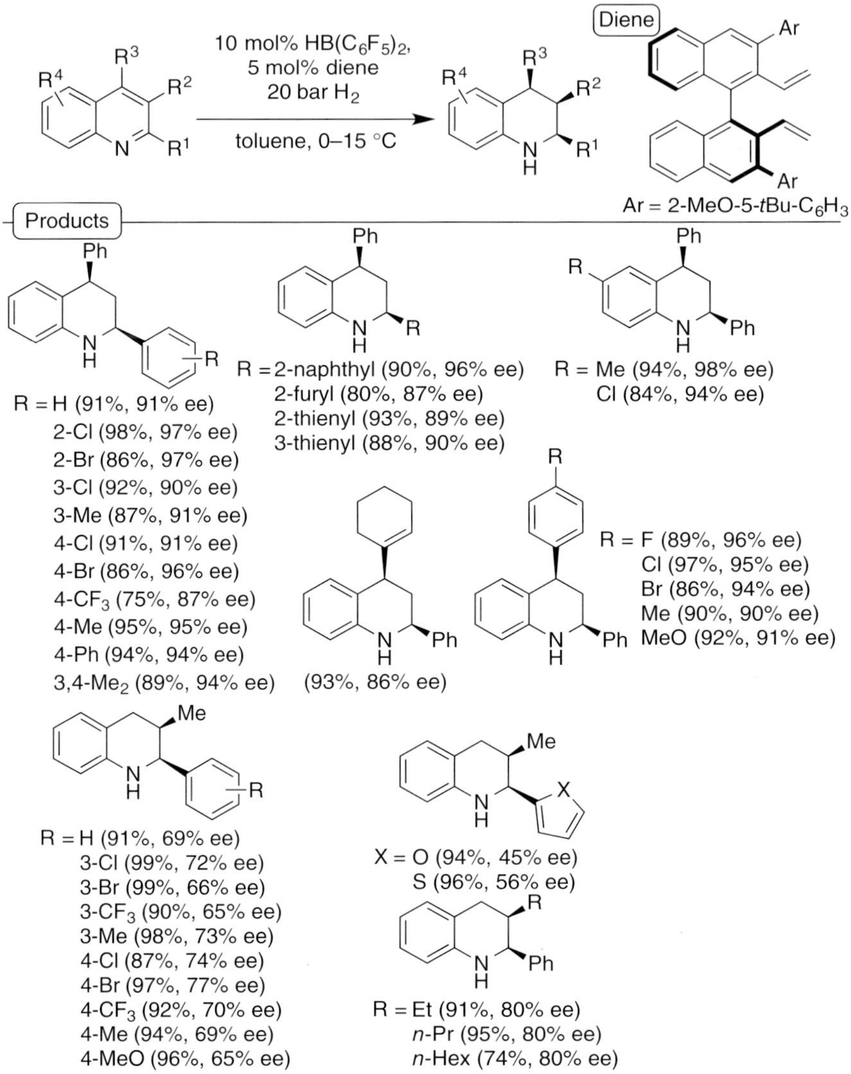

Figure 7.61 FLP-catalyzed enantioselective hydrogenation of disubstituted quinolines.

Electron-deficient allenes are susceptible to FLP-catalyzed hydrogenation in good to excellent yields to afford the corresponding substituted vinyl diethylmalonates. Here, a catalyst consisting of DABCO/$B(C_6F_5)_3$ (15 mol%, 60 bar H_2, 80 °C, 72 hours) proved as most reactive (Figure 7.66) [135].

The same catalytic system was even more reactive (10 mol%, 60 bar H_2, 80 °C, 24 hours) in the hydrogenation of alkylidene malonates, which gave access to 2-alkyl malonates in excellent yields (Figure 7.67) [135].

The electronic modification of the triaryl borane proved as very useful to achieve catalytic hydrogenations of densely functionalized substrates. The less Lewis acidic borane $B(2,4,6\text{-}F_3\text{–}C_6H_2)_3$ was highly competent in the reduction

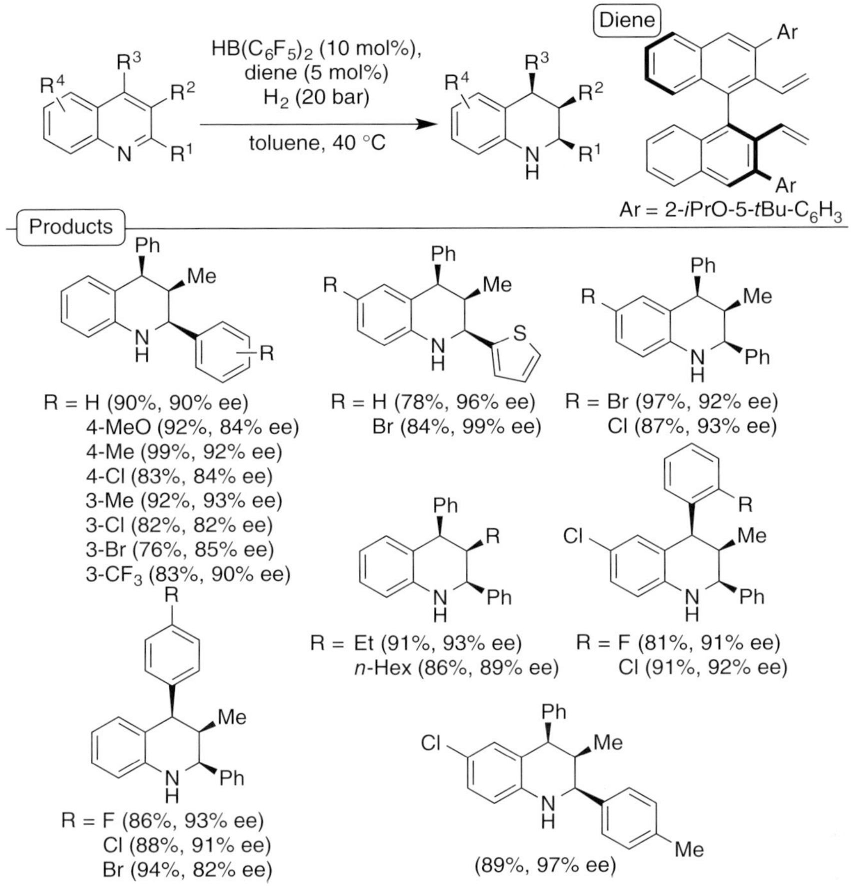

Figure 7.62 *cis*-Selective and enantioselective metal-free hydrogenation of 2,3,4-trisubstituted quinolines.

of functionalized, activated acrylates, sulfones, and nitroolefins [102]. The pronounced functional group tolerance of the borane is achieved by the reduced Lewis acidity (70% compared to $B(C_6F_5)_3$ (100%) according to Childs method) and permits milder reaction conditions (5 mol%, 10 bar H_2, 50 °C, 24 hours; Figure 7.68) [102].

The weaker Lewis acidic $B(2,6\text{-}F_2\text{–}C_6H_3)_3$ (56% according to Childs' method) could be directly utilized as THF adduct as obtained from its synthesis as catalyst for the hydrogenation of nitroolefins. A large selection of nitroolefins was reactive and were hydrogenated in excellent yields under mild reaction conditions (4 bar H_2, 40 °C, Figure 7.69) [103].

Particularly, the mixture of $B(2,6\text{-}F_2\text{–}C_6H_3)_3$ with 2,6-lutidine as Lewis base enabled the efficient hydrogenation of a number of nitroolefins. On the one hand, not only the nitro-functionality is well tolerated but on the other hand also even stronger donors such as furyl or thiophenyl. Substrates with substituents in α-position are highly challenging for metal-free Hantzsch's ester hydrogenations

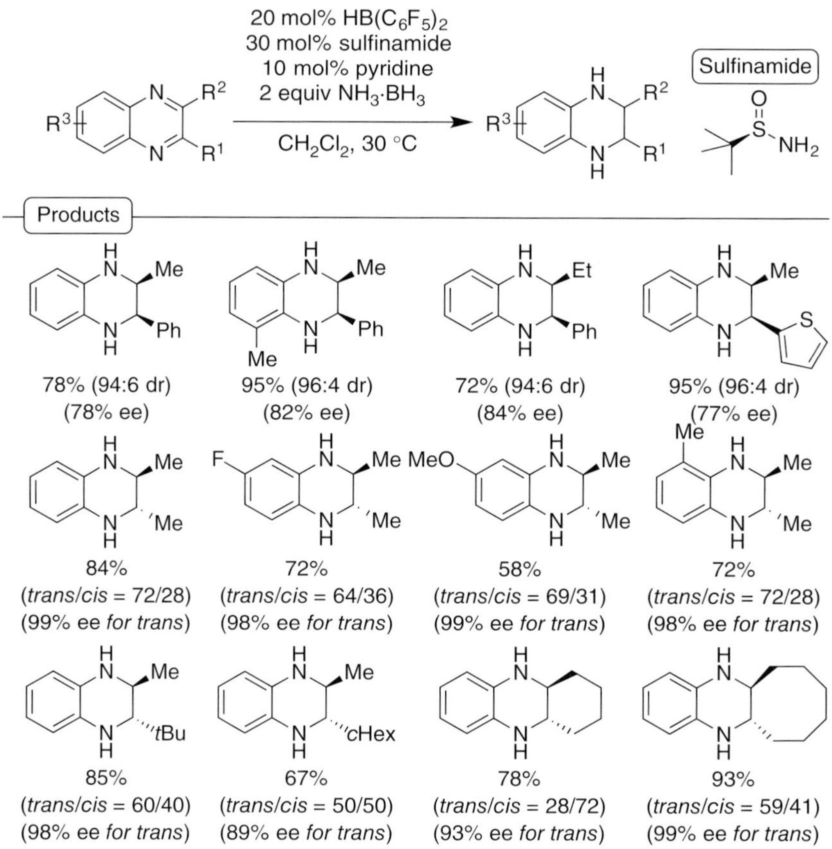

Figure 7.63 Asymmetric transfer hydrogenation using $HB(C_6F_5)_2$/*tert*-butyl sulfinimide as catalyst and ammonia-borane as hydrogen transfer reagent.

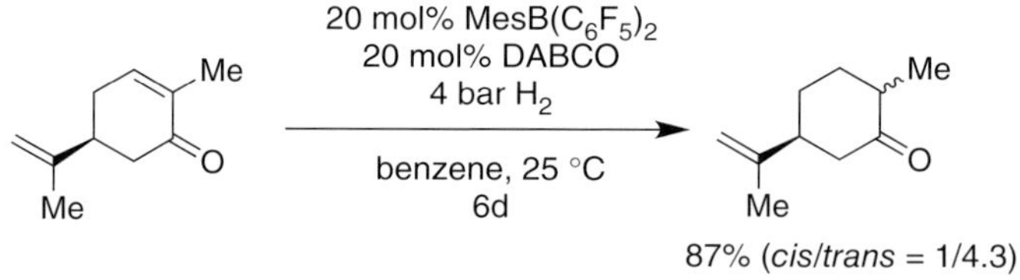

Figure 7.64 Hydrogenation of (*S*)-carvone.

catalyzed by thioureas or phosphoric acid esters. However, 2-nitro-2-methyl styrene was cleanly hydrogenated to the corresponding nitro alkane in 94% yield in only 24 hours. The combination of the Lewis acid $B(2,6\text{-}F_2{-}C_6H_3)_3$ with collidine was an efficient catalyst for the hydrogenation of alkylidene malonates and acrylates even at room temperature. Even the smallest α,β-unsaturated ester, methylacrylate, was hydrogenated in quantitative yield although prolonged reaction time is required.

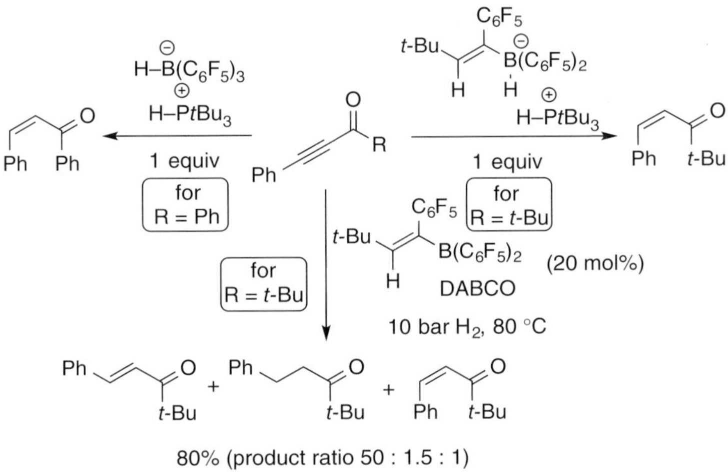

Figure 7.65 Hydrogenation of ynones.

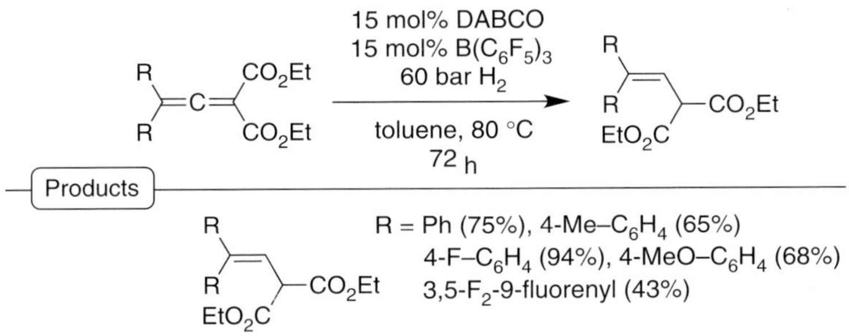

Figure 7.66 Hydrogenation of electron-deficient allenes.

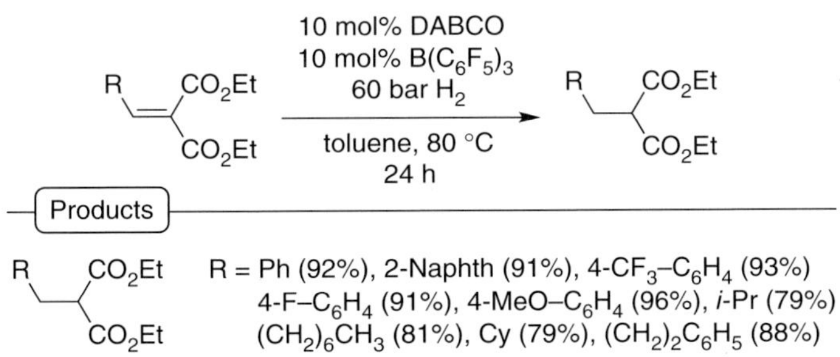

Figure 7.67 Catalytic hydrogenation of alkylidene malonates.

7.8 Hydrogenation of Unpolarized Olefins and Polycyclic Aromatic Hydrocarbons

One of the oldest but at the same time one of the most important catalytic hydrogenations is the conversion of unsaturated hydrocarbons to saturated ones. The

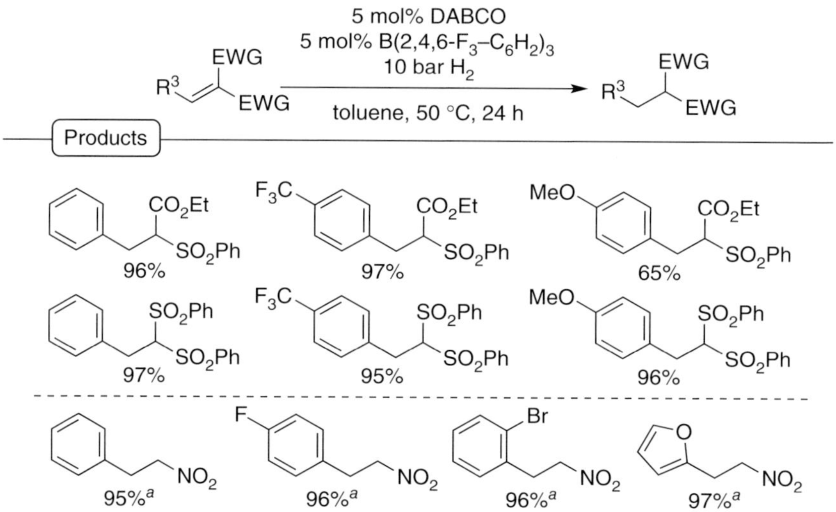

Figure 7.68 FLP-catalyzed hydrogenation of strongly activated acrylates, sulfones, and nitroolefins ([a]reactions were carried out at 5 bar H_2 pressure).

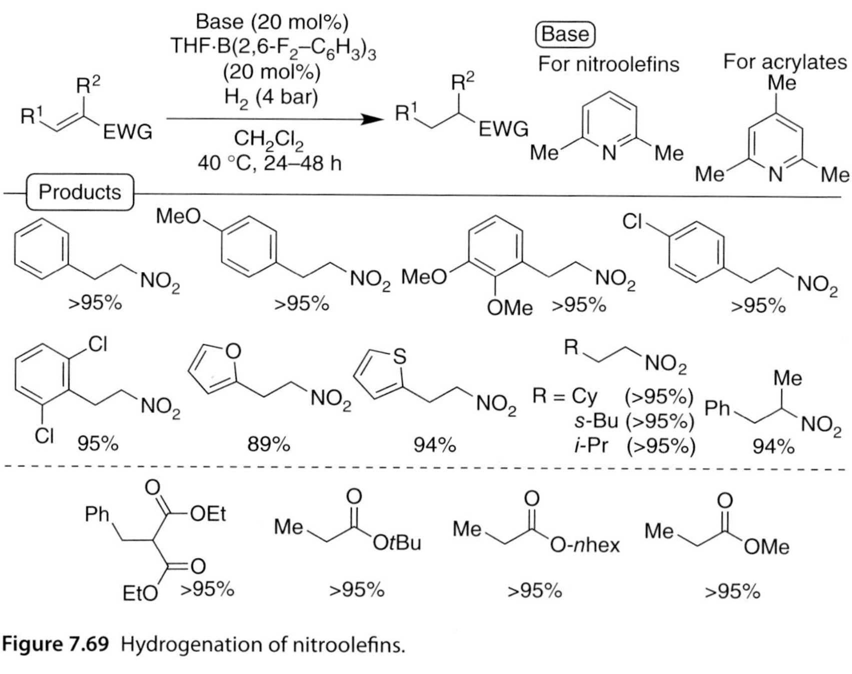

Figure 7.69 Hydrogenation of nitroolefins.

hydrogenation of unfunctionalized olefins is more challenging compared to the reduction of heteroatom-substituted double bonds, e.g. imines or heterocycles, due to the absence of activating heteroatomic sites. Generally, the activation of double bonds can be realized either by transition metal coordination or by reaction with strong electrophiles, e.g. hydrohalic acids. The conveyance of such

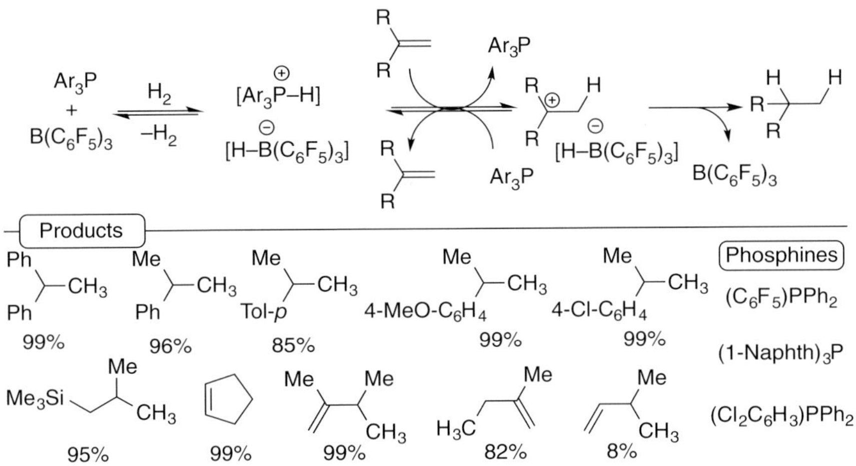

Figure 7.70 FLP-catalyzed hydrogenation of olefins.

a protonation/addition sequence to FLP chemistry requires the formation of a strong Brønsted acidic onium ion concomitant to the formation of the borohydride upon H_2 activation. The majority of the reported FLP systems derived from Lewis bases (phosphine, amine, pyridine, carbene, or phosphinimide) and boranes have demonstrated their capability to activate H_2, but none of the resulting Brønsted acids were potent to protonate an olefinic double bond.

Surprisingly, it was found that even weak donors, e.g. fluorinated phosphines, are useful Lewis bases for the FLP-mediated activation of H_2 in combination with $B(C_6F_5)_3$. The H_2 activation is reversible at room temperature and the products, $[Ar_3P{-}H][H{-}B(C_6F_5)_3]$, could only be detected by NMR spectroscopy below −60 °C. Particularly, these phosphines were applied in the first FLP-catalyzed hydrogenation of olefins [94]. The reaction proceeds via transient hydrogen activation by the FLP, proton transfer to the olefin, which results in the formation of a carbocation. This reactive intermediate is then irreversibly quenched by the corresponding borohydride (Figure 7.70) [94].

The hydrogenation proceeds smoothly with electron-rich olefins in excellent yield. The nature of the Lewis base influences the reaction rate and is responsible for the formation of by-products, e.g. from Friedel–Crafts dimerization [93]. Such effects are even more pronounced if less electron-donating Lewis bases are applied. Even diethyl ether, which was later found to be crucial for the hydrogenation of carbonyl groups (see section 7.5.3), can act as Lewis base in combination with $B(C_6F_5)_3$ as hydrogenation catalyst for olefins. The presence of 2 equiv of Et_2O relative to $B(C_6F_5)_3$ was found to facilitate the H_2 cleavage and was necessary to stabilize the diethyloxonium cation through hydrogen bonding (Figure 7.71) [79].

Experimental evidence for the H_2 activation was acquired from the isotope exchange reaction of a HD atmosphere. This mixture was catalytically converted to the statistical mixture of HD, H_2, and D_2 (2 : 1 : 1), which strongly corroborates the cleavage of the H=D bond by the FLP. Further hydrogenation experiments

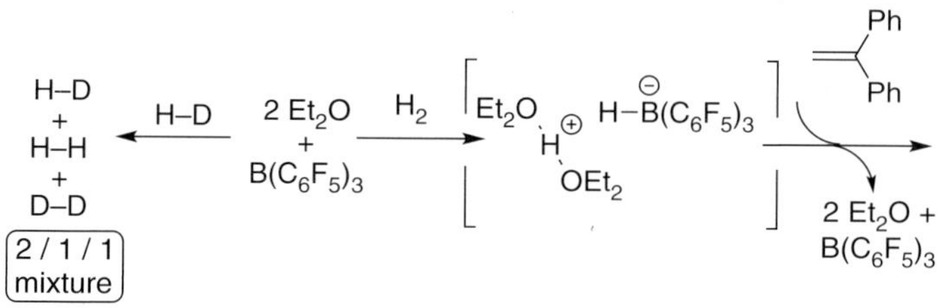

Figure 7.71 H_2 activation by $Et_2O/B(C_6F_5)_3$.

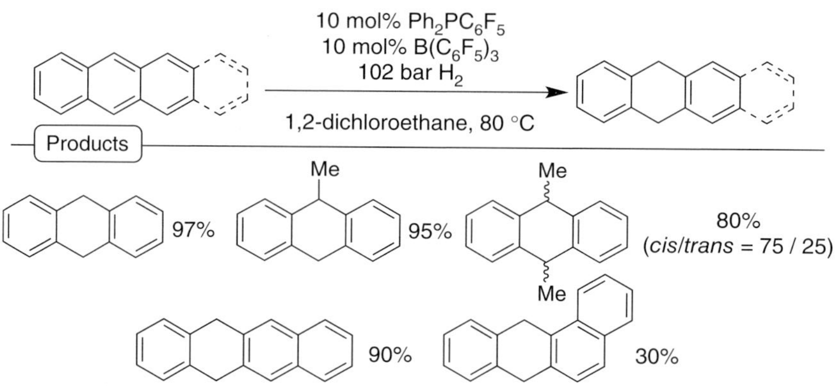

Figure 7.72 Hydrogenation of polycyclic aromatic hydrocarbons.

with other styrene derivatives resulted in dimerizations through Friedel–Crafts type reactions due to the high acidity of the diethyloxonium cation, which clearly underlines the importance of the Lewis base's nature in FLP-catalyzed reactions.

Similar to the reduction of olefins, also, fused, electron-rich aromatic hydrocarbons should be susceptible to FLP-catalyzed hydrogenation. Both the $(C_6F_5)PPh_2/B(C_6F_5)_3$ and the $Et_2O/B(C_6F_5)_3$ systems display reactivity in the reduction of polycyclic aromatic hydrocarbons (Figure 7.72) [79, 95]. The reaction required elevated temperature (80 °C) and 102 bar H_2 pressure to obtain the products within a reasonable amount of time and in high yields. The hydrogenation occurred selectively at the highest substituted aromatic ring, leading to isolated and thus thermodynamically more stable benzo or naphtho systems.

A conceptually different approach to the olefin hydrogenation shown above takes advantage of the intrinsic reactivity of hydroboranes toward olefins. An earlier study stressed on the reduction of triisobutylborane at 235 °C and 172 bar H_2 pressure [185, 186] releasing *iso*-butane and borane. Accordingly, the reaction of Piers' borane with olefins in the presence of H_2 was investigated. The drastic conditions could be optimized to comparably milder reaction conditions (6 bar H_2, 140 °C) and provided good to excellent yields for a broad selection of substrates (Figure 7.73) [187]. In particular, higher substituted olefins, e.g. methylcyclohexene or 2,3-dimethyl-but-2-ene, were converted in quantitative yields to the saturated products. The reaction proceeds via initial hydroboration

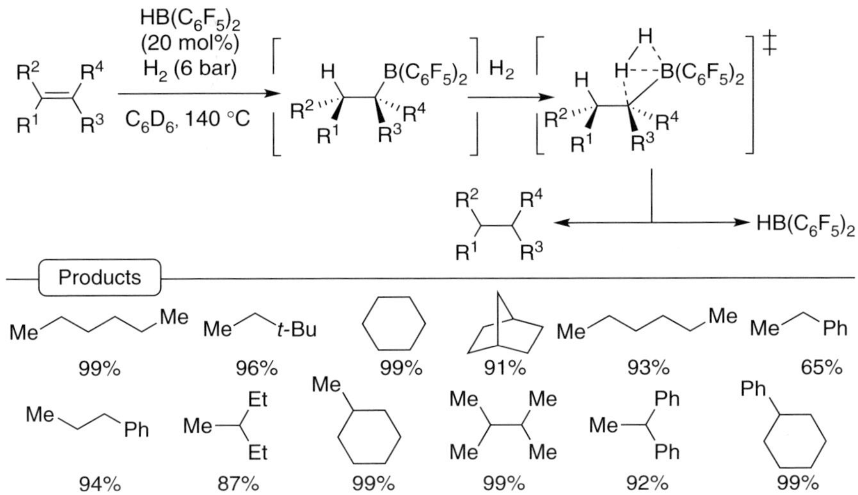

Figure 7.73 Metal-free hydrogenation of olefins involving σ-bond metathesis.

of the olefin to furnish the diarylalkylborane, which subsequently undergoes σ-bond metathesis with H_2. This liberates the saturated product together with the hydroborane catalyst.

The stereoselective *cis*-hydrogenation of alkynes was achieved by exploiting an alternative pathway to σ-bond metathesis: protodeborylation of sp^3 or sp^2 carbon–boron bonds results in the release of the corresponding alkyl or alkene/aryl molecule. The formed FLP catalyst consists of an *ansa*-amino hydroborane, which readily undergoes hydroboration of the alkyne (Figure 7.74) [124].

The H_2 activation is achieved by the intramolecular FLP resulting from hydroboration, which provides a perfect setup for the final protodeborylation. This step releases the *Z*-olefin with concomitant regeneration of the catalyst. A large number of alkynes were exclusively converted to *Z*-olefins with perfect diastereoselectivity. Remarkably, overreduction, a major problem in transition metal based hydrogenation, was not observed, and substrates bearing olefinic moieties were chemoselectively reduced at the alkyne. Furthermore, TMS-protected terminal alkynes and TMS-protected alcohols or esters bearing an alkyne moiety were efficiently converted to the alkenes under mild reaction conditions (2 bar, 80 to 120 °C) demonstrating the high preparative value for organic synthesis.

7.9 Electrophilic Phosphonium Cations (EPCs)

Electrophilic phosphonium cations (EPCs) are active in C=F bond activation [63–65] and are now found to activate molecular hydrogen [61]. Since EPCs are isoelectronic to boranes, a similar mode of H_2 activation was proposed on the basis of quantum mechanical investigations (compare Figure 7.2a and b) [61]. The Lewis acidity is strongly increased by the fluorine atoms in the pentafluorophenyl

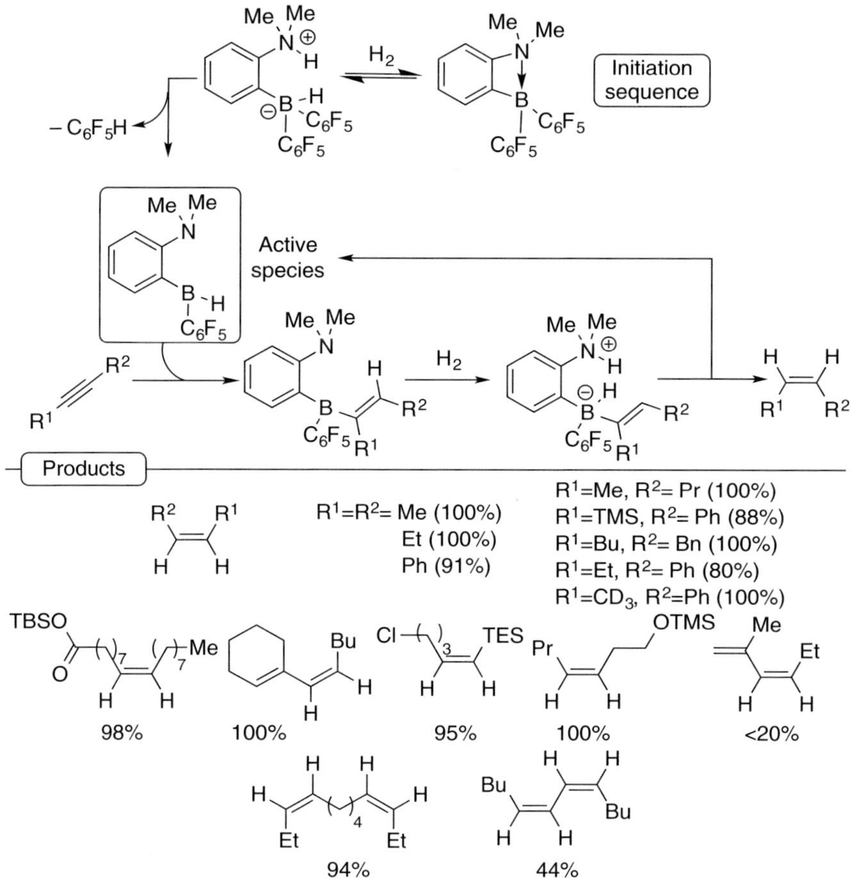

Figure 7.74 FLP-catalyzed diastereoselective hydrogenation of alkynes.

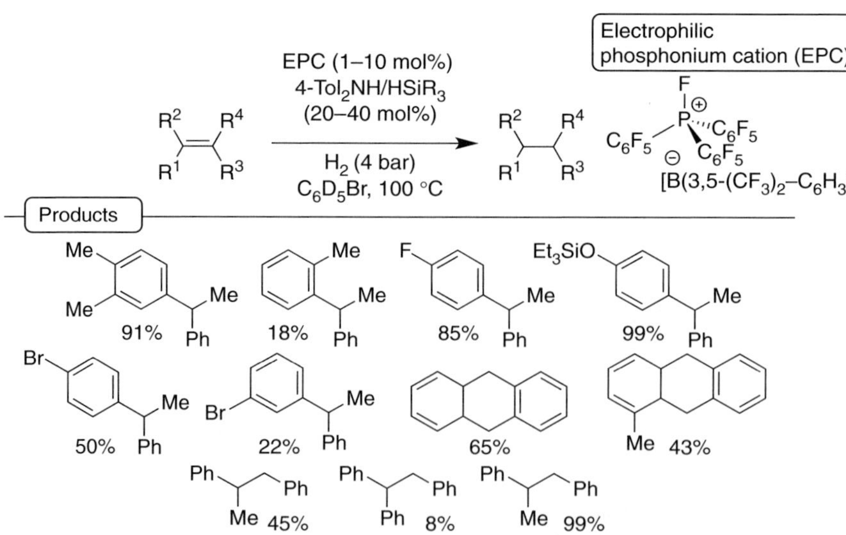

Figure 7.75 Catalytic hydrogenation with EPCs.

rings and on the phosphorus atom itself in $[(C_6F_5)_3PF][B(3,5\text{-}(CF_3)_2\text{–}C_6H_3)_4]$, which ultimately leads to the stabilization of the phosphorus(V) species.

This phosphonium species (1–10 mol%) was active in the catalytic hydrogenation of unfunctionalized olefins and polycyclic aromatic hydrocarbons. The system required the presence of a bis(4-toloyl)silyl amine as Lewis base, which was generated *in situ* by dehydrogenative Si–N coupling (Figure 7.75) [61]. The reactions proceeded in high yields for styrene-derived substrates. The significant dependency of the reaction on the nature of the Lewis base supports that the mechanism is similar to the borane-based hydrogenation of olefins (compare Figure 7.70), with the protonation of olefin as the rate-determining step.

7.10 Summary

FLP-catalyzed hydrogenation evolved from a laboratory curiosity to a fully developed methodology. Within only a very short period even highly enantioselective processes have been developed providing access to almost enantiopure products. Significant advances in the last years have provided access to even moisture-stable boranes or boranes with increased functional group tolerance for FLP-mediated H_2 activation, which now allows for practical transformations, e.g. reductive aminations or deoxygenative reduction of carboxylic amides. However, further efforts are necessary to advance this hydrogenation technology as the state of the art of transition metal catalysis. Mechanistic investigations now offer a structure–reactivity relationship for the targeted choice of an FLP for the selected application. New Lewis acids, e.g. on the basis of EPCs, reveal unprecedented opportunities and perspectives in main group element catalysis.

Abbreviations

2,6-Lutidine	2,6-dimethyl pyridine
2,4,6-Collidine	2,4,6-trimethyl pyridine
bar	pressure unit, 1.024 mbar, 760 Torr
Bn	benzyl
°C	degree Celsius
Cy	cyclohexyl
DABCO	1,4-diaza-bicyclo[2.2.2]octane
DCE	1,2-dichloroethane
DCM	dichloromethane
de	diastereomeric excess
d.r.	diastereomeric ratio
Dipp	2,6-diisopropylphenyl
ee	enantiomeric excess
Et	ethyl
FLP	frustrated Lewis pair
H_2	molecular hydrogen, dihydrogen
i-Pr	2-methylethyl
K	Kelvin

Me	methyl
Mes	mesitylene, 2,4,6-trimethylphenyl
MTBE	methyl-*tert*-butyl ether
n-Hex	*n*-hexyl
NMR	nuclear magnetic resonance
Ph	phenyl
PMP	*p*-methoxyphenyl
py	pyridine
RT	room temperature
s-Bu	*sec*-butyl, 1-methylpropyl
t-Bu	1,1-dimethylethyl
TBS	*tert*-butyldimethylsilyl
TES	triethylsilyl
TMS	trimethylsilyl
Tol	4-methyl-phenyl
THF	tetrahydrofurane
TMP	2,2,6,6-tetramethylpiperidine

References

1 G.C. Welch, R.R.S. Juan, J.D. Masuda, D.W. Stephan, *Science* 2006, 314, 1124–1126.
2 P. Spies, G. Erker, G. Kehr, K. Bergander, R. Fröhlich, S. Grimme, D.W. Stephan, *Chem. Commun.* 2007, 5072–5074.
3 G.C. Welch, D.W. Stephan, *J. Am. Chem. Soc.* 2007, 129, 1880–1881.
4 D.W. Stephan, G. Erker, *Angew. Chem. Int. Ed.* 2010, 49, 46–76.
5 G. Erker, *Pure Appl. Chem.* 2012, 84, 2203–2217.
6 D.W. Stephan, *Dalton Trans.* 2009, 3129–3136.
7 D.W. Stephan, *Org. Biomol. Chem.* 2008, 6, 1535–1539.
8 J. Paradies, *Synlett* 2013, 777–780.
9 J. Paradies, *Angew. Chem. Int. Ed.* 2014, 53, 3552–3557.
10 D.W. Stephan, G. Erker, *Angew. Chem. Int. Ed.* 2015, 54, 6400–6441.
11 D.W. Stephan, *J. Am. Chem. Soc.* 2015, 137, 10018–10032.
12 D.W. Stephan, *Org. Biomol. Chem.* 2012, 10, 5740–5746.
13 D.W. Stephan, *Acc. Chem. Res.* 2015, 48, 306–316.
14 J. Paradies, *Angew. Chem. Int. Ed.* 2014, 53, 3552–3557.
15 G. Erker, D.W. Stephan, Frustrated lewis pair mediated hydrogenations, *Topics in Current Chemistry: Frustrated Lewis Pairs*, vol. 1 (eds. G. Erker and D.W. Stephan) Berlin Heidelberg: Springer-Verlag, 2013, 85–110.
16 D. Stephan, G. Erker, *Isr. J. Chem.* 2015, 55, 133–133.
17 D.W. Stephan, G. Erker, *Philos. Trans. R. Soc. A* 2017, 375, 20170239.
18 F.G. Fontaine, D.W. Stephan, *Philos. Trans. R. Soc. A* 2017, 375, 20170004.
19 J. Paradies, *Coord. Chem. Rev.* 2019, 380, 170–183.
20 J. Paradies, Mechanisms in frustrated Lewis pair-catalyzed reactions. *Eur. J. Org. Chem.* 2019, 283–294. https://doi.org/10.1002/ejoc.201800944.

21 D.W. Stephan, G. Erker, *Frustrated Lewis Pairs I: Uncovering and Understanding*, vol. 332, 2013, 85–110.
22 *Frustrated Lewis Pairs II: Expanding the Scope* (eds. G. Erker and D.W. Stephan), Berlin Heidelberg: Springer-Verlag, vol. 334, 2013.
23 J. Paradies, Cooperative H_2 activation by borane-derived frustrated lewis pairs, *Cooperative Catalysis: Designing Efficient Catalysts for Synthesis* (ed. R. Peters). Weinheim, New York: Wiley-VCH, 2015, 263–291.
24 *Frustrated Lewis Pairs I: Uncovering and Understanding* (eds. G. Erker and D.W. Stephan), Berlin Heidelberg: Springer-Verlag, vol. 332, 2013.
25 J. Paradies, Chiral borane-based lewis acids for metal free hydrogenations, *Chiral Lewis Acids* (ed. K. Mikami) Cham, Switzerland: Springer, 2018, 193–216.
26 C. Appelt, H. Westenberg, F. Bertini, A.W. Ehlers, J.C. Slootweg, K. Lammertsma, W. Uhl, *Angew. Chem. Int. Ed.* 2011, 50, 3925–3928.
27 C. Appelt, J.C. Slootweg, K. Lammertsma, W. Uhl, *Angew. Chem. Int. Ed.* 2012, 51, 5911–5914.
28 S. Roters, C. Appelt, H. Westenberg, A. Hepp, J.C. Slootweg, K. Lammertsma, W. Uhl, *Dalton Trans.* 2012, 41, 9033–9045.
29 H. Westenberg, J.C. Slootweg, A. Hepp, J. Kösters, S. Roters, A.W. Ehlers, K. Lammertsma, W. Uhl, *Organometallics* 2010, 29, 1323–1330.
30 J. Boudreau, M.A. Courtemanche, F.G. Fontaine, *Chem. Commun.* 2011, 47, 11131–11133.
31 G. Menard, D.W. Stephan, *J. Am. Chem. Soc.* 2010, 132, 1796–1797.
32 G. Menard, D.W. Stephan, *Angew. Chem. Int. Ed.* 2011, 50, 8396–8399.
33 Y.T. Zhang, G.M. Miyake, E.Y.X. Chen, *Angew. Chem. Int. Ed.* 2010, 49, 10158–10162.
34 T. Holtrichter-Rossmann, C. Rosener, J. Hellmann, W. Uhl, E.-U. Würthwein, R. Fröhlich, B. Wibbeling, *Organometallics* 2012, 31, 3272–3283.
35 M.J. Sgro, D.W. Stephan, *Angew. Chem. Int. Ed.* 2012, 51, 11343–11345.
36 M.P. Boone, D.W. Stephan, *J. Am. Chem. Soc.* 2013, 135, 8508–8511.
37 M.P. Boone, D.W. Stephan, *Chem. Eur. J.* 2014, 20, 3333–3341.
38 A.M. Chapman, M.F. Haddow, D.F. Wass, *J. Am. Chem. Soc.* 2011, 133, 18463–18478.
39 A.M. Chapman, M.F. Haddow, D.F. Wass, *J. Am. Chem. Soc.* 2011, 133, 8826–8829.
40 D.F. Wass, A.M. Chapman, Frustrated lewis pairs beyond the main group: transition, *Frustrated Lewis Pairs II: Expanding the Scope*, vol. 334 (eds. D.W. Stephan and G. Erker) Berlin, Heidelberg: Springer, 2013, 261–280.
41 A.M. Chapman, M.F. Haddow, D.F. Wass, *Eur. J. Inorg. Chem.* 2012, 1546–1554.
42 A.M. Chapman, D.F. Wass, *Dalton Trans.* 2012, 41, 9067–9072.
43 S. Khan, M. Alcarazo, Carbon-based frustrated lewis pairs, *Frustrated Lewis Pairs II: Expanding the Scope*, vol. 334 (eds. D.W. Stephan and G. Erker) Berlin, Heidelberg: Springer, 2013, 157–170.
44 M. Alcarazo, C. Gomez, S. Holle, R. Goddard, *Angew. Chem. Int. Ed.* 2010, 49, 5788–5791.

45 L. Cabrera, G.C. Welch, J.D. Masuda, P. Wei, D.W. Stephan, *Inorg. Chim. Acta* 2006, 359, 3066–3071.
46 A. Schafer, M. Reissmann, A. Schafer, W. Saak, D. Haase, T. Müller, *Angew. Chem. Int. Ed.* 2011, 50, 12636–12638.
47 D.W. Stephan, G. Erker, *Chem. Sci.* 2014, 5, 2625–2641.
48 H.C. Brown, H.I. Schlesinger, S.Z. Cardon, *J. Am. Chem. Soc.* 1942, 64, 325–333.
49 T.A. Rokob, A. Hamza, A. Stirling, T. Soós, I. Pápai, *Angew. Chem. Int. Ed.* 2008, 47, 2435–2438.
50 A. Hamza, A. Stirling, T.A. Rokob, I. Pápai, *Int. J. Quantum Chem.* 2009, 109, 2416–2425.
51 T.A. Rokob, I. Bako, A. Stirling, A. Hamza, I. Pápai, *J. Am. Chem. Soc.* 2013, 135, 4425–4437.
52 T.A. Rokob, A. Hamza, A. Stirling, I. Pápai, *J. Am. Chem. Soc.* 2009, 131, 2029–2036.
53 S. Grimme, H. Kruse, L. Goerigk, G. Erker, *Angew. Chem. Int. Ed.* 2010, 49, 1402–1405.
54 C.M. Momming, S. Fromel, G. Kehr, R. Fröhlich, S. Grimme, G. Erker, *J. Am. Chem. Soc.* 2009, 131, 12280–12289.
55 C.M. Mömming, E. Otten, G. Kehr, R. Fröhlich, S. Grimme, D.W. Stephan, G. Erker, *Angew. Chem. Int. Ed.* 2009, 48, 6643–6646.
56 L. Rocchigiani, G. Ciancaleoni, C. Zuccaccia, A. Macchioni, *J. Am. Chem Soc.* 2014, 136, 112–115.
57 E.F. Pettersen, T.D. Goddard, C.C. Huang, G.S. Couch, D.M. Greenblatt, E.C. Meng, T.E. Ferrin, *J. Comput. Chem.* 2004, 25, 1605–1612.
58 B. Schirmer, S. Grimme, *Chem. Commun.* 2010, 46, 7942–7944.
59 T.A. Rokob, A. Hamza, I. Pápai, *J. Am. Chem. Soc.* 2009, 131, 10701–10710.
60 G. Skara, F. De Vleeschouwer, P. Geerlings, F. De Proft, B. Pinter, *Sci. Rep. UK* 2017, 7, 16024.
61 T. vom Stein, M. Perez, R. Dobrovetsky, D. Winkelhaus, C.B. Caputo, D.W. Stephan, *Angew. Chem. Int. Ed.* 2015, 54, 10178–10182.
62 P. Spies, S. Schwendemann, S. Lange, G. Kehr, R. Fröhlich, G. Erker, *Angew. Chem. Int. Ed.* 2008, 47, 7543–7546.
63 C.B. Caputo, L.J. Hounjet, R. Dobrovetsky, D.W. Stephan, *Science* 2013, 341, 1374–1377.
64 I. Mallov, T.C. Johnstone, D.C. Burns, D.W. Stephan, *Chem. Commun.* 2017, 53, 7529–7532.
65 M. Pérez, C.B. Caputo, R. Dobrovetsky, D.W. Stephan, *Proc. Natl. Acad. Sci. U.S.A.* 2014, 111, 10917.
66 S. Postle, V. Podgorny, D.W. Stephan, *Dalton Trans.* 2016, 45, 14651–14657.
67 A. Augurusa, M. Mehta, M. Perez, J. Zhu, D.W. Stephan, *Chem. Commun.* 2016, 52, 12195–12198.
68 E.R. Clark, M.J. Ingleson, *Angew. Chem. Int. Ed.* 2014, 53, 11306–11309.
69 A. Schäfer, M. Reißmann, A. Schäfer, W. Saak, D. Haase, T. Müller, *Angew. Chem. Int. Ed.* 2011, 50, 12636–12638.
70 H.D. Wang, R. Fröhlich, G. Kehr, G. Erker, *Chem. Commun.* 2008, 5966–5968.

71 L. Greb, P. Ona-Burgos, A. Kubas, F.C. Falk, F. Breher, K. Fink, J. Paradies, *Dalton Trans.* 2012, 41, 9056–9060.
72 C. Jiang, D.W. Stephan, *Dalton Trans.* 2013, 42, 630–637.
73 D.W. Stephan, S. Greenberg, T.W. Graham, P. Chase, J.J. Hastie, S.J. Geier, J.M. Farrell, C.C. Brown, Z.M. Heiden, G.C. Welch, M. Ullrich, *Inorg. Chem.* 2011, 50, 12338–12348.
74 S. Tussing, L. Greb, S. Tamke, B. Schirmer, C. Muhle-Goll, B. Luy, J. Paradies, *Chem. Eur. J.* 2015, 21, 8056–8059.
75 S. Tussing, K. Kaupmees, J. Paradies, *Chem. Eur. J.* 2016, 22, 7422–7426.
76 S. Tussing, J. Paradies, *Dalton Trans.* 2016, 45, 6124–6128.
77 V. Sumerin, F. Schulz, M. Atsumi, C. Wang, M. Nieger, M. Leskelä, T. Repo, P. Pyykkö, B. Rieger, *J. Am. Chem. Soc.* 2008, 130, 14117–14118.
78 V. Sumerin, F. Schulz, M. Nieger, M. Leskelä, T. Repo, B. Rieger, *Angew. Chem. Int. Ed.* 2008, 47, 6001–6003.
79 L.J. Hounjet, C. Bannwarth, C.N. Garon, C.B. Caputo, S. Grimme, D.W. Stephan, *Angew. Chem. Int. Ed.* 2013, 52, 7492–7495.
80 G.D. Frey, V. Lavallo, B. Donnadieu, W.W. Schoeller, G. Bertrand, *Science* 2007, 316, 439–441.
81 P.A. Chase, D.W. Stephan, *Angew. Chem. Int. Ed.* 2008, 47, 7433–7437.
82 D. Holschumacher, T. Bannenberg, C.G. Hrib, P.G. Jones, M. Tamm, *Angew. Chem. Int. Ed.* 2008, 47, 7428–7432.
83 P.A. Chase, A.L. Gille, T.M. Gilbert, D.W. Stephan, *Dalton Trans.* 2009, 7179–7188.
84 D. Holschumacher, C. Taouss, T. Bannenberg, C.G. Hrib, C.G. Daniliuc, P.G. Jones, M. Tamm, *Dalton Trans.* 2009, 6927–6929.
85 D. Holschumacher, C.G. Daniliuc, P.G. Jones, M. Tamm, *Z. Naturforsch. B* 2011, 66, 371–377.
86 S. Kronig, E. Theuergarten, D. Holschumacher, T. Bannenberg, C.G. Daniliuc, P.G. Jones, M. Tamm, *Inorg. Chem.* 2011, 50, 7344–7359.
87 H. Li, A.J.A. Aquino, D.B. Cordes, F. Hung-Low, W.L. Hase, C. Krempner, *J. Am. Chem. Soc.* 2013, 135, 16066–16069.
88 M. Periasamy, P. Thirumalaikumar, *J. Organomet. Chem.* 2000, 609, 137–151.
89 J. Seyden-Penne, *Reductions by the Alumino- and Borohydrides in Organic Synthesis.* Paris: VCH -Lavoisier, 1991.
90 P.A. Chase, G.C. Welch, T. Jurca, D.W. Stephan, *Angew. Chem. Int. Ed.* 2007, 46, 8050–8053.
91 P.A. Chase, T. Jurca, D.W. Stephan, *Chem. Commun.* 2008, 1701–1703.
92 D.J. Chen, J. Klankermayer, *Chem. Commun.* 2008, 2130–2131.
93 L. Greb, S. Tussing, B. Schirmer, P. Oña-Burgos, K. Kaupmees, M. Lõkov, I. Leito, S. Grimme, J. Paradies, *Chem. Sci.* 2013, 4, 2788–2796.
94 L. Greb, P. Ona-Burgos, B. Schirmer, S. Grimme, D.W. Stephan, J. Paradies, *Angew. Chem. Int. Ed.* 2012, 51, 10164–10168.
95 Y. Segawa, D.W. Stephan, *Chem. Commun.* 2012, 48, 11963–11965.
96 M.A. Beckett, D.S. Brassington, S.J. Coles, M.B. Hursthouse, *Inorg. Chem. Commun.* 2000, 3, 530–533.
97 M.A. Beckett, G.C. Strickland, J.R. Holland, K. Sukumar Varma, *Polymer* 1996, 37, 4629–4631.

98 V. Gutmann, *Coord. Chem. Rev.* 1976, 18, 225–255.

99 U. Mayer, V. Gutmann, W. Gerger, *Monatsh. Chem.* 1975, 106, 1235–1257.

100 R.F. Childs, D.L. Mulholland, A. Nixon, *Can. J. Chem.* 1982, 60, 801–808.

101 M. Ullrich, A.J. Lough, D.W. Stephan, *J. Am. Chem. Soc.* 2009, 131, 52–53.

102 J.A. Nicasio, S. Steinberg, B. Ines, M. Alcarazo, *Chem. Eur. J.* 2013, 19, 11016–11020.

103 L. Greb, C.G. Daniliuc, K. Bergander, J. Paradies, *Angew. Chem. Int. Ed.* 2013, 52, 5876–5879.

104 G. Eros, H. Mehdi, I. Pápai, T.A. Rokob, P. Király, G. Tárkányi, T. Soós, *Angew. Chem. Int. Ed.* 2010, 49, 6559–6563.

105 L. Greb, J. Paradies, Paracyclophane derivatives in frustrated lewis pairs, *Frustrated Lewis Pairs II: Expanding the Scope*, vol. 334 (eds. D.W. Stephan and G. Erker) Berlin, Heidelberg: Springer, 2013, 81–100.

106 S.C. Binding, H. Zaher, F. Mark Chadwick, D. O'Hare, *Dalton Trans.* 2012, 41, 9061–9066.

107 Z.P. Lu, Z.H. Cheng, Z.X. Chen, L. Weng, Z.H. Li, H. Wang, *Angew. Chem. Int. Ed.* 2011, 50, 12227–12231.

108 T.J. Herrington, A.J.W. Thom, A.J.P. White, A.E. Ashley, *Dalton Trans.* 2012, 41, 9019–9022.

109 J. Mohr, M. Durmaz, E. Irran, M. Oestreich, *Organometallics* 2014, 33, 1108–1111.

110 A.E. Ashley, T.J. Herrington, G.G. Wildgoose, H. Zaher, A.L. Thompson, N.H. Rees, T. Krämer, D. O'Hare, *J. Am. Chem. Soc.* 2011, 133, 14727–14740.

111 A. Gyomore, M. Bakos, T. Foldes, I. Pápai, A. Domján, T. Soós, *ACS Catal.* 2015, 5, 5366–5372.

112 É. Dorkó, M. Szabó, B. Kótai, I. Pápai, A. Domján, T. Soós, *Angew. Chem. Int. Ed.* 2017, 56, 9512–9516.

113 I. Kaljurand, A. Kütt, L. Sooväli, T. Rodima, V. Mäemets, I. Leito, I.A. Koppel, *J. Org. Chem.* 2005, 70, 1019–1028.

114 L. Greb, S. Tamke, J. Paradies, *Chem. Commun.* 2014, 50, 2318–2320.

115 A.F.G. Maier, S. Tussing, T. Schneider, U. Flörke, Z.-W. Qu, S. Grimme, J. Paradies *Angew. Chem. Int. Ed.* 2016, 55, 12219–12223.

116 D.J. Parks, R.E.V.H. Spence, W.E. Piers, *Angew. Chem. Int. Ed.* 1995, 34, 809–811.

117 D.J. Parks, W.E. Piers, *Tetrahedron* 1998, 54, 15469–15488.

118 D.J. Parks, W.E. Piers, G.P.A. Yap, *Organometallics* 1998, 17, 5492–5503.

119 S. Schwendemann, T.A. Tumay, K.V. Axenov, I. Peuser, G. Kehr, R. Fröhlich, G. Erker, *Organometallics* 2010, 29, 1067–1069.

120 K.V. Axenov, C.M. Momming, G. Kehr, R. Fröhlich, G. Erker, *Chem. Eur. J.* 2010, 16, 14069–14073.

121 S. Schwendemann, R. Fröhlich, G. Kehr, G. Erker, *Chem. Sci.* 2011, 2, 1842–1849.

122 V. Sumerin, K. Chernichenko, M. Nieger, M. Leskelä, B. Rieger, T. Repo, *Adv. Synth. Catal.* 2011, 353, 2093–2110.

123 K. Chernichenko, M. Nieger, M. Leskela, T. Repo, *Dalton Trans.* 2012, 41, 9029–9032.

124 K. Chernichenko, A. Madarasz, I. Pápai, M. Nieger, M. Leskelä, T. Repo, *Nat. Chem.* 2013, 5, 718–723.

125 M. Lindqvist, K. Borre, K. Axenov, B. Kótai, M. Nieger, M. Leskelä, I. Pápai, T. Repo, *J. Am. Chem. Soc.* 2015, 137, 4038–4041.

126 A. Stute, G. Kehr, R. Fröhlich, G. Erker, *Chem. Commun.* 2011, 47, 4288–4290.

127 A. Stute, L. Heletta, R. Fröhlich, C.G. Daniliuc, G. Kehr, G. Erker, *Chem. Commun.* 2012, 48, 11739–11741.

128 A. Stute, G. Kehr, C.G. Daniliuc, R. Fröhlich, G. Erker, *Dalton Trans.* 2013, 42, 4487–4499.

129 M. Sajid, A. Lawzer, W.S. Dong, C. Rosorius, W. Sander, B. Schirmer, S. Grimme, C.G. Daniliuc, G. Kehr, G. Erker, *J. Am. Chem. Soc.* 2013, 135, 18567–18574.

130 C. Rosorius, C.G. Daniliuc, R. Fröhlich, G. Kehr, G. Erker, *J. Organomet. Chem.* 2013, 744, 149–155.

131 O. Ekkert, G.G. Miera, T. Wiegand, H. Eckert, B. Schirmer, J.L. Petersen, C.G. Daniliuc, R. Fröhlich, S. Grimme, G. Kehr, G. Erker, *Chem. Sci.* 2013, 4, 2657–2664.

132 C. Rosorius, G. Kehr, R. Fröhlich, S. Grimme, G. Erker, *Organometallics* 2011, 30, 4211–4219.

133 G. Eros, K. Nagy, H. Mehdi, I. Pápai, P. Nagy, P. Király, G. Tárkányi, T. Soós, *Chem. Eur. J.* 2012, 18, 574–585.

134 T. Soós, *Pure Appl. Chem.* 2011, 83, 667–675.

135 B. Inés, D. Palomas, S. Holle, S. Steinberg, J.A. Nicasio, M. Alcarazo, *Angew. Chem. Int. Ed.* 2012, 51, 12367–12369.

136 D.J. Scott, M.J. Fuchter, A.E. Ashley, *Angew. Chem. Int. Ed.* 2014, 53, 10218–10222.

137 D.J. Scott, M.J. Fuchter, A.E. Ashley, *J. Am. Chem. Soc.* 2014, 136, 15813–15816.

138 D.J. Chen, Y.T. Wang, J. Klankermayer, *Angew. Chem. Int. Ed.* 2010, 49, 9475–9478.

139 D.J. Chen, V. Leich, F.F. Pan, J. Klankermayer, *Chem. Eur. J.* 2012, 18, 5184–5187.

140 G. Ghattas, D.J. Chen, F.F. Pan, J. Klankermayer, *Dalton Trans.* 2012, 41, 9026–9028.

141 M. Lindqvist, K. Borre, K. Axenov, B. Kótai, M. Nieger, M. Leskelä, I. Pápai, T. Repo, *J. Am. Chem. Soc.* 2015, 137, 4038–4041.

142 Y.B. Liu, H.F. Du, *J. Am. Chem. Soc.* 2013, 135, 6810–6813.

143 S. Li, G. Li, W. Meng, H. Du, *J. Am. Chem. Soc.* 2016, 138, 12956–12962.

144 S. Li, W. Meng, H. Du, *Org. Lett.* 2017, 19, 2604–2606.

145 J. Mohr, M. Oestreich, *Angew. Chem. Int. Ed.* 2014, 53, 13278–13281.

146 G. Erker, D.W. Stephan, Frustrated lewis pair mediated hydrogenations, *Topics in Current Chemistry: Frustrated Lewis Pairs*, vol. 1 (ed. D.W.S.G. Erker) Berlin, Heidelberg: Springer-Verlag, 2013, 85–110.

147 H. Pellissier, *Chem. Rev.* 2013, 113, 442–524.

148 L.F. Tietze, *Chem. Rev.* 1996, 96, 115–136.

149 J.M. Blackwell, D.J. Morrison, W.E. Piers, *Tetrahedron* 2002, 58, 8247–8254.

150 S. Wei, H. Du, *J. Am. Chem. Soc.* 2014, 136, 12261–12264.
151 T. Mahdi, D.W. Stephan, *J. Am. Chem. Soc.* 2014, 136, 15809–15812.
152 H. Babad, A.G. Zeiler, *Chem. Rev.* 1973, 73, 75–91.
153 R.D. Howells, J.D. McCown, *Chem. Rev.* 1977, 77, 69–92.
154 R. Salmon, *Encyclopedia of Reagents for Organic Synthesis.* New York: Wiley, 2001. https://doi.org/10.1002/047084289X.ro047084015.
155 M. Eggersdorfer, D. Laudert, U. Letinois, T. McClymont, J. Medlock, T. Netscher, W. Bonrath, *Angew. Chem. Int. Ed.* 2012, 51, 12960–12990.
156 L. Horner, W.D. Balzer, *Tetrahedron Lett.* 1965, 6, 1157–1162.
157 N.L. Bauld, F. Farr, *J. Am. Chem. Soc.* 1969, 91, 2788–2789.
158 R. Tang, K. Mislow, *J. Am. Chem. Soc.* 1969, 91, 5644–5645.
159 K. Naumann, G. Zon, K. Mislow, *J. Am. Chem. Soc.* 1969, 91, 7012–7023.
160 K.L. Marsi, *J. Org. Chem.* 1974, 39, 265–267.
161 T. Coumbe, N.J. Lawrence, F. Muhammad, *Tetrahedron Lett.* 1994, 35, 625–628.
162 H.C. Wu, J.Q. Yu, J.B. Spencer, *Org. Lett.* 2004, 6, 4675–4678.
163 M. Lemaire, M. Berthod, A. Favre-Réguillon, J. Mohamad, G. Mignani, G. Docherty, *Synlett* 2007, 1545–1548.
164 C. Petit, A. Favre-Reguillon, B. Albela, L. Bonneviot, G. Mignani, M. Lemaire, *Organometallics* 2009, 28, 6379–6382.
165 C.J. O'Brien, J.L. Tellez, Z.S. Nixon, L.J. Kang, A.L. Carter, S.R. Kunkel, K.C. Przeworski, G.A. Chass, *Angew. Chem. Int. Ed.* 2009, 48, 6836–6839.
166 L. Horner, H. Hoffmann, P. Beck, *Chem. Ber.* 1958, 91, 1583–1588.
167 T. Yano, M. Hoshino, M. Kuroboshi, H. Tanaka, *Synlett* 2010, 801–803.
168 T. Yano, M. Kuroboshi, H. Tanaka, *Tetrahedron Lett.* 2010, 51, 698–701.
169 R. Salmon, I.V. Efremov, Oxalyl chloride. *In Encyclopedia of Reagents for Organic Synthesis,* 2008. https://doi.org/10.1002/047084289X.ro015.pub2.
170 K.V. Rajendran, D.G. Gilheany, *Chem. Commun.* 2012, 48, 817–819.
171 J. Paradies, A.J. Stepen, D.W. Stephan, *Angew. Chem. Int. Ed.* 2018. https://doi.org/10.1002/anie.201809275\ignorespacesand\ignorespaces201809210.201801002/ange.201809275.
172 C.B. Caputo, D. Winkelhaus, R. Dobrovetsky, L.J. Hounjet, D.W. Stephan, *Dalton Trans.* 2015, 44, 12256–12264.
173 J.-W. Wang, L.-Y. Ding, B.-Q. Wang, Y.-Y. He, Y. Guo, X.-F. Jia, W. Zheng, *J. Mol. Struct.* 2014, 1058, 62–70.
174 R. Zelli, W. Zeinyeh, R. Haudecoeur, J. Alliot, B. Boucherle, I. Callebaut, J.-L. Décout, *Org. Lett.* 2017, 19, 6360–6363.
175 Q. Liu, J. Wu, J. Li, J. Wang, W. Zheng, H.W. Roesky, *Phosphorus, Sulfur Silicon Relat. Elem.* 2014, 189, 1455–1466.
176 N.A. Sitte, M. Bursch, S. Grimme, J. Paradies, *Angew. Chem. Int. Ed..* 2018, 57, 15253–15256.
177 S.J. Geier, P.A. Chase, D.W. Stephan, *Chem. Commun.* 2010, 46, 4884–4886.
178 T. Mahdi, J.N. del Castillo, D.W. Stephan, *Organometallics* 2013, 32, 1971–1978.
179 Y. Liu, H. Du, *J. Am. Chem. Soc.* 2013, 135, 12968–12971.
180 Z.H. Zhang, H.F. Du, *Angew. Chem. Int. Ed.* 2015, 54, 623–626.
181 Z.H. Zhang, H.F. Du, *Org. Lett.* 2015, 17, 6266–6269.

182 Z.H. Zhang, H.F. Du, *Org. Lett.* 2015, 17, 2816–2819.
183 J.S. Reddy, B.-H. Xu, T. Mahdi, R. Fröhlich, G. Kehr, D.W. Stephan, G. Erker, *Organometallics* 2012, 31, 5638–5649.
184 B.H. Xu, G. Kehr, R. Fröhlich, B. Wibbeling, B. Schirmer, S. Grimme, G. Erker, *Angew. Chem. Int. Ed.* 2011, 50, 7183–7186.
185 E.J. Dewitt, L.E. Trapasso, F.L. Ramp, *J. Am. Chem. Soc.* 1961, 83, 4672.
186 F.L. Ramp, E.J. Dewitt, L.E. Trapasso, *J. Org. Chem.* 1962, 27, 4368.
187 Y. Wang, W. Chen, Z. Lu, Z.H. Li, H. Wang, *Angew. Chem. Int. Ed.* 2013, 52, 7496–7499.

8

Recent Advances in Selective Biocatalytic (Hydrogen Transfer) Reductions

Gonzalo de Gonzalo[1] *and Iván Lavandera*[2]

[1] *Universidad de Sevilla, Departamento de Química Orgánica, c/Profesor García González, Sevilla, 41012, Spain*
[2] *University of Oviedo, Organic and Inorganic Chemistry Department, Avenida Julián Clavería 8, Oviedo, 33006, Spain*

8.1 Introduction

The use of enzymes (also called biocatalysts) in the synthesis of organic compounds is the core of biocatalysis [1–4]. Since the discovery of Zaks and Klibanov regarding the applicability of these catalysts in organic solvents [5], the implementation of this technique has largely increased, even at industrial scale [6, 7].

Among the different advantages that biocatalysis can offer in a chemical transformation [8], it can be highlighted that the biodegradability of the catalyst and the mildness of the reaction conditions are advantageous from an environmental point of view. Furthermore, catalyst immobilization is also possible, increasing its stability and simplifying handling and the downstream process. Enzymes are extremely efficient, enhancing the reaction rate by several orders of magnitude, making the implementation of the catalyzed process feasible at larger scale. But probably the more valued aspect of biocatalysts is their excellent selectivity for the synthesis of the desired compound with high purity, shortening previously established synthetic routes, thus diminishing costs and the waste/by-products generated. Among the different types of selectivities that enzymes can show, chemo-, regio-, and stereoselectivity can be mentioned. Moreover, in the last two decades, the modification of biological catalysts can be easily approached via molecular biology tools, providing novel enzymes with improved activities and stabilities [9].

Biocatalytic (hydrogen transfer) reductions are effected by enzymes belonging to the oxidoreductase class. These catalysts transfer the required protons and hydrides from/to the substrates by the mediation of some specific molecules called cofactors/coenzymes. Among the various naturally existing cofactors and coenzymes, the ones most often found are those that depend on nicotinamide cofactor NAD or its phosphorylated analog NADP. This chapter will deal with recent synthetic applications of different oxidoreductases to perform selective reductions, resembling (metal- or organo-catalyzed) hydrogenations (Scheme 8.1). The different transformations that will be mentioned are stereoselective ketone/aldehyde reductions to obtain chiral alcohols mediated by alcohol

Homogeneous Hydrogenation with Non-Precious Catalysts, First Edition. Edited by Johannes F. Teichert.

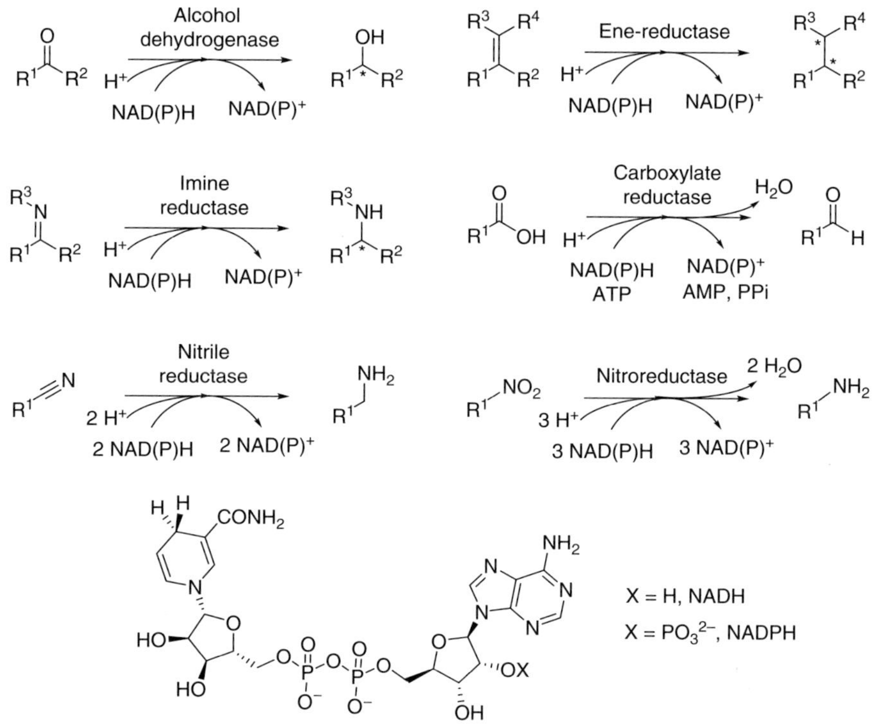

Scheme 8.1 Examples of biocatalytic reductions discussed in this chapter and structure of nicotinamide cofactors.

dehydrogenases (ADHs), hydrogenations of C=C double bonds catalyzed by ene-reductases (EREDs), reductions of imines to synthesize the corresponding amines by the action of imine reductases (IREDs), transformations of carboxylic acids into aldehydes mediated by carboxylic acid reductases, and biocatalytic reductions of nitrile and nitro derivatives.

Beyond the scope of this chapter are the hydrogenases [10], enzymes responsible for the formation/rupture of molecular hydrogen from/into protons and electrons. These biocatalysts are an important factor for adjusting the redox balance in green algae, bacteria, and archaea in H_2-based respiration processes, and while they have demonstrated an unsurpassed H_2 release activity, they have not been successfully applied to conventional hydrogenation chemical reactions due to their relatively low stability under nonoptimal conditions.

8.2 Ketoreductases

The most exploited group of biocatalysts in asymmetric hydrogenations are alcohol dehydrogenases (ADHs; E.C. 1.1.1.x), also called ketoreductases (KREDs) or carbonyl reductases (CRs). These enzymes are oxidoreductases that selectively catalyze the reversible transformation of carbonyl compounds into the corresponding alcohols, requiring the presence of nicotinamide cofactors (NADH or NADPH) to perform their activity (Scheme 8.1) [11–14]. During the course of the reduction, the enzyme delivers a hydride from the cofactor C4 to the *Si*- (anti-Prelog ADHs) or to the *Re*-face (Prelog enzymes) of the carbonyl compound, yielding (*R*)- or (*S*)-alcohols, respectively. Especially when working with purified enzymes, efficient cofactor regeneration strategies have to be established due to the high costs of these compounds. In most of these, the desired bioreduction is coupled with a chemical, photochemical, electrochemical, or enzymatic reaction to regenerate the nicotinamide cofactor. The enzymatic reaction is in general most compatible with the reaction conditions for the dehydrogenase and can be developed by two approaches (and both can be seen as hydrogen transfer protocols):

- Coupled enzyme approach, in which a second enzymatic system is employed to regenerate the NAD(P)H. Typical examples are the combination of glucose with glucose dehydrogenase (GDH) or glucose-6-phosphate with glucose-6-phosphate dehydrogenase (G6PDH).
- Coupled substrate approach, in which the ADH catalyzes the desired bioreduction and the oxidation of a cosubstrate, in general isopropanol (IPA). The presence of IPA and acetone as coproduct facilitates the substrate solubilization, but it has to be taken into account that the ketone reduction is under thermodynamic control and a large excess of IPA is required.

Asymmetric hydrogenations catalyzed by ADHs can be performed using whole cell systems or isolated enzymes with different degrees of purification. The use of whole cells facilitates cofactor regeneration, since glucose can be employed as cosubstrate and hydrogen donor, and the cofactors are in the cell, but when using these systems, the ketone concentration must be kept low during the bioreduction to avoid inhibition processes. In contrast, the use of isolated enzymes requires the presence of an additional cofactor and a system for its regeneration. A study to determine whether whole cells or crude enzymes were more effective in the catalyzed bioreduction of three model ketones at large scale has been recently performed [15]. Results obtained demonstrate that both preparations can be employed to develop simple and feasible reactions. When the ADH accepts IPA as cosubstrate for cofactor regeneration, crude or purified enzymes can be easily employed.

8.2.1 Alcohol Dehydrogenases in "Nonconventional" Media

Enzymatic reactions are usually performed in aqueous medium. The use of "nonconventional media" included the development of asymmetric biocatalyzed processes in different non-aqueous solvents. The use of ADHs in reaction media containing organic solvents, even in the so-called "micro-aqueous" medium (with 99% v/v of organic solvent), or ionic liquids (ILs) has been well documented [16–18]. In recent times, deep eutectic solvents (DESs) have appeared as useful cosolvents in many biocatalytic applications. These nontoxic and biodegradable compounds show a high solvation capacity and are formed by the combination of a quaternary ammonium salt and a hydrogen bond donor molecule, which leads to a liquid at room temperature [19]. In 2015, overexpressed ADH from *Ralstonia* sp. (RasADH) was studied as biocatalyst in the reduction of propiophenone in the presence of different mixtures of buffer–DES (choline chloride and glycerol 1 : 2 v/v) using IPA as cosubstrate [20]. Conversions varied at different DES proportions, but an increase in the optical purity of the final alcohol was observed at high DES proportions (ee >90% at DES 80–95% v/v). The selective reduction of various ketones employing liophilized *Escherichia coli* whole cells containing the *Candida parapsilosis* carbonyl reductase (CPCR) has been performed in neat substrates with a defined activity of water [21]. The absence of solvents for these reactions presents high cost effectiveness and an environmentally friendly operation mode. This methodology can also be employed for water-unstable compounds. Thus, bioreduction of 3-butyn-2-one to enantiopure (*S*)-3-butyn-2-ol was carried out in an IPA/substrate ratio 9 : 1 with 67.6% conversion after 24 hours, which corresponds to 57.4 g/l of the optically pure (*S*)-alcohol. This product concentration is twice higher than the reported value in plain buffer.

8.2.2 Dynamic Processes Employing Ketoreductases

KREDs are valuable tools for the production of chiral alcohols with one or more stereocenters. In some cases, the racemization of a nonreactive stereocenter is possible in order to obtain multiple chiral centers in only one process. The epimerizable stereogenic center is located in an adjacent position to the carbonyl moiety, containing an acidic proton that facilitates the racemization, in the so-called dynamic reductive kinetic resolutions (DYRKRs) [22].

Horse liver alcohol dehydrogenase (HLDH) is a commercially available NADH-dependent KRED. This biocatalyst has been employed in the enantioselective synthesis of a set of (*S*)-2-arylpropanols (**2**), precursors in the preparation of profens, starting from 2-arylpropanals (**1**, Scheme 8.2) [23]. These compounds can be obtained with high optical purities and yields in a dynamic process in which the enzymatic bioreduction is coupled with the base-catalyzed racemization of the substrate under the reaction conditions. A set of α-alkyl β-keto

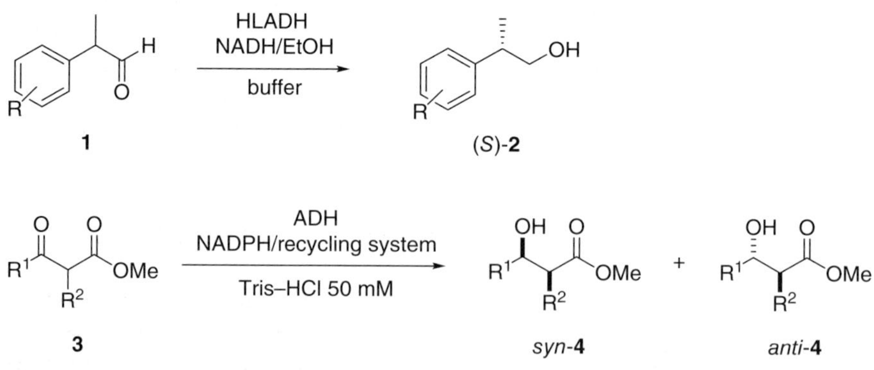

Scheme 8.2 Dynamic kinetic resolution of 2-arylpropanals (**1**) and α-alkyl β-keto esters (**3**) employing alcohol dehydrogenases.

esters (**3**) has been reduced in a DYRKR catalyzed by various KREDs to afford optically active α-alkyl β-hydroxy esters (**4**), building blocks of several bioactive compounds, with excellent selectivities and high conversions (Scheme 8.2) [24]. Owing to the structure of the substrates, a fast racemization was possible at neutral pH. The use of Prelog KREDs as ADH-A from *Rhodococcus ruber*, CPADH from *C. parapsilosis*, and TeSADH from *Thermoanaerobacter ethanolicus* afforded the *syn*-(2*R*,3*S*)-β-hydroxy esters with high diastereoselectivities for small substrates, while bulkier keto esters were reduced by the ADHs from *Sphingobium yanoikuyae* (SyADH) to the *syn*-(2*R*,3*S*)-hydroxy ester, and RasADH to the *anti*-(2*S*,3*S*) with good selectivities. The anti-Prelog ADHs from *Lactobacillus brevis* (LBADH) and from *Lactobacillus kefir* (LKADH) led to the *syn*-(2*S*,3*R*) alcohols with high yields and generally high diastereoselectivities.

The biocatalyzed reduction of methyl 2-benzamidomethyl-3-oxobutanonate (**5**) by the anti-Prelog ADH from *L. brevis* (*Bg*ADH2) expressed in *E. coli* afforded methyl (2*S*,3*R*)-2-benzamidomethyl-3-hydroxybutyrate (**6**), a precursor in the synthesis of carbapenem antibiotics (Scheme 8.3) [25]. When the reaction was carried out in buffer pH 6.5 and 30 °C, a fast substrate racemization was observed, leading to (2*S*,3*R*)-**6** with 99% conversion and complete enantio- and diastereoselectivity in a DYRKR. The study of various cosolvents at 50% v/v showed higher productivity when using a biphasic system buffer/toluene. Thus, 60 mM of **5** afforded a 91% yield after two hours, recovering (2*S*,3*R*)-**6** with 99% enantiomeric excess (ee) and 99.2 : 0.8 diastereomeric ratio (d.r.). A DYRKR has been developed for the synthesis of a set of chiral 3,4-dialkyl-3,4-dihydroisocoumarins starting from 2-(3-oxoalkyl)benzonitriles through a simultaneous biocatalytic reduction combined with the substrate racemization (Scheme 8.3) [26]. Initial studies on 2-(3-oxobutan-2-yl)benzonitrile (**7**) were carried out at 30 °C in tris(hydroxymethyl)aminomethane (Tris)/HCl buffer pH 7.5 containing 5% v/v IPA and 5% v/v hexane. The bioreduction catalyzed by ADH-A afforded chiral (*S*,*S*)-**8** with an excellent selectivity and 56% conversion after 24 hours. As racemization catalysts, Et_3N or the exchange resin Dowex MWA-1 were studied. This procedure was extended to other ketones bearing different

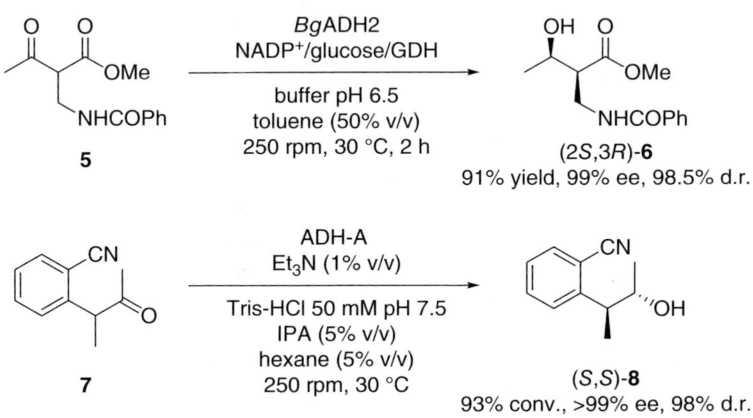

Scheme 8.3 Examples of DYRKR catalyzed by alcohol dehydrogenases.

substituents at the aromatic ring or the stereocenter, leading to the final enantiopure alcohols in good to excellent yields and complete diastereoselectivities. The (*S*,*S*)-alcohols were subjected to acid-catalyzed cyclization, leading to (*S*,*S*)-3-dialkyl-3,4-dihydroisocoumarins in good yields in most of the cases. A similar methodology has also been applied for the synthesis of optically active 3-arylpropan-2-ols starting from 3-arylalkan-2-ones. Bioreduction of racemic 3-phenylbutan-2-one catalyzed by *E. coli*/ADH-A led to *syn*-(*S*,*S*)-alcohol with good selectivity (83 : 17 d.r.) and a conversion of 50% at short reaction times [27]. This compound was transformed into an (*S*,*S*)-isochroman by treatment with zinc chloride at room temperature (RT). The substrate racemization was studied in the presence of anion exchange resins, but biocatalyst inactivation was observed at the reaction conditions. Thus, a higher temperature was used and the biocatalyst was added stepwise to the reaction medium. The bioreductions were then performed at 30 °C in buffer pH 10.0 using an exchange resin, while ADH-A was added in portions over 3–4 days, obtaining excellent enantio- and diastereoselectivities.

8.2.3 Alcohol Dehydrogenases in Multicatalytic Processes

The combination of multiple biocatalysts might be very plausible, since enzymes usually operate under similar reaction conditions concerning pH, temperature, and the solvent medium (i.e. water) [28, 29]. Also, the high specificity frequently shown by enzymes leads to a high control of reactivities, allowing the presence of different substrates, cofactors, intermediates, and catalysts in the same medium. Likewise, the great efforts that are also being made in order to combine enzymes with other chemical catalysts such as organo- and metal-based ones [30, 31], opening the door to possible applications of biocatalysts to completely unexplored transformations, are worth mentioning. In many cases, the product of one reaction acts as the substrate for the next one. This allows rapid generation of complexity, the shifting of reaction equilibria, the elimination of inhibition problems, or the minimization of the decomposition of unstable intermediates. Obviously, from an operational point of view, the application of these methodologies can present several advantages over the classic step-by-step design. As they are usually performed in the same vessel ("one-pot"), there is no need for intermediate purification, and therefore costs, operating time, and waste are reduced, improving atom economy and the overall process yield. ADH-catalyzed reactions can also be coupled with other catalytic procedures in order to design multicatalytic protocols, which can be performed stepwise or in a cascade fashion.

Levulinic acid (**9**) is a synthon for numerous chemical compounds such as (*S*)-γ-valerolactone [32]. For this synthesis, levulinic acid was esterified over an anion exchange resin to ethyl levulinate (**10**). This keto ester was reduced to (*S*)-ethyl-4-hydroxypentanoate (**11**) by CPADH at 30 °C in triethanolamine (TEA) buffer pH 7.0 containing 0.1 mM NAD^+ and a 20-fold excess of IPA, obtaining 95% of the enantiopure alcohol. The bioreduction was also performed at continuous conditions, immobilizing the catalyst in a polypropylene enzyme membrane reaction. After 24 hours, a productivity of 5.6 mg of chiral alcohol/μg of ADH was reached. The obtained (*S*)-ethyl-4-hydroxypentanoate was

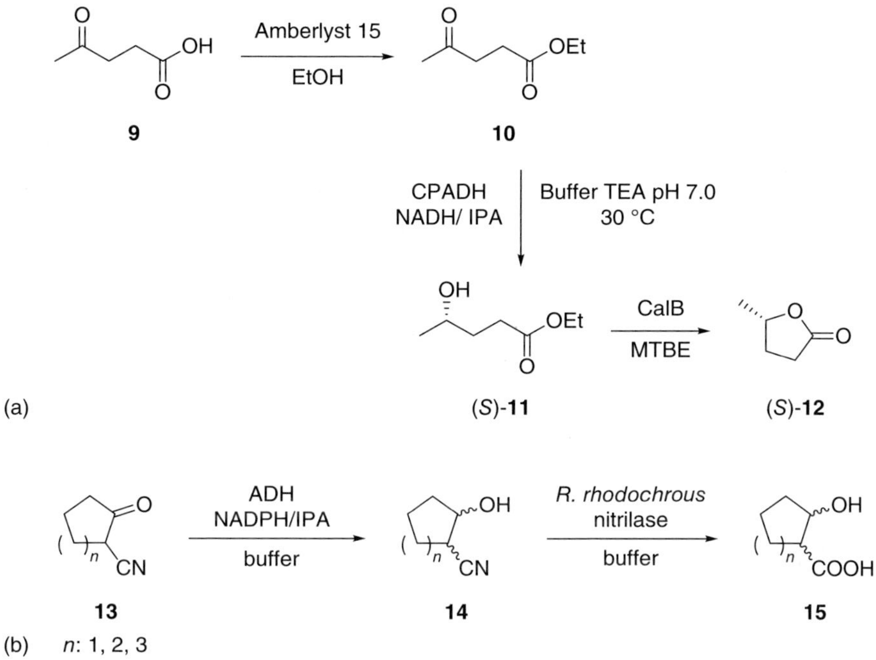

Scheme 8.4 Enzymatic biocascades for the synthesis of (*S*)-γ-valerolactone (**12**) using ADHs and CalB (a) and for the preparation of optically active cyclic β-hydroxy acids (**15**) by combining ADHs and the nitrilase from *Rhodococcus rhodochrous* (b).

converted into (*S*)-γ-valerolactone **12** by *Candida antarctica* lipase B-catalyzed lactonization in *tert*-butyl methyl ether (MTBE) (Scheme 8.4). A quantitative reaction was measured after two hours, with the lipase catalyst being recycled by filtration. An overall yield of 90% was achieved in the three chemoenzymatic steps in order to obtain the enantiopure lactone from levulinic acid.

A biocascade has been recently proposed for the conversion of β-keto nitriles (**13**) into β-hydroxy acids (**15**) in a two-step procedure [33], combining the ADH-catalyzed reduction of the β-keto nitriles to the β-hydroxy nitriles (**14**), which yielded the final products by the action of a nitrilase (Scheme 8.4). Bioreductions were performed using IPA as cosubstrate and all the tested purified ADHs led to complete conversion after 24 hours in a DYRKR process, as the starting materials can be racemized at the reaction conditions. The syntheses, with excellent selectivities, of three out of four of the diastereomers of 2-hydroxycyclopentanecarbonitrile were carried out in buffer pH 5.0 when using KRED-P2-D11 for the (1*R*,2*S*)-*trans*-diastereomer, with KRED-P1-B12 for the (1*R*,2*R*)-*cis*-diastereomer, and with KRED-NADH101 for the (1*S*,2*S*)-*cis*-alcohol. Bioreductions on the cyclohexane and cycloheptane derivatives were carried out at pH 7.0, affording both enantiomers of the *cis*-diastereomer with excellent selectivities. Hydroxy nitriles were completely converted into the hydroxy acids in the presence of the nitrilase from *Rhodococcus rhodochrous* IFO 15564 after 24 hours. Once each biocatalytic step was optimized, a cascade process with both enzymes was developed in phosphate buffer pH 7.0, IPA 5% v/v, and the

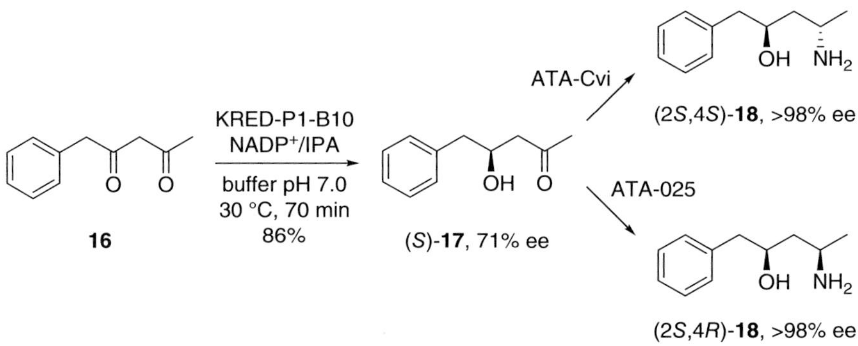

Scheme 8.5 Bienzymatic approach to obtain chiral 1,3-amino alcohols.

biocatalysts. Both cyclohexyl and cycloheptyl keto nitriles were converted to the corresponding hydroxy acids in high yields (>95%) and stereoselectivities. Regarding the cyclopentyl analog, a stepwise process was developed. Thus, the keto nitrile was reduced by KRED-P2-D11 at pH 5.0, and after completion, pH was set to 7.0 and *R. rhodochrous* was added. The (1*S*,2*S*)-hydroxy acid was isolated with 19 : 1 d.r. and 99% ee, but in moderate yield (52%).

Chiral 1,3-amino alcohols are present as structural motif of several biologically active compounds. Recently, a method involving two enzymes (Scheme 8.5) has been described combining KREDs and amine transaminases (ATAs) [34]. The preparation of the four diastereoisomers of 4-amino-1-phenylpentan-2-ol (**18**) was performed by stepwise biocatalytic reactions using this methodology starting from 4-hydroxy-5-phenylpentan-2-one (**17**). After enzymatic screening, (*S*)-**17** was selectively oxidized via kinetic resolution catalyzed by KRED-P1-B10 to the diketone 5-phenylpentan-2,4-dione **16**, remaining (*R*)-**17** with 89% ee at 50% conversion. In order to obtain (*S*)-**17**, the diketone can selectively be reduced with KRED-P1-B10. After 24 hours, 100 mM of **16** led to the final product with complete conversion (86% yield) and 71% ee. Once both enantiomers of **17** were obtained, the treatment of each of them with ATA-Cvi or ATA-025, which showed different stereopreference, afforded the four diastereomers of **18** with high enantioselectivities (>98%).

In 2010, a one-pot procedure for the synthesis of β-hydroxytriazoles from α-haloketones in a cascade fashion with four biocatalytic processes and a subsequent click reaction was described [35]. In the cascade reaction, α-haloketones were reduced to α-halohydrins in a process catalyzed by the ADH from *Thermoanaerobacter* sp., which was expressed in *E. coli* together with a halohydrin dehalogenase, which mediated the ring closure of the halohydrin to an epoxide and the ring opening by azide to form a β-azido alcohol, which reacted in a click-type reaction with an alkyne in a copper-catalyzed cycloaddition. The NADPH was regenerated by IPA–substrate coupling, with the final products being obtained with good yields and excellent enantioselectivities.

The Pd-catalyzed isomerization of alkynols in water led to 2,3-dihydrofurans, which suffered the nucleophilic attack of water to obtain hydroxy ketones such as **20**. The best conversions were achieved after short times using a Pd(II) dimer. Under the reaction conditions, the isomerization can be coupled

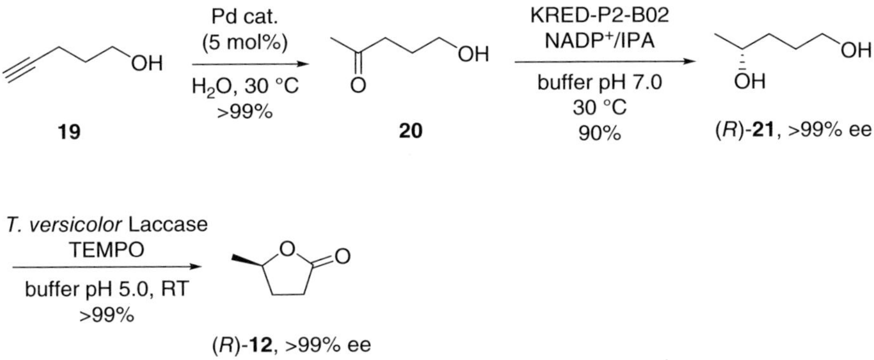

Scheme 8.6 Synthesis of γ-valerolactones combining Pd- and enzyme-catalyzed processes.

with a bioreduction of the hydroxy ketone using KREDs from a commercial source to yield optically active diols in a one-pot procedure [36]. Thus, starting from both pent-4-yn-1-ol (**19**) and 3-pentyn-1-ol, it was possible to obtain both enantiopure (*S*)-pentane-1,4-diol (**21**) using KRED-P2-B03 and (*R*)-**21** (KRED-P2-B02) with yields higher than 90%. A further biocatalyzed step can be included using the laccase from *Trametes versicolor* in the presence of TEMPO as mediator, which oxidized the primary alcohol leading to chiral γ-valerolactone **12** after cyclization. The γ-valerolactones were accessed with excellent optical purities (Scheme 8.6). This methodology was extended to other substrates, such as pent-4-ynoic acid. Levulinic acid obtained in this process was selectively reduced by KRED P1-A04 or KRED P3-H12 to enantiopure (*R*)- or (*S*)-γ-hydroxyvaleric acid, respectively, with excellent conversions.

α-Amino-γ-butyrolactones have been prepared by combining a proline-catalyzed Mannich reaction of an aldimine with acetone, leading to chiral amino keto esters that were reduced to all possible diastereomers of the amino alcohol by properly choosing the ADH [37]. The best results were obtained with ADH-A, yielding the (*S*)-stereoisomers, and with evo-1.1.200, affording the (*R*)-diastereoisomers with ee >99%. Both enzymes required IPA for cofactor recycling (10% v/v). This synthesis was developed at 300-mg scale, yielding the *cis/trans*-lactones in good yields (72–82%) and excellent selectivities (d.r. >98 : 2; ee >99%). The process has also been developed in a one-pot fashion. The starting aldimine was synthesized by mixing *p*-anisidine with ethyl glyoxylate in IPA, which also served as solvent for the Mannich reaction, and as cosubstrate for the reduction of the formed (*S*)-ketone to the corresponding (2*S*,4*R*)-amino alcohol catalyzed by evo-1.1.200, after dilution of the medium with buffer. Transesterification in the presence of HCl/MeOH led to the enantiopure *cis*-(3*S*,5*R*)-lactone in 47% yield and 86 : 14 d.r. In another one-pot approach, the aldimine was employed in the Mannich reaction with acetone as solvent and L-proline as organocatalyst. After 16 hours, the solvent was evaporated and the (*S*)-ketone was selectively reduced by evo-1.1.200 in buffer/IPA to the (2*S*,4*R*)-amino alcohol. Further hydrolysis afforded the enantiopure *cis*-(3*S*,5*R*)-amino lactone in 51% yield.

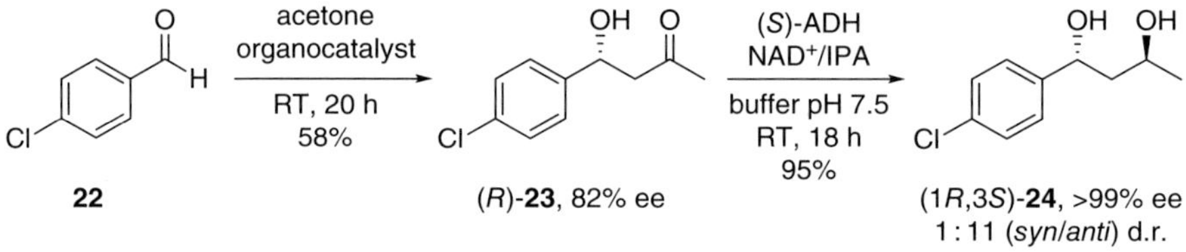

Scheme 8.7 Combination of an organocatalyst with an ADH to obtain chiral 1,3-diols.

In 2009, the sequential one-pot synthesis in aqueous media of chiral 1,3-diols combining a proline-catalyzed organocatalytic aldol reaction between several benzaldehydes and acetone, followed by the ADH-catalyzed bioreduction of the optically active β-hydroxy ketones formed was described (Scheme 8.7) [38]. Initial studies on the β-hydroxy ketone reduction showed that the use of both the (*S*)-ADH from *Rhodococcus* sp. and the (*R*)-LKADH afforded the four diastereoisomers of the 1,3-diol in high yields and excellent selectivities. This system was employed in the one-pot two-step process starting from *p*-chlorobenzaldehyde (**22**) and acetone. (*R*)-**23** was attained using a proline-based catalyst after 20 hours at RT (82% ee). The biocatalyzed reduction by the (*S*)-ADH from *Rhodococcus* sp. was carried out in phosphate buffer pH 7.0 containing 25% v/v of IPA at RT, affording (1*R*,3*S*)-**24** in 95% conversion and >99% ee, 1 : 10 *syn*/*anti* d.r. after 18 hours. This procedure was extended to *meta*-substituted aromatic aldehydes with similar results [39]. In a further achievement, organic media was used in a similar one-pot process for an alternative preparation of chiral 1,3-diols [40]. The synthesis of (1*R*,3*S*)-**24** has recently been performed by combining the (*S*,*S*)-proline organocatalyst and the (*S*)-ADH from *Rhodococcus* sp. in a simultaneous way adding both catalysts at the beginning of the reaction [41]. Conditions of both the aldol reaction and the enzymatic reduction were optimized to develop an efficient process and to minimize side reactions.

8.2.4 Application of Ketoreductases to the Synthesis of Valuable Compounds

Owing to the recent advances in biocatalysis and molecular biology, ADHs can now be considered complementary for the synthesis of valuable molecules, in terms of both costs and operation, to the "classical" catalysts such as boranes or metal complexes bearing phosphine ligands [42].

α-Halohydrins, key synthons in organic synthesis, have been synthesized using ADHs as catalysts in the asymmetric transfer hydrogenation of prochiral α-haloketones. For example, a set of α-fluoroketones have been recently prepared and selectively reduced in the presence of ADHs [43]. Bioreductions were performed in Tris/SO_4 buffer pH 7.5, using IPA or glucose/GDH in order to avoid undesired S_N2 reactions. Overexpressed ADH-A in *E. coli* and LBADH led to the best activities and selectivities. In addition, the biocatalyzed preparation of enantiopure (*R*)-2-bromo-1-(4-fluorophenyl)ethanol (**25**, Scheme 8.8), a valuable precursor in the synthesis of cholesterol absorption inhibitors, was

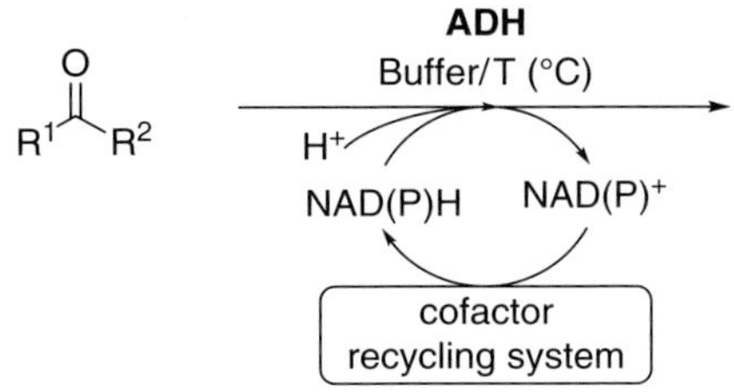

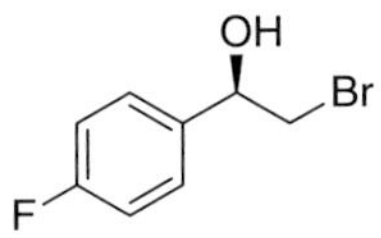

(*R*)-**25**
ADH-A
70% yield, >99% ee

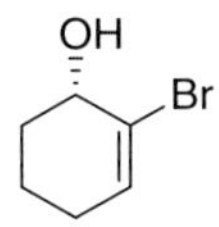

(*S*)-**26**
CRED A-601
88% yield, 99.8% ee

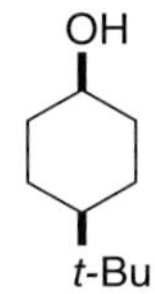

27
CRED A-161
91% yield, >99% d.r.

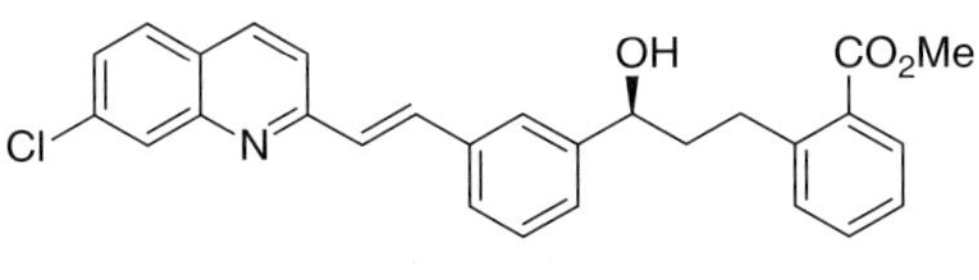

(*S*)-**28**
KRED CDX-026
95% yield, >99% ee

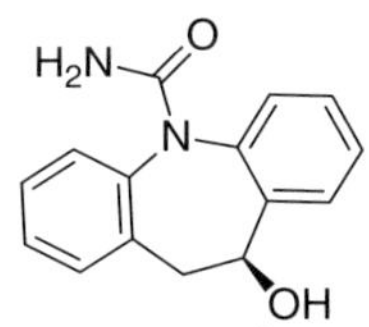

(*S*)-**29**
KRED CDX-021
96% yield, >99% ee

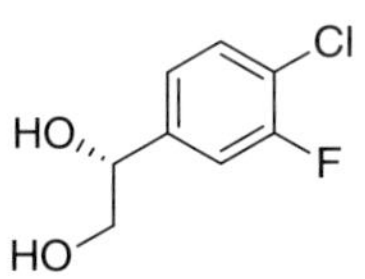

(*R*)-**30**
KRED NADH-112
98% yield, >99% ee

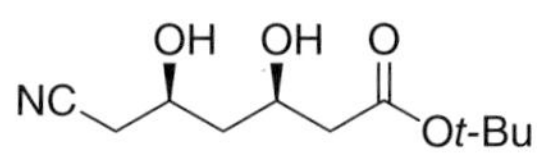

(3*R*,5*R*)-**31**
*Lb*CR
99% conv., >99% ee, >99% d.r.

Scheme 8.8 Examples of chiral alcohols obtained through ADH-mediated reductions.

developed using *E. coli*/ADH-A cells at 260-mg scale with 0.1 M substrate concentration. After purification, the enantiopure bromohydrin was recovered in 70% yield. Some antiretroviral drugs include hydroxyethylene or hydroxyethylamine moieties. ADH-catalyzed bioreduction of N-protected chloroketones to obtain *threo-* or *erythro-*N-protected chlorohydrins was recently described [44]. The best results were achieved with RasADH and SyADH. Both enzymes use IPA for cofactor recycling and are tolerant to organic solvents. After optimization, bioreductions were performed on a 50-mg scale to obtain both isomers of the chlorohydrins in good yields (70–84%) and high d.r. (from 92 : 7 to 95 : 5). (*S*)-Bromo-2-cyclohexen-1-ol (**26**, Scheme 8.8) is an α-halogenated α,β-unsaturated chiral alcohol key intermediate for the synthesis of various biologically active compounds. This derivative was obtained by selective reduction of the prochiral ketone using ADH CRED A-601 (Almac) with complete conversion and 99.8% ee when using glucose/GDH and with slightly lower conversion (95.7%) using IPA as hydrogen donor [45]. The bioreduction was scaled up to 100-g scale by performing an *in situ product removal* (ISPR) approach, in order to overcome the equilibrium problem associated with the coupled substrate approach. Thus, the acetone formed by IPA oxidation was continuously removed from the reactor, making it possible to isolate the chiral alcohol in 88% yield and 99.8% ee after 24 hours.

The O-acetylated derivative of *cis*-4-*tert*-butylcyclohexanol (**27**, Scheme 8.8) is a perfume employed for cosmetics. This alcohol was obtained by the asymmetric reduction of 4-*tert*-butylcyclohexanone employing different KREDs [46]. The best results were attained with the NADH-dependent CRED A-161, which employs IPA for cofactor regeneration. Reaction optimization led to 91% isolated yield (>98% purity) of **27** after 22 hours. To ensure these results, the prochiral ketone (500 g) was dissolved in phosphate buffer pH 7.0 at 45 °C containing 1.5 volumes of MTBE, 30% v/v IPA, 50 g of A161 cell paste, and 1.0 g of NAD^+. Montelukast sodium (Merck) is a leukotriene receptor agonist [47]. The chirality in this compound was introduced by selective hydrogen transfer of a bulky ketone to form the (*S*)-alcohol **28** (Scheme 8.8). As the starting ketone was a hydrophobic compound, high amounts of organic solvent contents were required to perform the bioreduction. Alcohol **28** crystallized in the medium, making the process irreversible. Biocatalyst screening provided some candidates and after rounds of evolution, KRED CDX-026 showed a high activity, which can be increased using toluene as cosolvent. After optimization, the ketone bioreduction was carried out at 100 g/l in a mixture 1 : 5 : 3 v/v/v toluene/IPA/Tris–HCl buffer pH 8.0 at 45 °C in the presence of 3.0 g/l of the ketone for 22–24 hours. The process could be scaled up to a 230-kg scale, allowing for obtaining (*S*)-**28** in yields higher than 95%, 99.9% ee, and 98.5% chemical purity.

Ethyl (*S*)-4-chloro-3-hydroxybutanoate [(*S*)-CHBE] is a valuable intermediate in the synthesis of atorvastatin, a cholesterol-lowering drug. This compound has been prepared from ethyl 4-chloro-3-oxobutanoate (COBE) by biocatalyzed asymmetric reduction using different KREDs. An NADH-dependent CR obtained from *Streptomyces coelicolor* (ScCR) was able to selectively reduce COBE with high activity using IPA for cofactor regeneration in a hydrogen transfer mode [48]. The preparation of (*S*)-CHBE using *E. coli* cells expressing ScCR was implemented at kilogram scale [49]. After optimization of the reaction parameters, the ketone concentration was established at 100 g/l, with an NAD^+ concentration of 0.1 M. The reduction was carried out in a biphasic mixture buffer/toluene 1 : 1 v/v, at 25 °C and pH 6.5, in the presence of IPA (1.5 equiv) as cosubstrate. The bioreduction of COBE catalyzed by ScCR was implemented in a 50-l thermostated stirred-tank reactor, affording (*S*)-CHBE after separation of the aqueous phase and vacuum distillation in 85.4% yield and 99.9% ee.

Wet cells from *Candida krusei* ZJB-09162 were employed for the reduction of 4-hydroxybutan-2-one to (*R*)-butane-1,3-diol [50], a valuable building block for the preparation of pheromones, insecticides, and fragrances. They contained an NADH-dependent CR that was able to produce 19.8 g/l of the diol in 96.6% conversion and 99% ee when working in phosphate buffer 0.1 M, pH 8.5, and 35 °C, using glucose as cosubstrate for cofactor regeneration. Eslicarbazepine acetate is a sodium channel inhibitor used for the treatment of epileptic seizures. It is prepared by O-acetylation of the prodrug (*S*)-licarbazepine (**29**, Scheme 8.8), which is obtained by asymmetric reduction of oxacarbazepine [51]. Several mutants of the KRED from *L. kefir* were able to convert oxacarbazepine into (*S*)-**29** with excellent stereoselectivity. One of the variants was improved by directed evolution in order to find novel mutants that accepted low cofactor loading, high IPA concentrations, and high temperatures. Once the optimal conditions were identified (100 g/l ketone, 1 g/l of CDX-021, 0.1 g/l $NADP^+$, 60% v/v IPA, TEA buffer 0.1 M pH 10.0, and 55 °C), a reaction at 500-ml scale was performed. In this manner, compound **29** was obtained in 96% yield, 98.7% purity, and >99% ee.

Several pharmaceutically relevant intermediates are synthesized from (*S*)-*N*-Boc-3-hydroxypiperidine, which can be prepared from *N*-Boc-piperidin-3-one [52]. Enzyme screening showed that KR-110, an NAD^+-dependent KRED, was able to catalyze the preparation of the alcohol with 99.3% ee. Process optimization led to the enantiopure product in 98.6% conversion by adding the substrate in two 50 g/l batches after four hours, carrying out the bioreduction in buffer pH 7.0 at 25 °C. A 5-g scale reaction under these conditions led to 99.8% conversion in 24 hours, in 97.6% yield and 93% chemical purity. Enantiomers of 1-(3,5-bis(trifluoromethyl)phenyl)ethanol are building blocks for the synthesis of NK1 receptor antagonists. The (*S*)-alcohol was prepared in enantiopure form and high conversion by reduction of the corresponding ketone using an ADH from *Rhodococcus erythropolis* [53], whereas in 2013 the preparation of the (*R*)-enantiomer in the presence of whole cells of *Tricodermia asperellum* ZJPH0810, containing an NADH-selective KRED was described [54]. After optimization of the reaction parameters, the best results were obtained employing two cosubstrates for the cofactor regeneration: IPA (6.0% v/v) and glycerol (0.5% v/v). The bioreduction can be performed in pure distilled water, which represents an advantage for the industrial development of this procedure. A 50 mM concentration of the starting ketone was converted into the (*R*)-alcohol in 93.4% yield and 98% ee. This derivative has also been prepared employing KRED P1B2 from Codexis, using 150 g/l of the ketone in the presence of 1.0% weight of the lyophilized enzyme, 1.0% weight of $NADP^+$, and 30% v/v of IPA in phosphate buffer pH 7.0 at 35 °C. The chiral alcohol was recovered in 99% conversion and ee, but the product extraction from the reaction medium led to enzymatic deactivation. In order to circumvent this drawback, the lyophilized ADH was supported on an amino-epoxy functionalized polymethacrylate resin [55]. The bioreduction was performed at 50-g scale using 50 g/l of ketone and 100 g/l of enzyme in 50/40/10 v/v hexane/IPA/water at 40 °C, leading to the enantiopure (*R*)-alcohol in 98% yield. The enzyme retained 94% of its activity after seven days at 40 °C.

Both *cis*- and *trans*-4-substituted-4-aminocyclohexanols are valuable building blocks in organic and pharmaceutical chemistry, but unfortunately to date no effective routes for their preparation have been described. Therefore, KREDs have been tested in order to synthesize these molecules from the corresponding prochiral ketones [56]. A set of 24 commercial KREDs were employed in the selective reduction of allyl *N*-(1-methyl-4-oxocyclohexyl)-carbamate, observing for most of the enzymes high conversions and complete diastereoselectivities for the *cis* isomer. After optimization, five enzymes were employed for the preparation of *cis*-allyl *N*-(4-hydroxy-1-methyl-cyclohexyl)-carbamate at 100-mg scale, with KRED-P1-B10, a biocatalyst that tolerated high IPA concentrations, being the best candidate. The bioreduction was carried out at 30-g scale, achieving 95% conversion after 24 hours with complete diastereoselectivity. The reaction was further scaled up to 1.0 kg of substrate in a 20-l reactor. After 24 hours, the *cis*-alcohol was recovered by extraction in 99% yield and 98% purity. The study was extended to other analogs, and KRED-P1-B10, KRED-P1-B12, and KRED-P2-D11 showed the best performance for the *cis* isomers. In order to obtain the *trans*-alcohols, both KRED-130 and KRED-NADH-110 afforded good selectivities depending on the substrate structure.

The preparation of the extracellular-regulated kinase inhibitor GDC-0994 has been recently developed on multikilogram scale [57]. This process consisted of a seven-step synthesis starting from 4-chloro-3-fluorostyrene, which was selectively dihydroxylated in 94% ee. In order to improve the optical purity of diol (*R*)-**30** (Scheme 8.8), a novel approach was adopted by carrying out the bioreduction of an α-hydroxy ketone using KREDs. KRED NADH-112, using the glucose/GDH recycling system, produced (*R*)-**30** with complete conversion and >99% ee. Further process optimization at 10% w/v substrate concentration in phosphate buffer pH 7.2 at 30 °C led to 98% yield of the enantiopure diol after 24 hours. (*S*)-2-Chloro-1-(3,4-difluorophenyl)ethanol is a key intermediate for the synthesis of Ticagrelor, employed for the treatment of coronary syndromes. In 2017, a biocatalytic approach for its synthesis was described testing 13 KREDs in the bioreduction of 2-chloro-1-(3,4-difluorophenyl)acetone [58]. KR-01 showed the best catalytic performance, obtaining a complete conversion after 18 hours in a 200-ml scale with a substrate concentration of 500 g/l, 0.05 g/l NAD^+, 50 g/l of KRED, 3.5 equiv of IPA in phosphate buffer 0.1 M pH 6.0. After removing IPA and acetone by evaporation and product separation by extraction, 96% yield of the product was achieved with a purity of 98.9% and 99.9% ee. *tert*-butyl (5*R*)-6-cyano-5-hydroxy-3-oxohexanoate can be selectively reduced into the corresponding (3*R*,5*R*)-diol (**31**, Scheme 8.8) [59], employed in the synthesis of the side chain of atorvastatin, a cholesterol-lowering drug. Different KREDs were evaluated, obtaining the highest activities with the KRED *Lb*CR from *L. brevis* expressed on *E. coli*. This biocatalyst accepted a broad range of bulky substrates and was employed at gram scale for the asymmetric reduction of the keto ester using lyophilized *E. coli*/p*Lb*CR cells coupled with GDH/glucose. Reductions were carried out in phosphate buffer pH 6.0. After 16 hours, 300 g/l of substrate was completely reduced in >99 : 1 d.r.

(*S*)-5-Fluoro-3-methylisobenzofuran-1(3*H*)-one (**36**, Scheme 8.9) is a valuable intermediate in the preparation of lorlatinib, a biologically active compound with anticancer properties. This compound can be obtained via carbonylation reaction or lactonization procedure [60]. Regarding the first procedure, the key step was the selective reduction of a prochiral ketone **32**, and it was found that the 2,4-diketogluconic acid (DkgA) enzyme was able to reduce the ketone in 98% extent with complete selectivity. The NADPH cofactor was recycled using IPA as cosubstrate and LBADH as secondary enzyme. The optimal conditions were established in enzyme/substrate ratio 0.06/0.07 w/w, pH 6.5–7.0, 35–40 °C,

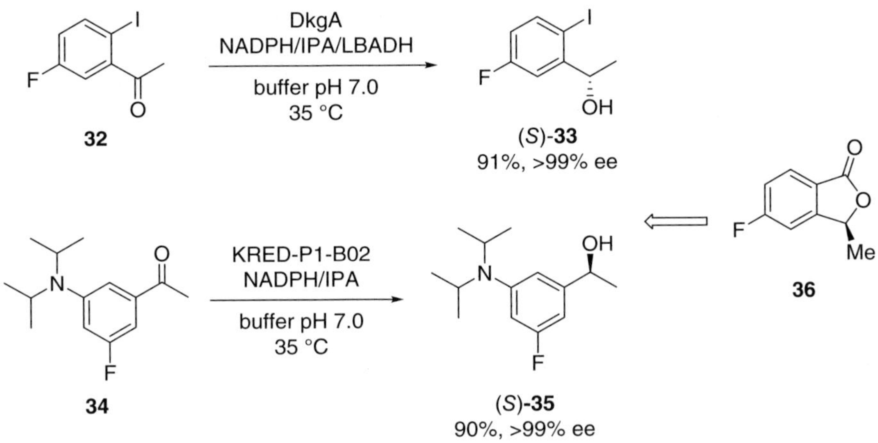

Scheme 8.9 Biocatalytic hydrogenations of (*S*)-5-fluoro-3-methylisobenzofuran-1(3*H*)-one (**36**) precursors employing alcohol dehydrogenases.

and IPA/substrate 1.0–1.2 w/w and were applied to prepare 350 kg of the product (*S*)-**33** with 91% yield. The reaction reached complete conversion and the product was easily isolated by precipitation and recrystallization. Regarding the lactonization procedure, the reduction of the intermediate ketone was performed using different ADHs, obtaining the best results with KRED-P1-B02 (Codexis) and evo 1.1.440 (Evocatal). Chiral alcohol (*S*)-**35** (Scheme 8.9) was obtained in 90% isolated yield and 99% ee from ketone **34**.

8.3 Ene-Reductases

These enzymes (EREDs, E.C. 1.6.99.1) are able to catalyze the asymmetric hydrogenation of C=C double bonds, thus affording the reduced compounds with up to two new chiral centers (Scheme 8.1) [61, 62]. In most cases, they are flavin-dependent proteins, and due to this fact they belong to the Old Yellow Enzyme family, as flavin provides this color to the biocatalyst. While the addition of the hydride to the alkene comes from this molecule, it must be regenerated by the action of, e.g. a nicotinamide cofactor. Therefore, as in the case of ADHs, efficient cofactor recycling systems have been developed to use it in catalytic amounts. The most widely employed methodology is the coupled enzyme approach, using glucose/GDH, formate/formate dehydrogenase, or IPA/alcohol dehydrogenase to recycle the reduced form of the nicotinamide cofactor. However, there are examples of coupled substrate approach where the ERED by itself is able to perform the reduction of the alkene compound at the expense of the oxidation of a cosubstrate such as 3-methylcyclohex-2-enone [63] or *N*-Boc-pyrrolidinone [64]. All these examples are hydrogen transfer reductions. More recently, light-driven oxidation of alternate hydride donors for flavin mononucleotide (FMN) regeneration has emerged as an alternative to nicotinamide-dependent protocols. For instance, the oxidation of water using

TiO_2-semiconductor as photocatalyst [65], or the employment of cyanobacteria whole cells, based on natural photosynthesis [66], can recycle NADPH. Finally, inexpensive nicotinamide biomimetics (mNADHs) also can be used as redox equivalents for the development of various ERED-mediated transformations [67, 68].

8.3.1 Substrate Scope of Ene-Reductases

EREDs can promote the reduction of alkenes (usually in a *trans* fashion) that are conjugated with activating (electron-withdrawing) groups. Thus, among the different substrates converted at small scale, (cyclic) enones [69–71], α,β-unsaturated (di)esters [72–74], β-cyano-α,β-unsaturated esters [75], nitroalkenes [76–78], α,β-unsaturated nitriles [79, 80], α,β-unsaturated lactones [81], α,β-unsaturated carboxylic acids [82], and β-activated vinylphosphonates [83] can be mentioned. In these contributions, EREDs from the Old Yellow Enzyme family were used, utilizing stoichiometric amounts of cofactor or the coupled enzyme approach to regenerate NAD(P)H. The transformations were usually performed at temperatures close to RT and at pHs close to the neutral value, demonstrating the mildness of this methodology. In some specific examples the addition of an organic cosolvent [72, 82] or an ionic liquid [79] was studied in order to improve the solubility of hydrophobic substrates, and for one substrate even a dynamic kinetic protocol could be successfully achieved [81].

This enzymatic approach has already been applied to the selective synthesis of highly valuable compounds at preparative scale. (*R*)-3-Hydroxy-2-methylpropanoate, commonly denoted as the "Roche ester," is a popular chiral building block for the synthesis of vitamins, fragrances, and antibiotics. In 2010, the synthesis of this compound was described via ERED-catalyzed reduction of several O-protected derivatives of methyl 2-hydroxymethylacrylate. After enzymatic screening, xenobiotic reductase A (XenA) from *Pseudomonas putida* afforded complete conversion toward the enantiopure compound with the hydroxyl group protected with the *tert*-butyldimethylsilyloxy (TBDMS) group (**37**, Scheme 8.10). Using glucose and GDH as cofactor recycling system in Tris/HCl buffer 50 mM pH 7.5 at 30 °C, after 24 hours, 10 mM of the substrate was fully converted [84].

Lilial™ and Helional™ (**38**, Scheme 8.10) are olfactory aldehydes present in fragrances. Faber and coworkers designed an approach from different cinnamaldehyde derivatives performing their bioreduction mediated by different EREDs [85]. After optimization of the enzymatic conditions, it was found that OYE3 from *Saccharomyces cerevisiae* could afford the aldehyde (*S*)-**38** (10 mM) with complete conversion and ee in a mixture of Tris/HCl buffer 50 mM pH 7.5 with MTBE (80 : 20 v/v) at 30 °C after 24 hours in the presence of glucose/GDH as cofactor recycling system.

2-Arylpropionic acids are an important class of nonsteroidal anti-inflammatory drugs. In 2012, an enzymatic approach to synthesize enantiomerically pure (*R*)-profen derivatives was achieved using YqjM, an ERED from *Bacillus subtilis*. After synthesizing the α,β-unsaturated ester precursors, enzymatic transformations were performed at 5 mM substrate concentration in a mixture of

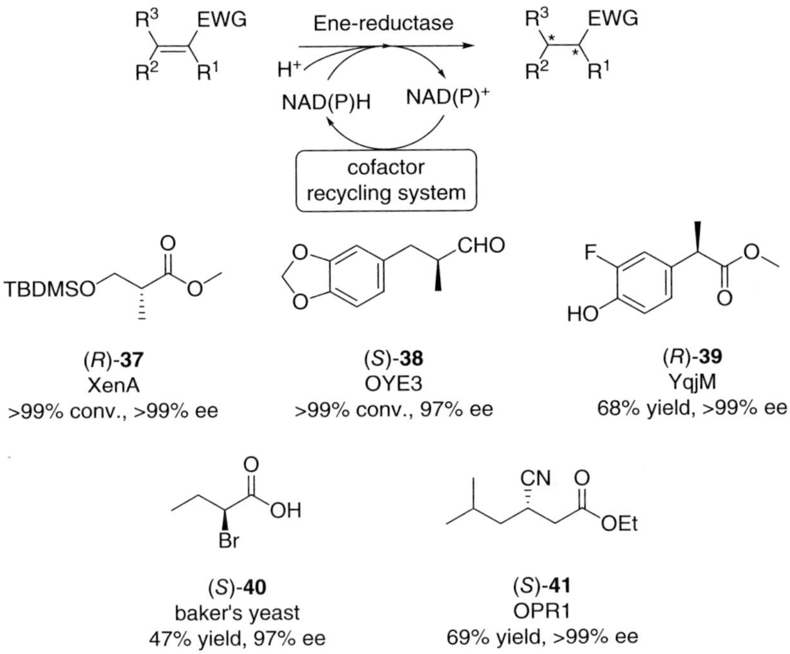

Scheme 8.10 Examples of chiral compounds obtained through ERED-mediated reductions.

phosphate buffer 50 mM pH 7.0 and 2-methyltetrahydrofuran (2-MeTHF, 1% v/v) using glucose/GDH to recycle the cofactor at 25 °C. In some cases, after 24 hours the reduced (*R*)-derivatives were obtained with complete conversion and in enantiopure form. Even a transformation at 50-mg scale to obtain a precursor (**39**, Scheme 8.10) of (*R*)-flurbiprofen was accomplished, obtaining the enantiopure compound in 68% yield [86].

ERED-catalyzed reactions have also been utilized to produce methyl (*S*)-2-bromobutanoate used as building block for several active pharmaceutical ingredients [87]. Starting from the corresponding crotonate derivatives, baker's yeast and OYEs1–3 from *Saccharomyces pastorianus* (OYE1) and *S. cerevisiae* (OYEs2–3) afforded the desired compound in excellent conversions (up to >99%) and ee (97%). Depending on the *Z*- or *E*-alkene stereochemistry different results were observed. The reaction with baker's yeast could be scaled up at 0.05 mol of the substrate (25 mM) using tap water with ethanol (0.5% v/v) and adding glucose to regenerate the cofactor. After 72 hours at 30 °C the final products could be obtained in moderate yields (30–48%) and high ee (85–97%) including the carboxylic acid precursor (**40**, Scheme 8.10) of methyl (*S*)-2-bromobutanoate.

Pregabalin, [(*S*)-3-aminomethyl-5-methylhexanoic acid], belongs to the class of drugs known as γ-aminobutyric acid analogs. A library of β-cyanoacrylate esters were reduced using different EREDs. After screening, various reactions were achieved at 10-mmol scale (100 mM) using OYE2 overexpressed in *E. coli* and IPA and LBADH to recycle NADPH. Using phosphate buffer 100 mM pH 7 at 30 °C after 24 hours, various derivatives were isolated in moderate to high yields (60–83%) and excellent ee (>99%) [88]. In a related approach, a precursor of pregabalin (**41**, Scheme 8.10) was obtained using OPR1 from *Lycopersicon esculentum* at 5-mmol scale in 69% yield and in enantiopure form [89].

8.3.2 Ene-Reductases in Multicatalytic Processes

It has been envisaged that the combination of an ERED and an ADH should lead to both the reduction of an α,β-unsaturated carbonyl derivative and the theoretical introduction of chirality in up to three contiguous stereogenic centers at the same time in a cascade fashion (Scheme 8.11). The main problem of these protocols is the competition for the ADH between the unsaturated substrate and the saturated one, which can provide the unsaturated alcohol as a by-product that cannot be reduced by the ERED. Among the various examples that can be found in Refs. [61, 62], the reduction of several α-substituted cinnamaldehyde derivatives combining baker's yeast and HLDH provided the final alcohol

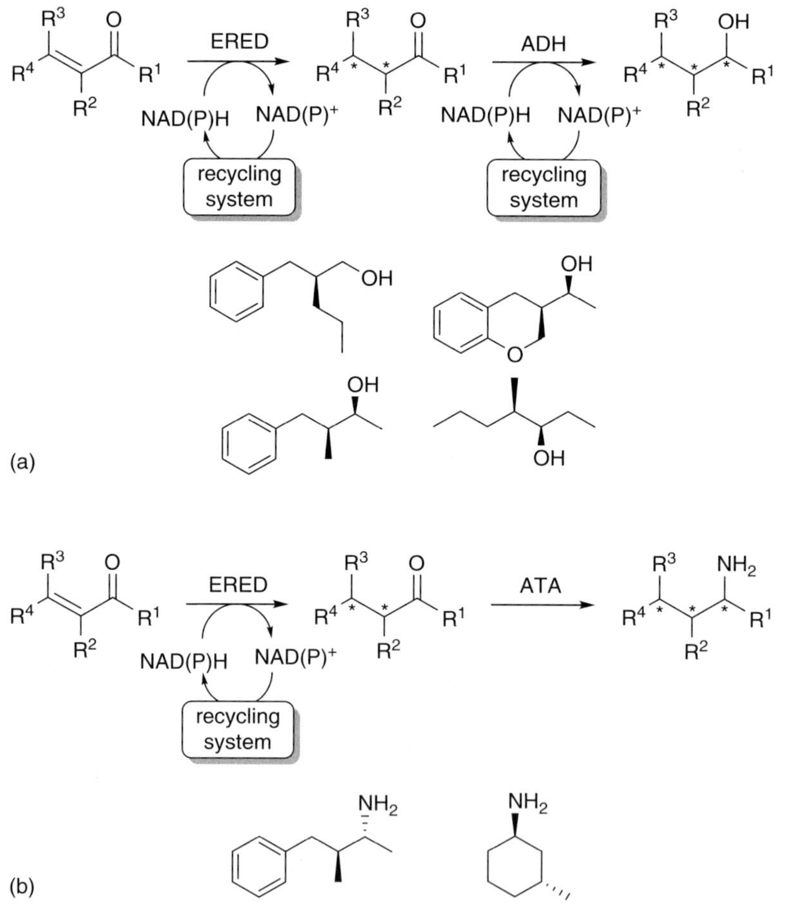

Scheme 8.11 Combination of an ERED and an ADH to obtain chiral alcohols (a) and combination of an ERED and an ATA to obtain chiral amines (b).

derivatives (2.0 g/l) at high yield (>80%) and high ee values (>90%) after 12 hours at 30 °C [90]. In a subsequent contribution, starting from the corresponding α,β-unsaturated aldehydes or ketones (5.0 g/l), and combining EREDs with different ADHs, the final alcohols, precursors of tetralin and chroman-based drugs such as Robalzotan, Ebalzotan, and Rotigotine, could be attained with usually very high conversions (>85%) and selectivities (ee >90%) after 12 hours at 30 °C [91]. This cascade was also applied to obtain two out of four isomers of odorant Muguesia®, using OYE3 and two stereocomplementary ADHs from *R. erythropolis* and *Parvibaculum lavamentivorans* [92]. Recently, this methodology was also applied to synthesize the four isomers of 4-methylheptan-3-ol, an insect pheromone, using complementary EREDs (OYE2.6 from *Pichia stipitis* and variant OYE1-W116V) and commercial ADHs, obtaining the final compounds (8 mM) in high yields (72–83%) and excellent ee (>98%) and d.r. (>96 : 4) [93]. The application of this cascade protocol to obtain a series of 2-methyl-3-substituted tetrahydrofurans starting from the corresponding α-bromo-α,β-unsaturated ketones has also been demonstrated [94]. After the synthesis of the bromohydrin intermediates, two different chemical approaches afforded the final compounds. In all these examples, the system glucose/GDH was selected to recycle the nicotinamide cofactor employed by both the ERED and the ADH.

This combination was also utilized to synthesize chiral lactones. Starting from linear α,β-unsaturated γ- or δ-keto esters, the sequential selective bioreduction of the double bond mediated by an ERED, followed by the transformation into the hydroxy ester using an ADH, was envisaged. Then, the spontaneous (or acid-catalyzed) intramolecular attack of the alcohol to the ester was highly favored, providing the final derivatives. Among the different compounds that were accessed through this methodology, 3,4-, 3,5-, and 4,5-disubstituted γ-butyrolactones [95], and all isomers of *Nicotiana tabacum* lactone can be underlined [96].

The combination of transaminases with EREDs (Scheme 8.11) starting from α,β unsaturated carbonyl derivatives has also been investigated to obtain enantioenriched amines with several chiral centers. Various ATAs and EREDs were used to selectively reduce a series of α,β-unsaturated ketones (10 mM). These transformations were done at pH 7.0 and 30 °C, and after 18 hours the saturated amines were produced in high conversions (>70%) and d.r. (>92.5 : 7.5). To recycle the cofactors, glucose/GDH (for ERED) and isopropylamine (for ATA) were utilized [97]. Bornscheuer and coworkers have designed a similar cascade to obtain the different diastereoisomers of 1-amino-3-methylcyclohexane combining ERED from *Saccharomyces carlsbergensis* with different variants from *Vibrio fluvialis* ATA (Vf-TA). Two out of four [(1*R*,3*S*) and (1*S*,3*S*)] stereoisomers could be obtained with high diastereomeric excess (>80%) in the presence of *N*,*N*-dimethylformamide (DMF) or dimethylsulfoxide (DMSO) as organic cosolvents [98]. In a recent update, the (1*R*,3*R*)-isomer of this compound could be synthesized by a combination of a variant of ERED from *B. subtilis* and a variant of ATA from Vf-TA under similar reaction conditions [99].

Redox neutral cascades have been designed linking two oxidoreductases, one in an oxidation mode and the other in a reduction manner, just requiring catalytic amounts of NAD(P) as hydride shuttle between both processes. Hence,

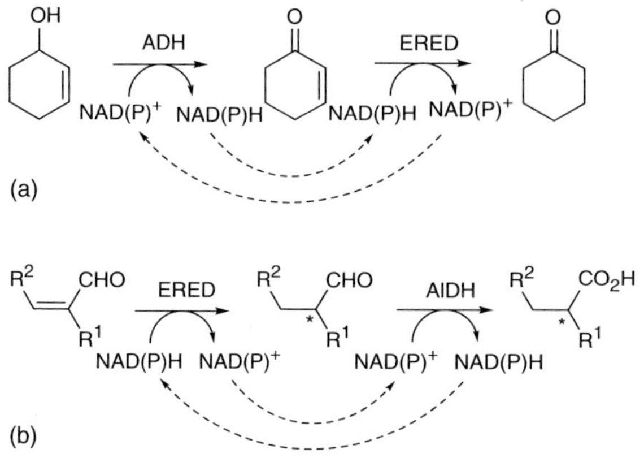

Scheme 8.12 Redox neutral cascades combining an ERED and an ADH to isomerize allylic alcohols (a) and an ERED and an AlDH to obtain chiral carboxylic acids (b).

the isomerization of allylic alcohols to saturated ketones employing an ERED in combination with an ADH to obtain cyclohexanone from cyclohex-2-enol has been described (Scheme 8.12) [100]. This transformation was further extended by adding a Baeyer–Villiger monooxygenase (BVMO), which oxidized the cyclic ketone to form the corresponding lactone [101]. Another example was developed by Scrutton and coworkers, coupling an ERED with an aldehyde dehydrogenase (AlDH, Scheme 8.12). Starting from substituted α,β-unsaturated aldehydes (5 mM), the ERED-catalyzed stereoselective bioreduction of the double bond was achieved, followed by the oxidation of the aldehyde into the corresponding carboxylic acid mediated by the AlDH. The reactions were carried out at pH 7.0 and 30 °C, and the final compounds were obtained in excellent conversions (>80%) and ee (>95%) [102].

Very recently, the isomerization of allylic alcohols to form (chiral) saturated ketones has also been possible, combining a chemoenzymatic oxidative system (a laccase with the radical oxidant TEMPO) with an ERED [103, 104]. Owing to the different reaction conditions, these protocols were designed in a sequential mode in the same pot.

In 2011, a Wittig reaction between aromatic aldehydes and ylides was coupled to an ERED-mediated transformation to obtain several saturated ketone derivatives. The enzyme employed was ERED from *Gluconobacter oxydans* and the synthesis of 4-(4-nitrophenyl)butan-2-one could be performed in a cascade mode at preparative scale (0.5 mmol) in phosphate buffer 50 mM pH 7.0 using glucose/GDH to recycle the cofactor. After 67.5 hours, this compound was isolated in 89% yield [105]. Later, some contributions related to the combination of EREDs with organo- or metal-catalyzed processes have also been described as for the case of ADHs. A novel synthetic approach toward Guerbet alcohols, which are used as plasticizers, lubricants, or surfactants, was demonstrated coupling organo- and biocatalyzed steps [106]. Starting from hexan-1-ol, a chemical oxidation (NaOCl, NaBr, and TEMPO) quantitatively afforded hexanal, which in a subsequent aldol reaction promoted by lysine as organocatalyst gave the corresponding enal. Later, ERED from *G. oxydans* reduced the C=C double

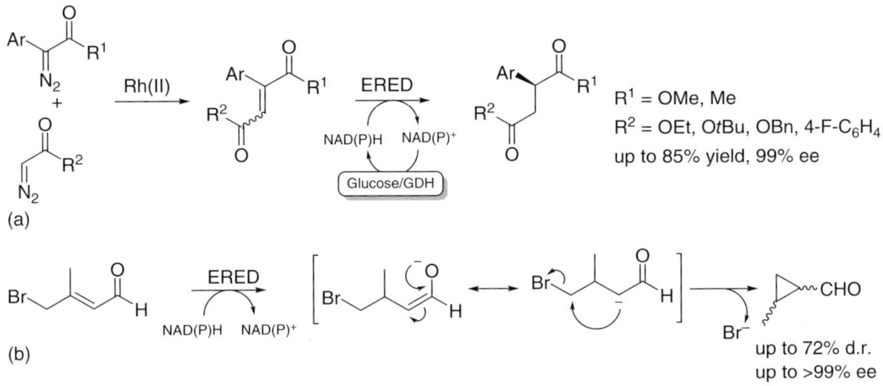

Scheme 8.13 Combination of a Rh(II)-catalyzed cross-coupling of diazo compounds with an ERED (a) and asymmetric reductive cyclopropanation promoted by an ERED (b).

bond providing a saturated aldehyde, which in the last step was reduced by ADH from *Rhodococcus* sp. to yield the final Guerbet alcohol. Owing to the different reaction conditions, the overall transformation was achieved in a sequential mode but in the same reactor. This synthetic route could be scaled up at 8 mmol of hexan-1-ol, delivering the final derivative in 62% yield. A combination between the Rh(II)-catalyzed cross-coupling of two diazo compounds and an ERED hydrogenation was recently demonstrated (Scheme 8.13) [107]. Owing to the different noncompatible reaction conditions, the protocol must be run sequentially. In the first step, several α,β-unsaturated diesters or keto esters were obtained with high *E*-selectivity using the complexes $[Rh_2(Oct)_4]$ or $[Rh_2(OPiv)_4]$ in CH_2Cl_2 at −78 °C. After one hour, the solvent was evaporated and then the ERED, the cofactor, glucose, and GDH were added in phosphate buffer 200 mM pH 7.5 in the presence of DMSO (2.5% v/v). YersER from *Yersinia bercovieri*, OPR1 from *L. esculentum*, and OYE2 from *S. cerevisiae* usually showed the best results in terms of activity and stereoselectivity toward the final 1,4-diesters or 1,4-keto esters.

Special mention needs to be made of a very recent contribution from Breinbauer and coworkers that shows a cascade initiated by the action of an ERED (Scheme 8.13) [108]. In this case, several 4-halobut-2-enals were synthesized and tried as possible substrates for a series of EREDs. As predicted by the mechanism known for these enzymes, the formation of an enolate intermediate could afford the internal cyclopropanation by nucleophilic displacement of the halogen atom. This example can be envisaged as a biocatalytic example of a Michael-initiated ring closure (MIRC) reaction, a general strategy in organic synthesis to construct (hetero)cycles [109, 110]. Moreover, after protein engineering in the enzyme-active site, different variants of OPR3 and YqjM from *B. subtilis* produced the chiral cyclopropane derivatives as the only product. While ee and d.r. values were still moderate, this article shows the tremendous potential that engineered biocatalysts can function as initiators for C–C forming reactions.

8.4 Imine Reductases

These enzymes (IREDs, E.C. 1.5.1.x) are responsible for the transfer hydrogenation of imines to the corresponding amines (Scheme 8.1) [111–114]. They are involved in many natural processes such as the biosynthesis of cofactors, alkaloids, and cyclic amino acids. IREDs are NAD(P)-dependent enzymes. Obviously, the already mentioned cofactor regeneration systems for ADHs and EREDs are also applicable for these biocatalysts. They work in aqueous medium at neutral or basic pHs under mild conditions, but they can also be active in the presence of high extents of organic solvents [115].

Owing to the hydrolytic instability of imines in water, the first applications using IREDs were described with cyclic imines, which are more stable than the acyclic analogs. Thus, among the typical products obtained (Figure 8.1), pyrrolidines [116], β-tetrahydrocarbolines [115, 117], 1,2,3,4-tetrahydroisoquinolines [115, 117–121], piperidines [118, 119, 122], azepanes [119, 122], indolines [123], dibenz[c,e]azepines [124], 3,4-dihydro-2*H*-1,4-benzothiazines [125], and 3-thiazolidines [125] can be mentioned. These biotransformations are usually performed at 5–100 mM substrate concentration, at pH 7.0–7.5, and 25–30 °C, using the glucose/GDH system to recycle the nicotinamide cofactor, providing in most cases the reduced products in excellent conversions and ee. Moreover, these studies showed that it was possible to find two stereocomplementary IREDs to access both product enantiomers. In some cases [119, 120, 122–124], scale-up protocols (from 100-mg up to 1-g scale) were successfully developed, demonstrating the tremendous potential of this enzymatic approach.

It has also been demonstrated that these biocatalysts can perform reductive aminations in the presence of a huge excess of the amine regarding the carbonyl

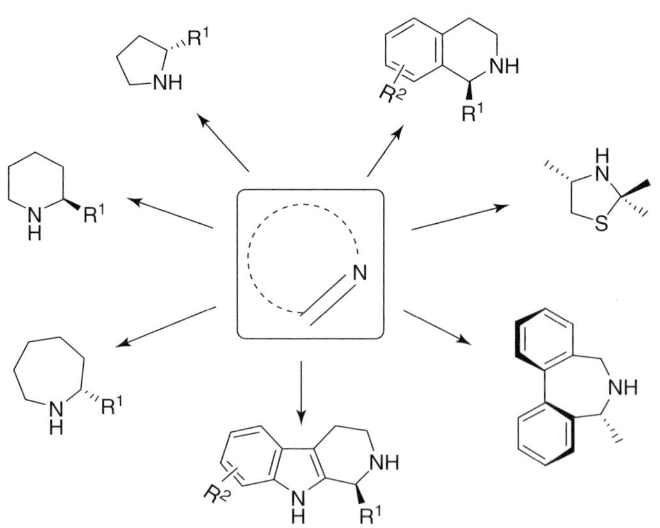

Figure 8.1 Examples of chiral cyclic amines that can be obtained using IREDs.

compound at basic pHs in order to form unstable imines in relatively high concentrations [126, 127]. Very recently, it has been discovered that a subclass of these enzymes, the so-called reductive aminases, were not only able to catalyze the imine reduction but also the imine formation [128]. In these cases it was possible to form the final amines even employing equimolar amounts of the carbonyl compound and the amine, leading to an overall reductive amination protocol, which is a powerful synthetic transformation. This methodology opens the door to synthesizing chiral secondary and tertiary amines from easily accessible ketone substrates.

8.5 Carboxylic Acid Reductases

These biocatalysts (CARs, E.C. 1.2.1.30), also called carboxylate reductases, can catalyze the reduction of carboxylic acids to form aldehydes under very mild conditions (Scheme 8.1) [129, 130]. This transformation is very useful as in classic organic synthesis, the use of reducing agents, even at stoichiometric amounts, leads to the undesired overreduction of the formed aldehyde into the corresponding alcohol. Moreover, extreme reaction conditions such as very low temperatures are necessary, and even under those conditions by-products are still obtained. While the reducing equivalents come from the nicotinamide cofactor, the mechanism is more complicated and still is not fully understood. In this case, adenosine triphosphate (ATP) is also involved, releasing after the end of the reaction adenosine monophosphate (AMP) and pyrophosphate (PPi).

While they were discovered a few decades ago, CARs were not used for synthetic applications until very recently. Thus, they can reduce benzoic acid and heteroaromatic derivatives, e.g. to provide vanillin [131], but also aliphatic carboxylic acids to afford the corresponding aldehydes. In some cases, when these enzymes are overexpressed in a host microorganism, the synthesis of the final alcohols is observed due to the action of endogenous alcohol dehydrogenases that can reduce the aldehyde [132, 133]. Overexpressing in *E. coli* the CAR from *Mycobacterium avium* K-10 and an aldehyde reductase from *E. coli*, a series of diols were obtained from the corresponding diacids or hydroxy acids in high yield. Propane-1,2-diol (**42**, Scheme 8.14), a bulk industrial chemical, was synthesized at 7.0–9.6 mM concentration using these engineered cells [134].

CARs have also been applied in sequential or cascade protocols. For instance, in 2015 the bioreduction of a series of aromatic and aliphatic carboxylic acids mediated by CAR from *Mycobacterium* sp. was coupled to a Wittig transformation to synthesize various α,β-unsaturated esters in a two-step reaction [135]. Also, the application of a CAR combined with a phenyl ammonia lyase and an ADH allowed the synthesis of cinnamyl alcohol starting from easily accessible L-phenylalanine [136]. The synthesis of enantiomerically pure mono- and disubstituted piperidines and pyrrolidines was achieved using a biocatalytic cascade involving a CAR, an ATA, and an IRED (Scheme 8.15) [137]. Starting from δ-keto acids **43**, the final compounds were accessed with high enantio- and diastereoselectivities by (i) formation of the corresponding δ-keto aldehydes **44** catalyzed by CAR; (ii) amination mediated by ATA into the amino ketones that spontaneously

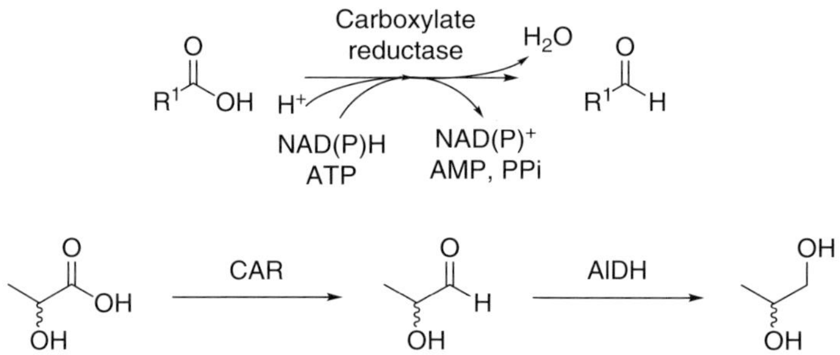

42

Scheme 8.14 Synthesis of propane-1,2-diol overexpressing a carboxylic acid reductase and an aldehyde reductase.

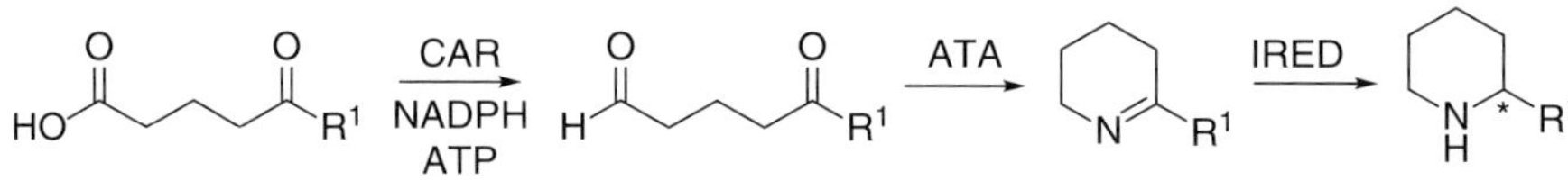

43 44 45 46

Scheme 8.15 Combination of a CAR, an ATA, and an IRED to obtain chiral 2-substituted piperidines **46**.

cyclized into the cyclic imines **45**; and (iii) stereoselective reduction of the imines catalyzed by IRED. Using 5 mM substrate concentration, different 2-substituted piperidines (**46**) were obtained with excellent conversions (>90%) and high selectivities (>90%) after 24 hours at pH 7 and 30 °C.

8.6 Emerging Enzymes: Nitrile Reductases and Nitroreductases

Nitrile reductases or nitrile oxidoreductases (NReds) catalyze the four-electron reduction of nitriles to afford primary amines (Scheme 8.1) [138]. For instance, 7-cyano-7-deazaguanine reductase is involved in the biosynthesis of the hypermodified nucleosides present in transfer RNA. Typical synthetic approaches for nitrile reduction involve either the use of various transition metal catalysts or the addition of stoichiometric amounts of metal hydride; therefore, the development of a biocatalytic alternative may certainly bring about new synthetic approaches either in regioselective reductions or in asymmetric synthesis. NReds from *Geobacillus kaustophilus* and several variants were used with the natural substrate 7-cyano-7-deazaguanine (**47**, Figure 8.2) and with other nitrile derivatives. Among them, substrates **48** and **49** (Figure 8.2) could also be accepted by this enzyme [139].

The reduction of nitro derivatives is usually accomplished employing metals under acidic conditions; hence, the use of enzymes can also be advantageous. The biodegradation of nitroaromatics usually proceeds via nitro group reduction, catalyzed by flavin-dependent nitroreductases [140]. Type I enzymes mediate several sequential two-electron reductions at the expense of NAD(P)H. In the case that three additions are performed, the nitro compound will be

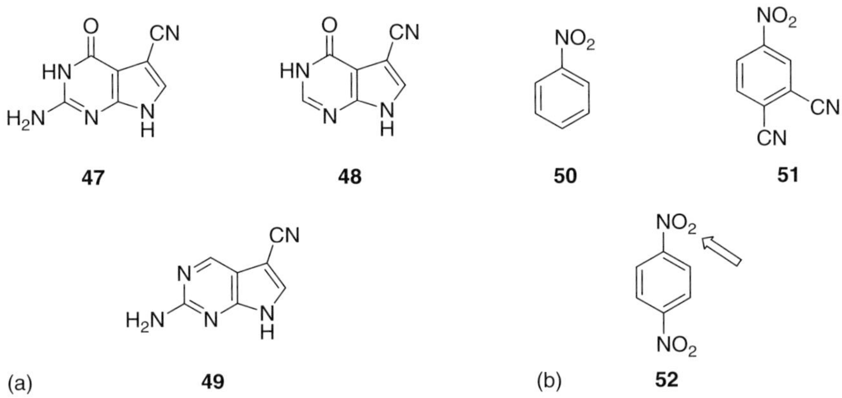

Figure 8.2 Examples of substrates for nitrile reductases (a) and nitroreductases (b).

completely reduced to form the corresponding amine, as in the case of nitrobenzene (**50**, Figure 8.2) to form aniline employing nitroreductase from *Salmonella typhimurium* [141]. However, this transformation delivered only low conversions due to incomplete reduction (giving the corresponding nitroso and hydroxylamine derivatives), and also the azoxybenzene originating from the coupling of the last two compounds. Using baker's yeast, the chemoselective reduction of aromatic nitro compounds bearing electron-withdrawing groups (e.g. derivative **51**, Figure 8.2) gave the corresponding hydroxylamines with good to excellent conversions in aqueous medium at 30 °C and utilizing glucose to recycle the nicotinamide cofactor, although the corresponding anilines were usually detected at only small amounts (<30%) [142]. More recently, a nitroreductase from *Bacillus amyloliquefaciens* (*Ba*NTR1) was also employed in the same transformation affording the hydroxylamines in excellent conversions under mild conditions. Even in the case of dinitro compounds such as **52** (Figure 8.2), the only product observed was the mono-reduced hydroxylamine [143].

8.7 Summary and Outlook

Among the various chemical protocols that have been developed to selectively reduce chemical compounds in a catalytic manner, reductive transformations such as metal-catalyzed hydrogenations, hydrogen transfer reductions, or hydrosilylation reactions are among the best studied. These methodologies have been developed to a great extent and, in many cases, offer very good results in terms of selectivity and catalytic efficiency. However, these strategies still present some drawbacks mainly from an environmental point of view due to the employment of harmful reagents and harsh reaction conditions.

Biocatalysis can be an appealing alternative due to the exquisite chemo-, regio-, and stereoselectivities obtained through mild and environmentally friendly protocols. Apart from catalyst engineering using molecular biology tools, it is undeniable that tremendous progress has been made to produce more robust biocatalytic processes enhancing their efficiency, thereby minimizing waste and costs. Also, the design of suitable reaction conditions (e.g.

adding organic cosolvents or developing cofactor recycling systems), bioprocess engineering (e.g. type of reactor or stirring), and novel catalyst formulations (e.g. immobilization) can suppress the formation of by-products, maximizing the biocatalyst stability and productivity. The use of genetically modified over-expressed biocatalysts is also enhancing the production of the enzymes, thus greatly improving their availability. Hence, the knowledge accumulated so far is resulting in a true alternative for industry where chemicals can be prepared not only in the laboratory scale (milligram/gram) but also at industrial scale (kilogram/ton), using a biocatalyzed process as key step.

On the other hand, the need for more sustainable processes based on highly efficient transformations has turned the attention toward the development of systems where the combination of several (bio)catalytic steps can be performed in the same vessel. These strategies provide multistep catalytic cascade processes that avoid the isolation of the corresponding intermediates, shorten reaction times, and reduce the energy required and waste generated. Moreover, the concentration of unstable intermediates and the formation of secondary reactions are also decreased in these protocols. Especially, the concurrent employment of different enzymes in cascade processes seems to be straightforward as they can perform their natural reactions under very similar conditions.

In the case of biocatalytic (hydrogen transfer) hydrogenations, the bioreduction of carbonyl and alkene derivatives has already been implemented at industrial scale in several cases using ADHs and EREDs, respectively. It is envisaged that IREDs and CARs will be applied at large scale in the next few years, while nitrile reductases and nitroreductases still need to be further developed.

Abbreviations

2-MeTHF	2-methyltetrahydrofuran
ADH	alcohol dehydrogenase
ADH-A	alcohol dehydrogenase from *Rhodococcus ruber*
AlDH	aldehyde dehydrogenase
AMP	adenosine monophosphate
ATA	amine transaminase
ATP	adenosine triphosphate
BVMO	Baeyer–Villiger monooxygenase
CHBE	ethyl 4-chloro-3-hydroxybutanoate
COBE	ethyl 4-chloro-3-oxobutanoate
CPADH	alcohol dehydrogenase from *Candida parapsilosis*
CR	carbonyl reductase
DES	deep eutectic solvent
DMF	*N,N*-dimethylformamide
DMSO	dimethylsulfoxide
d.r.	diastereomeric ratio
DYRKR	dynamic reductive kinetic resolution
E.C.	enzyme commission number
ee	enantiomeric excess
ERED	ene-reductase

FMN	flavin mononucleotide
GDH	glucose dehydrogenase
HLDH	horse liver alcohol dehydrogenase
IL	ionic liquid
IPA	isopropanol
IRED	imine reductase
KRED	ketoreductase
LBADH	alcohol dehydrogenase from *Lactobacillus brevis*
LKADH	alcohol dehydrogenase from *Lactobacillus kefir*
mNADH	nicotinamide biomimetic
MTBE	*tert*-butyl methyl ether
NRed	nitrile reductase
Oct	octanoate
OPiv	pivalate
PPi	pyrophosphate
RasADH	alcohol dehydrogenase from *Ralstonia* sp.
RT	room temperature
SyADH	alcohol dehydrogenase from *Sphingobium yanoikuyae*
TBDMS	*tert*-butyldimethylsilyloxy
TEA	triethanolamine
TEMPO	2,2,6,6-tetramethylpiperidinyloxy radical
TeSADH	alcohol dehydrogenase from *Thermoanaerobacter ethanolicus*
Tris	tris(hydroxymethyl)aminomethane

References

1 G. Williams, M. Hall (eds.) *Advances Towards Synthetic Biological Systems.* Cambridge: Royal Society of Chemistry, 2018.

2 K. Faber, W.-D. Fessner, N.J. Turner (eds.) *Science of Synthesis: Biocatalysis in Organic Synthesis.* Stuttgart: Thieme Verlag, 2015.

3 K. Drauz, H. Groger, O. May (eds.) *Enzyme Catalysis in Organic Synthesis,* 3e. Weinheim: Wiley-VCH, 2012.

4 V. Gotor, I. Alfonso, E. Garcia-Urdiales (eds.) *Asymmetric Organic Synthesis with Enzymes.* Weinheim: Wiley-VCH, 2008.

5 A. Zaks, A.M. Klibanov, *Proc. Natl. Acad. Sci. U.S.A.* 1985, 82, 3192–3196.

6 G. de Gonzalo, P. Dominguez de Maria (eds.) *Biocatalysis: An Industrial Perspective.* Cambridge: Royal Society of Chemistry, 2018.

7 B.M. Nestl, S.C. Hammer, B.A. Nebel, B. Hauer, *Angew. Chem. Int. Ed.* 2014, 53, 3070–3095.

8 K. Faber, *Biotransformations in Organic Chemistry,* 6e. Berlin, Springer, 2011.

9 G.A. Behrens, A. Hummel, S.K. Padhi, S. Schätzle, U.T. Bornscheuer, *Adv. Synth. Catal.* 2011, 353, 2191–2215.

10 F. Wittkamp, M. Senger, S.T. Stripp, U.-P. Apfel, *Chem. Commun.* 2018, 54, 5934–5942.

11 Y.-G. Zheng, H.-H. Yin, D.-F. Yu, X. Chen, X.-L. Tang, X.-J. Zhang, Y.-P. Xue, Y.-J. Wang, Z.-Q. Liu, *Appl. Microbiol. Biotechnol.* 2017, 101, 987–1001.

12 R. Kratzer, J.M. Woodley, B. Nidetzky, *Biotechnol. Adv.* 2015, 33, 1641–1652.

13 M.M. Musa, R.S. Phillips, *Catal. Sci. Technol.* 2011, 1, 1311–1323.

14 T. Matsuda, R. Yamanaka, K. Nakamura, *Tetrahedron, Asymmetry* 2009, 20, 513–557.

15 D. Dascier, S. Kamburaki, L. Hua, J.D. Rozzell, J.D. Stewart, *Org. Process Res. Dev.* 2014, 18, 793–800.

16 I. Lavandera, A. Kern, M. Schaffenberger, J. Gross, A. Glieder, S. de Wildeman, W. Kroutil, *ChemSusChem* 2008, 1, 431–436.

17 G. de Gonzalo, I. Lavandera, K. Faber, W. Kroutil, *Org. Lett.* 2007, 9, 2163–2166.

18 G. de Gonzalo, I. Lavandera, K. Durchschein, D. Wurm, K. Faber, W. Kroutil, *Tetrahedron, Asymmetry* 2007, 18, 2541–2546.

19 Y.P. Mbous, M. Hayyan, A. Hayyan, W.F. Wong, M.A. Hashim, C.Y. Looi, *Biotechnol. Adv.* 2017, 35, 105–134.

20 C.T. Muller, I. Lavandera, V. Gotor-Fernandez, P. Dominguez de Maria, *ChemCatChem* 2015, 7, 2654–2659.

21 A. Jakobinnert, R. Mladenov, A. Paul, F. Sibilla, U. Schwaneberg, M.B. Ansorge-Schumacher, P.D. de María, *Chem. Commun.* 2011, 47, 12230–12232.

22 J.A. Friest, Y. Maezato, S. Broussy, P. Blum, D.B. Berkowitz, *J. Am. Chem. Soc.* 2010, 132, 5930–5931.

23 P. Galleti, E. Emer, G. Gucciardo, A. Quintavalla, M. Pori, D. Giacomini, *Org. Biomol. Chem.* 2010, 8, 4117–4123.

24 A. Cuetos, A. Rioz-Martinez, F.R. Bisogno, B. Grischek, I. Lavandera, G. de Gonzalo, W. Kroutil, V. Gotor, *Adv. Synth. Catal.* 2012, 354, 1743–1749.

25 X. Chen, Z.-Q. Liu, J.-F. Huang, C.-P. Lin, Y.-G. Zheng, *Chem. Commun.* 2015, 51, 12328–12331.

26 J. Mangas-Sanchez, E. Busto, V. Gotor, V. Gotor-Fernandez, *Org. Lett.* 2013, 15, 3872–3875.

27 D. Mendez-Sanchez, J. Mangas-Sanchez, E. Busto, V. Gotor, V. Gotor-Fernández, *Adv. Synth. Catal.* 2016, 358, 122–131.

28 J.H. Schrittwieser, S. Velikogne, M. Hall, W. Kroutil, *Chem. Rev.* 2018, 118, 270–348.

29 S. Riva, W.-D. Fessner (eds.) *Cascade Biocatalysis, Integrating Stereoselective and Environmentally Friendly Reactions.* Weinheim, Wiley-VCH, 2014.

30 F.R. Bisogno, M.G. Lopez-Vidal, G. de Gonzalo, *Adv. Synth. Catal.* 2017, 359, 2026–2049.

31 C.A. Denard, J.F. Hartwig, H. Zhao, *ACS Catal.* 2013, 3, 2856–2864.

32 K. Gotz, A. Liese, M. Ansorge-Schumacher, L. Hilterhaus, *Appl. Microbiol. Biotechnol.* 2013, 97, 3865–3873.

33 E. Liardo, N. Rios-Lombardia, F. Moris, J. González-Sabín, F. Rebolledo, *Org. Lett.* 2016, 18, 3366–3369.

34 H. Kohls, M. Anderson, J. Dickerhoff, K. Weisz, A. Córdova, P. Berglund, H. Brundiek, U.T. Bornscheuer, M. Höhne, *Adv. Synth. Catal.* 2015, 357, 1808–1814.

35 W. Szymanski, C.P. Postema, C. Tarabiono, F. Berthiol, L. Campbell-Verduyn, S. de Wildeman, J.G. de Vries, B.L. Feringa, D.B. Janssen, *Adv. Synth. Catal.* 2010, 352, 2111–2115.

36 M.J. Rodriguez-Alvarez, N. Rios-Lombardia, S. Schumacher, D. Pérez-Iglesias, F. Morís, V. Cadierno, J. García-Álvarez, J. González-Sabín, *ACS Catal.* 2017, 7, 7753–7759.

37 R.C. Simon, E. Busto, J.H. Schrittwieser, J.H. Sattler, J. Pietruszka, K. Faber, W. Kroutil, *Chem. Commun.* 2014, 50, 15669–15672.

38 K. Baer, M. Krau*β*er, E. Burda, K. Baer, M. Kraußer, E. Burda, W. Hummel, A. Berkessel, H. Gröger, *Angew. Chem. Int. Ed.* 2009, 48, 9355–9358.

39 G. Rulli, N. Duangdee, K. Baer, W. Hummel, A. Berkessel, H. Gröger, *Angew. Chem. Int. Ed.* 2011, 50, 7944–7947.

40 M. Heidlindemann, G. Rulli, A. Berkessel, W. Hummel, H. Gröger, *ACS Catal.* 2014, 4, 1099–1103.

41 G. Rulli, N. Duangdee, W. Hummel, A. Berkessel, H. Gröger, *Eur. J. Org. Chem.* 2017, 812–817.

42 G.W. Huisman, J. Liang, A. Kreber, *Curr. Opin. Chem. Biol.* 2010, 14, 122–129.

43 W. Borzęcka, I. Lavandera, V. Gotor, *J. Org. Chem.* 2013, 78, 7312–7317.

44 A.S. de Miranda, R.C. Simon, B. Grischek, G.C. de Paula, B.A.C. Horta, L.S.M. de Miranda, W. Kroutil, C.O. Kappe, R.O.M.A. de Souza, *ChemCatChem* 2015, 7, 984–992.

45 S.J. Calvin, D. Mangan, I. Miskelly, T.S. Moody, P.J. Stevenson, *Org. Process Res. Dev.* 2012, 16, 82–86.

46 G. Brown, D. Mangan, I. Miskelly, T.S. Moody, *Org. Process Res. Dev.* 2011, 15, 1036–1039.

47 J. Liang, J. Lalonde, B. Borup, V. Mitchell, E. Mundorff, N. Trinh, D.A. Kochrekar, R.N. Cherat, G.G. Pai, *Org. Process Res. Dev.* 2010, 14, 193–198.

48 L.J. Wang, C.X. Li, Y. Ni, J. Zhang, X. Liu, J.H. Xu, *Bioresour. Technol.* 2011, 102, 7023.

49 J. Pan, G.-W. Zheng, Q. Ye, J.-H. Xu, *Org. Process Res. Dev.* 2014, 18, 739–743.

50 R.-C. Zheng, Z. Ge, Z.-K. Qiu, Y.-S. Wang, Y.-G. Zheng, *Appl. Microbiol. Biotechnol.* 2012, 94, 969–976.

51 N.K. Modukuru, J. Sukumaran, S.J. Collier, A.S. Chan, A. Gohel, G.W. Huisman, R. Keledjian, K. Narayanaswamy, S.J. Novick, S.M. Palanivel, D. Smith, Z. Wei, B. Wong, W.L. Yeo, D.A. Entwistle, *Org. Process Res. Dev.* 2014, 18, 810–815.

52 X. Ju, Y. Tang, X. Liang, M. Hou, Z. Wan, J. Tao, *Org. Process Res. Dev.* 2014, 18, 827–830.

53 D. Pollard, M. Truppo, J. Pollard, C.-Y. Chen, J. Moore, *Tetrahedron: Asymmetry* 2006, 17, 554–559.

54 J. Li, P. Wang, J. Hung, J. Tang, *Appl. Microbiol. Biotechnol.* 2013, 97, 6685–6692.

55 H. Li, J. Moncecchi, M.D. Truppo, *Org. Process Res. Dev.* 2015, 19, 695–700.

56 S. Garcia-Cerrada, L. Redondo-Gallego, F. Martinez-Olid, J.A. Rincón, P. García-Losada, *Org. Process Res. Dev.* 2017, 21, 779–784.

57 X. Linghu, N. Wong, H. Iding, V. Jost, H. Zhang, S.G. Koenig, C.G. Sowell, F. Gosselin, *Org. Process Res. Dev.* 2017, 21, 387–398.
58 X. Guo, J.-W. Tang, J.-T. Yang, G.-W. Ni, F.-L. Zhang, S.-X. Chen, *Org. Process Res. Dev.* 2017, 21, 1595–1601.
59 X.-M. Gong, G.-W. Zheng, Y.-Y. Liu, J.-H. Xu, *Org. Process Res. Dev.* 2017, 21, 1349–1354.
60 S. Duan, B. Li, R.W. Dugger, B. Conway, R. Kumar, C. Martínez, R. Makowski, R. Pearson, M. Olivier, R. Colon-Cruz, *Org. Process Res. Dev.* 2017, 21, 1340–1348.
61 H.S. Toogood, N.S. Scrutton, *ACS Catal.* 2018, 8, 3532–3549.
62 C.K. Winkler, K. Faber, M. Hall, *Curr. Opin. Chem. Biol.* 2018, 43, 97–105.
63 C. Stueckler, T.C. Reiter, N. Baudendistel, K. Faber, *Tetrahedron* 2010, 66, 663–667.
64 C.K. Winkler, D. Clay, M. Entner, M. Plank, K. Faber, *Chem. Eur. J.* 2014, 20, 1403–1409.
65 M. Mifsud, S. Gargiulo, S. Iborra, I.W.C.E. Arends, F. Hollmann, A. Corma, *Nat. Commun.* 2014, 5, 3145.
66 K. Koninger, A. Gomez Baraibar, C. Mugge, C.E. Paul, F. Hollmann, M.M. Nowaczyk, R. Kourist, *Angew. Chem. Int. Ed.* 2016, 55, 5582–5585.
67 T. Knaus, C.E. Paul, C.W. Levy, S. de Vries, F.G. Mutti, F. Hollmann, N.S. Scrutton, *J. Am. Chem. Soc.* 2016, 138, 1033–1039.
68 C.E. Paul, S. Gargiulo, D.J. Opperman, I. Lavandera, V. Gotor-Fernández, V. Gotor, A. Taglieber, I.W.C.E. Arends, F. Hollmann, *Org. Lett.* 2013, 15, 180–183.
69 C.K. Winkler, C. Stueckler, N.J. Mueller, D. Pressnitz, K. Faber, *Eur. J. Org. Chem.* 2010, 6354–6358.
70 E. Brenna, S.L. Cosi, E.E. Ferrandi, F.G. Gatti, D. Monti, F. Parmeggiani, A. Sacchetti, *Org. Biomol. Chem.* 2013, 11, 2988–2996.
71 N.G. Turrini, R.C. Cioc, D.J.H. van der Niet, E. Ruijter, R.V.A. Orru, M. Hall, K. Faber, *Green Chem.* 2017, 19, 511–518.
72 Y. Yanto, C.K. Winkler, S. Lohr, M. Hall, K. Faber, A.S. Bommarius, *Org. Lett.* 2011, 13, 2540–2543.
73 G. Tasnadi, C.K. Winkler, D. Clay, N. Sultana, W.M.F. Fabian, M. Hall, K. Ditrich, K. Faber, *Chem. Eur. J.* 2012, 18, 10362–10367.
74 N. Iqbal, F. Rudroff, A. Brige, J. Van Beeumen, M.D. Mihovilovic, *Tetrahedron* 2012, 68, 7619–7623.
75 E. Brenna, F.G. Gatti, A. Manfredi, D. Monti, F. Parmeggiani, *Catal. Sci. Technol.* 2013, 3, 1136–1146.
76 E. Burda, T. Reß, T. Winkler, C. Giese, X. Kostrov, T. Huber, W. Hummel, H. Gröger, *Angew. Chem. Int. Ed.* 2013, 52, 9323–9326.
77 M. Bertolotti, E. Brenna, M. Crotti, F.G. Gatti, D. Monti, F. Parmeggiani, S. Santangelo, *ChemCatChem* 2016, 8, 577–583.
78 E. Brenna, M. Crotti, F.G. Gatti, D. Monti, F. Parmeggiani, S. Santangelo, *ChemCatChem* 2017, 9, 2480–2487.
79 E. Brenna, M. Crotti, F.G. Gatti, A. Manfredi, D. Monti, F. Parmeggiani, S. Santangelo, D. Zampieri, *ChemCatChem* 2014, 6, 2425–2431.

80 C.K. Winkler, D. Clay, N.G. Turrini, H. Lechner, W. Kroutil, S. Davies, S. Debarge, P. O'Neill, J. Steflik, M. Karmilowicz, J.W. Wong, K. Faber, *Adv. Synth. Catal.* 2014, 356, 1878–1882.
81 N.G. Turrini, M. Hall, K. Faber, *Adv. Synth. Catal.* 2015, 357, 1861–1871.
82 T. Reß, W. Hummel, S.P. Hanlon, H. Iding, H. Gröger, *ChemCatChem* 2015, 7, 1302–1311.
83 I. Janicki, P. Kiełbasiński, N.G. Turrini, K. Faber, M. Hall, *Adv. Synth. Catal.* 2017, 359, 4190–4196.
84 C. Stueckler, C.K. Winkler, M. Bonnekessel, K. Faber, *Adv. Synth. Catal.* 2010, 352, 2663–2666.
85 C. Stueckler, N.J. Mueller, C.K. Winkler, S.M. Glueck, K. Gruber, G. Steinkellner, K. Faber, *Dalton Trans.* 2010, 39, 8472–8476.
86 J. Pietruszka, M. Scholzel, *Adv. Synth. Catal.* 2012, 354, 751–756.
87 E. Brenna, F.G. Gatti, A. Manfredi, D. Monti, F. Parmeggiani, *Org. Process Res. Dev.* 2012, 16, 262–268.
88 C.K. Winkler, D. Clay, S. Davies, P. O'Neill, P. McDaid, S. Debarge, J. Steflik, M. Karmilowicz, J.W. Wong, K. Faber, *J. Org. Chem.* 2013, 78, 1525–1533.
89 S. Debarge, P. McDaid, P.O. Neill, J. Frahill, J.W. Wong, D. Carr, A. Burrell, S. Davies, M. Karmilowicz, J. Steflik, *Org. Process Res. Dev.* 2014, 18, 109–121.
90 E. Brenna, F.G. Gatti, D. Monti, F. Parmeggiani, A. Sacchetti, *ChemCatChem* 2012, 4, 653–659.
91 E. Brenna, F.G. Gatti, L. Malpezzi, D. Monti, F. Parmeggiani, A. Sacchetti, *J. Org. Chem.* 2013, 78, 4811–4822.
92 E. Brenna, M. Crotti, F.G. Gatti, D. Monti, F. Parmeggiani, A. Pugliese, S. Santangelo, *J. Mol. Catal. B, Enzym.* 2015, 114, 37–41.
93 E. Brenna, M. Crotti, F.G. Gatti, D. Monti, F. Parmeggiani, A. Pugliese, *Molecules* 2017, 22, 1591.
94 E. Brenna, M. Crotti, F.G. Gatti, L. Marinoni, D. Monti, S. Quaiato, *J. Org. Chem.* 2017, 82, 2114–2122.
95 T. Classen, M. Korpak, M. Scholzel, J. Pietruszka, *ACS Catal.* 2014, 4, 1321–1331.
96 E. Brenna, F.G. Gatti, D. Monti, F. Parmeggiani, A. Sacchetti, J. Valoti, *J. Mol. Catal. B, Enzym.* 2015, 114, 77–85.
97 D. Monti, M.C. Forchin, M. Crotti, F. Parmeggiani, F.G. Gatti, E. Brenna, S. Riva, *ChemCatChem* 2015, 7, 3106–3109.
98 L. Skalden, C. Peters, J. Dickerhoff, A. Nobili, H.-J. Joosten, K. Weisz, M. Höhne, U.T. Bornscheuer, *ChemBioChem* 2015, 16, 1041–1045.
99 L. Skalden, C. Peters, L. Ratz, U.T. Bornscheuer, *Tetrahedron* 2016, 72, 7207–7211.
100 S. Gargiulo, D.J. Opperman, U. Hanefeld, I.W.C.E. Arends, F. Hollmann, *Chem. Commun.* 2012, 48, 6630–6632.
101 N. Oberleitner, C. Peters, J. Muschiol, M. Kadow, S. Saß, T. Bayer, P. Schaaf, N. Iqbal, F. Rudroff, M.D. Mihovilovic, U.T. Bornscheuer, *ChemCatChem* 2013, 5, 3524–3528.
102 T. Knaus, F.G. Mutti, L.D. Humphreys, N.J. Turner, N.S. Scrutton, *Org. Biomol. Chem.* 2015, 13, 223–233.

103 E. Brenna, M. Crotti, F.G. Gatti, D. Monti, F. Parmeggiani, A. Pugliese, F. Tentori, *Green Chem.* 2017, 19, 5122–5130.
104 L. Martinez-Montero, V. Gotor, V. Gotor-Fernandez, I. Lavandera, *ACS Catal.* 2018, 8, 2413–2419.
105 M. Kraußer, T. Winkler, N. Richter, S. Dommer, A. Fingerhut, W. Hummel, H. Gröger, *ChemCatChem* 2011, 3, 293–296.
106 M. Biermann, H. Grus, W. Hummel, H. Groger, *ChemCatChem* 2016, 8, 895–899.
107 Y. Wang, M.J. Bartlett, C.A. Denard, J.F. Hartwig, H. Zhao, *ACS Catal.* 2017, 7, 2548–2552.
108 K. Heckenbichler, A. Schweiger, L.A. Brandner, A. Binter, M. Toplak, P. Macheroux, K. Gruber, R. Breinbauer, *Angew. Chem. Int. Ed.* 2018, 57, 7240–7244.
109 D. Caine, *Tetrahedron* 2001, 57, 2643–2684.
110 R.D. Little, J.R. Dawson, *Tetrahedron Lett.* 1980, 21, 2609–2612.
111 M. Lenz, N. Borlinghaus, L. Weinmann, B.M. Nestl, *World J Microbiol. Biotechnol.* 2017, 33, 199.
112 J. Mangas-Sanchez, S.P. France, S.L. Montgomery, G.A. Aleku, H. Man, M. Sharma, J.I. Ramsden, G. Grogan, N.J. Turner, *Curr. Opin. Chem. Biol.* 2017, 37, 19–25.
113 G. Grogan, N.J. Turner, *Chem. Eur. J.* 2016, 22, 1900–1907.
114 J.H. Schrittwieser, S. Velikogne, W. Kroutil, *Adv. Synth. Catal.* 2015, 357, 1655–1685.
115 Z. Maugeri, D. Rother, *Adv. Synth. Catal.* 2016, 358, 2745–2750.
116 P.N. Scheller, S. Fademrecht, S. Hofelzer, J. Pleiss, F. Leipold, N.J. Turner, B.M. Nestl, B. Hauer, *ChemBioChem* 2014, 15, 2201–2204.
117 T. Huber, L. Schneider, A. Prag, S. Gerhardt, O. Einsle, M. Müller, *ChemCatChem* 2014, 6, 2248–2252.
118 H. Man, E. Wells, S. Hussain, F. Leipold, S. Hart, J.P. Turkenburg, N.J. Turner, G. Grogan, *ChemBioChem* 2015, 16, 1052–1059.
119 D. Wetzl, M. Berrera, N. Sandon, D. Fishlock, M. Ebeling, M. Müller, S. Hanlon, B. Wirz, H. Iding, *ChemBioChem* 2015, 16, 1749–1756.
120 H. Li, P. Tian, J.-H. Xu, G.-W. Zheng, *Org. Lett.* 2017, 19, 3151–3154.
121 J. Zhu, H. Tan, L. Yang, Z. Dai, L. Zhu, H. Ma, Z. Deng, Z. Tian, X. Qu, *ACS Catal.* 2017, 7, 7003–7007.
122 S. Hussain, F. Leipold, H. Man, E. Wells, S.P. France, K.R. Mulholland, G. Grogan, N.J. Turner, *ChemCatChem* 2015, 7, 579–583.
123 H. Li, Z.-J. Luan, G.-W. Zheng, J.-H. Xu, *Adv. Synth. Catal.* 2015, 357, 1692–1696.
124 S.P. France, G.A. Aleku, M. Sharma, J. Mangas-Sanchez, R.M. Howard, J. Steflik, R. Kumar, R.W. Adams, I. Slabu, R. Crook, G. Grogan, T.W. Wallace, N.J. Turner, *Angew. Chem. Int. Ed.* 2017, 56, 15589–15593.
125 N. Zumbragel, C. Merten, S.M. Huber, H. Groger, *Nat. Commun.* 2018, 9, 1949.
126 P.N. Scheller, M. Lenz, S.C. Hammer, B. Hauer, B.M. Nestl, *ChemCatChem* 2015, 7, 3239–3242.

127 D. Wetzl, M. Gand, A. Ross, H. Müller, P. Matzel, S.P. Hanlon, M. Müller, B. Wirz, M. Höhne, H. Iding, *ChemCatChem* 2016, 8, 2023–2026.

128 G.A. Aleku, S.P. France, H. Man, J. Mangas-Sanchez, S.L. Montgomery, M. Sharma, F. Leipold, S. Hussain, G. Grogan, N.J. Turner, *Nat. Chem.* 2017, 9, 961–969.

129 M. Winkler, *Curr. Opin. Chem. Biol.* 2018, 43, 23–29.

130 G. Qu, J. Guo, D. Yang, Z. Sun, *Green Chem.* 2018, 20, 777–792.

131 E.H. Hansen, B.L. Møller, G.R. Kock, C.M. Bünner, C. Kristensen, O.R. Jensen, F.T. Okkels, C.E. Olsen, M.S. Motawia, J. Hansen, *Appl. Environ. Microbiol.* 2009, 75, 2765–2774.

132 M.K. Akhtara, N.J. Turner, P.R. Jones, *Proc. Natl. Acad. Sci. U.S.A.* 2013, 110, 87–92.

133 Y. Duan, P. Yao, X. Chen, X. Liu, R. Zhang, J. Feng, Q. Wu, D. Zhu, *J. Mol. Catal. B, Enzym.* 2015, 115, 1–7.

134 L. Kramer, E.D. Hankore, Y. Liu, K. Liu, E. Jimenez, J. Guo, W. Niu, *ChemBioChem* 2018, 19, 1452–1460.

135 Y. Duan, P. Yao, Y. Du, J. Feng, Q. Wu, D. Zhu, *Beilstein J. Org. Chem.* 2015, 11, 2245–2251.

136 E. Klumbys, Z. Zebec, N.J. Weise, N.J. Turner, N.S. Scrutton, *Green Chem.* 2018, 20, 658–663.

137 S.P. France, S. Hussain, A.M. Hill, L.J. Hepworth, R.M. Howard, K.R. Mulholland, S.L. Flitsch, N.J. Turner, *ACS Catal.* 2016, 6, 3753–3759.

138 P. Dominguez de Maria, *ChemCatChem* 2011, 3, 1683–1685.

139 B. Wilding, M. Winkler, B. Petschacher, R. Kratzer, A. Glieder, N. Klempier, *Adv. Synth. Catal.* 2012, 354, 2191–2198.

140 K. Durchschein, M. Hall, K. Faber, *Green Chem.* 2013, 15, 1764–1772.

141 Y. Yanto, M. Hall, A.S. Bommarius, *Org. Biomol. Chem.* 2010, 8, 1826–1832.

142 F. Li, J. Cui, X. Qian, R. Zhang, *Chem. Commun.* 2004, 2338–2339.

143 H.-H. Nguyen-Tran, G.-W. Zheng, X.-H. Qian, J.-H. Xu, *Chem. Commun.* 2014, 50, 2861–2864.

9

Organocatalytic Transfer Hydrogenation

Jie Wang and Yong-Gui Zhou

Dalian Institute of Chemical Physics, Chinese Academy of Sciences, 457 Zhongshan Road, Dalian 116023, China

9.1 Introduction

The asymmetric hydrogenation of unsaturated double bonds presents one of the most versatile and efficient methodologies to access a broad range of biologically active and pharmacological compounds [1]. Therefore, two prominent scientists Knowles and Noyori were awarded the Nobel Prize in Chemistry in 2001 for their pioneering work in catalytic asymmetric hydrogenation [2]. In the traditional hydrogenation process, asymmetric reductions focused on the combination of transition metals and chiral ligands, applying dihydrogen, isopropanol, or formic acid as hydrogen source [3]. Although great progress has been made in the field of transition metal catalyzed hydrogenation, some limitations exist in substrate scope, metal residue, and high dihydrogen pressure.

Based on the reductive transformations in biological systems, taking advantage of enzymes and cofactors (NADH, NADPH, $FADH_2$, and others) [4], organocatalytic transfer hydrogenation gradually became a "hot" field, and an impressive number of organocatalytic reductions have been reported (Figure 9.1). A few classes of organocatalysts have been successfully developed into asymmetric transfer hydrogenations [5], particularly chiral phosphoric acids (CPAs) **5** , secondary amines, and thioureas, which were successful at controlling the stereoselectivity and were suitable for various unsaturated compounds containing C=C, C=N, C=O bonds as well as heteroaromatics. With respect to hydrogen source, Hantzsch esters (HEHs) **3** are the most typical representatives, containing a 1,4-dihydropyridine structure capable of delivering two hydrogen atoms to the unsaturated bonds. In addition, benzothiazolines **4** also could be regarded as hydrogen source due to their reducing ability by releasing molecular hydrogen to form the more stable aromatic benzothiazole [6]. In this way, organocatalytic transfer hydrogenation was successfully performed without metal residue and away from the danger of high pressure.

This chapter aims to give an up-to-date, comprehensive and concise introduction to recent developments in organocatalytic asymmetric transfer hydrogenation of C=C, C=N, C=O and heteroaromatics using CPAs, secondary amines, or thioureas as the catalyst and HEHs or benzothiazoline and their

Homogeneous Hydrogenation with Non-Precious Catalysts, First Edition. Edited by Johannes F. Teichert.

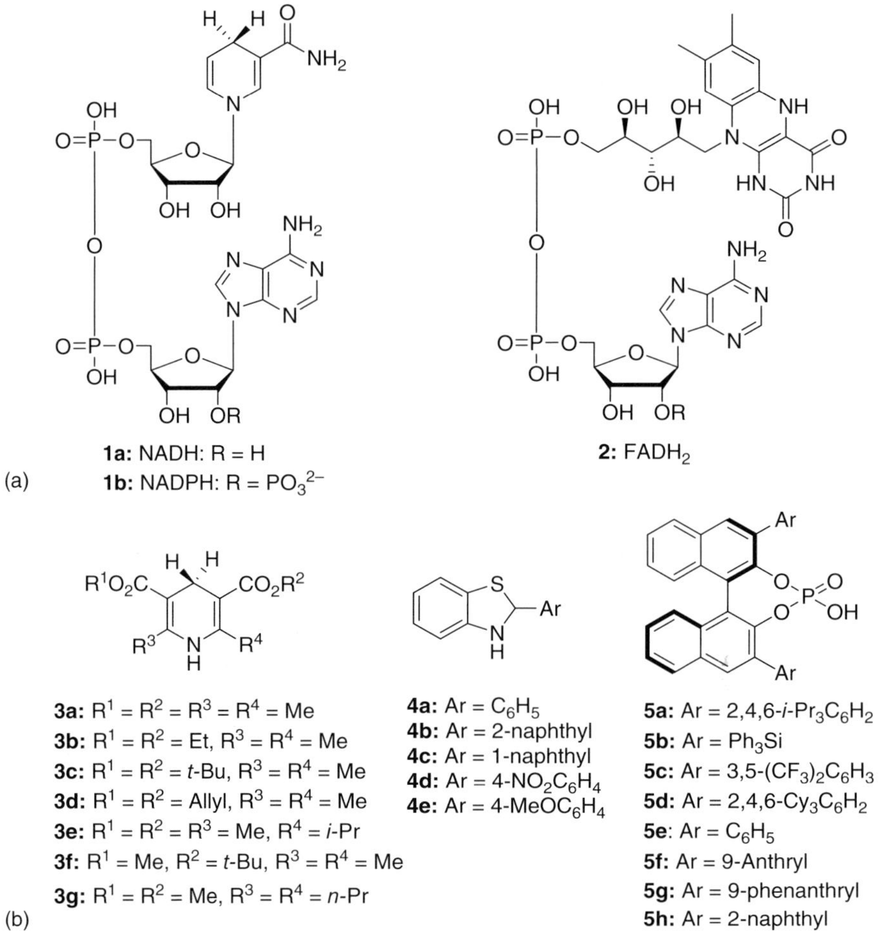

Figure 9.1 (a) The structures of cofactors for reductions in biological systems. (b) The commonly employed hydrogen sources and chiral phosphoric acids.

derivatives as hydrogen sources. Cascade reactions [7] involving organocatalytic transfer hydrogenation are not included.

9.2 Asymmetric Transfer Hydrogenation of C=C Bonds

Although the cofactor NAD(P)H-mediated reduction process has been known for a long time, organocatalytic asymmetric transfer hydrogenation did not develop until the List group first realized the asymmetric reduction of α,β-unsaturated aldehydes in 2004 [8]. Using chiral amine **6a** and trichloroacetic acid (TCA) as the catalyst and HEH **3e** as hydride source, the trisubstituted enals **7** could be reduced to chiral aldehydes **8** with up to 96% ee [9]. Nearly at the same time, the MacMillan group reported a similar strategy to realize the asymmetric transfer hydrogenation of α,β-unsaturated aldehydes **9**, combining

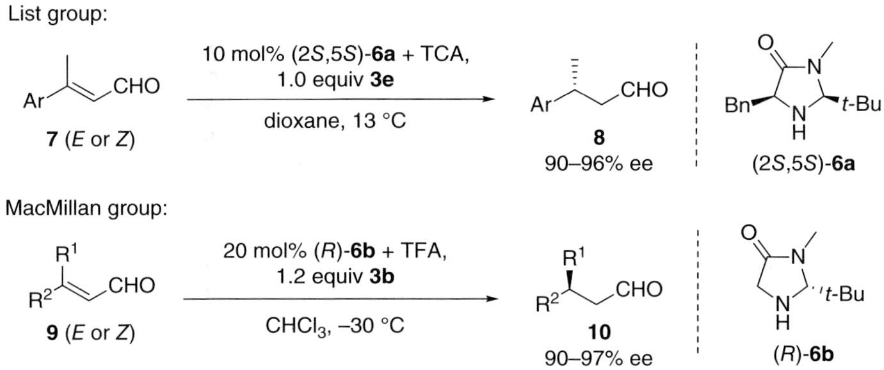

Scheme 9.1 Asymmetric transfer hydrogenation of enals catalyzed by chiral secondary amines.

chiral amine **6b** and trifluoroacetic acid (TFA) as the catalyst (Scheme 9.1) [10]. This approach was widely employed in the total synthesis of natural products and pharmaceutical intermediates such as marine macrolide (+)-neopeltolide, side chain of α-tocopherol, (+)-ricciocarpin A, and others [11].

It is worth noting that the excellent enantioselectivity and the same absolute configuration of the products were obtained when starting with either *E*- or *Z*-unsaturated aldehydes in a stereoconvergent process [9]. It was proposed that the *E*- and *Z*-α,β-unsaturated iminium ion intermediates were in a fast equilibrium *via* a conjugate dienamine species. The following rate-determining hydride transfer from dihydropyridine to *E*-enal proceeded faster than *Z*-enal [$k(E) > k(Z)$], resulting in high enantioselectivity (Scheme 9.2).

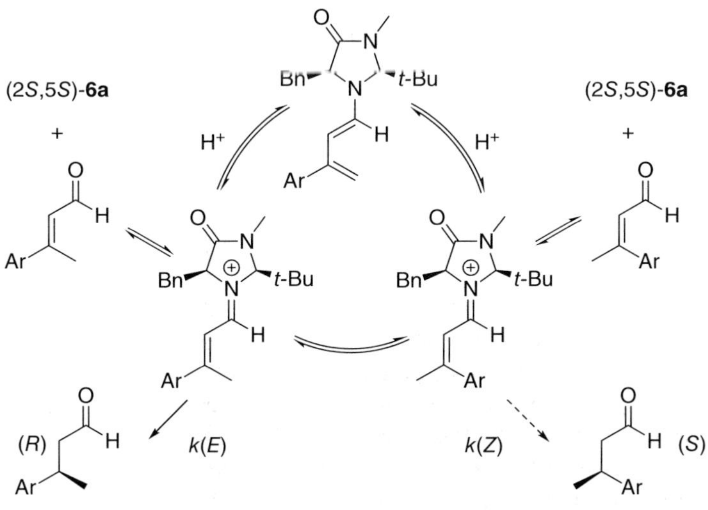

Scheme 9.2 Proposed mechanism for asymmetric transfer hydrogenation of enals.

Inspired by these results, the groups of Cossy, Arseniyadis, and coworkers successfully applied this strategy in the enantioselective organocatalytic conjugate reduction of β-azole-containing α,β-unsaturated aldehydes with up to 94% ee and provided an efficient method to synthesize the C7–C14 fragment of ulapualide A, which exhibited promising antitumor activity [12].

In 2008, the Kudo group reported that a resin-supported N-terminal prolyl peptide with a β-turn motif and hydrophobic polyleucine chain could efficiently catalyze asymmetric transfer hydrogenation of the α,β-unsaturated aldehydes in aqueous media (THF/H_2O) with up to 76% yield and 96% ee. The polyleucine tether provided a hydrophobic cavity that brought about a remarkable acceleration of the reaction [13].

In parallel with the development in iminium catalysis, a different type of catalysis became available to reduce α,β-unsaturated aldehydes in high enantioselectivity: the concept of asymmetric counteranion directed catalysis (ACDC) developed by List group, utilizing catalytic amounts of achiral ammonium salts and in combination with CPAs to induce asymmetry in the process [14]. After screening various commercially available ammonium salts and known chiral binaphthol-derived phosphoric acids, the combination of the sterically hindered TRIP anion and morpholinium ion **13** gave the best result with up to 98% ee. This strategy widened the substrate scope of the iminium-based catalytic transfer hydrogenation by allowing the use of sterically nonhindered α,β-unsaturated aldehydes. For example, (*R*)-citronellal and (*R*)-dihydrofarnesal could be reduced from the corresponding α,β-unsaturated aldehydes with 90% and 92% ee employing the ACDC strategy, but only moderate 40% ee could be obtained while using iminium catalysts (Scheme 9.3).

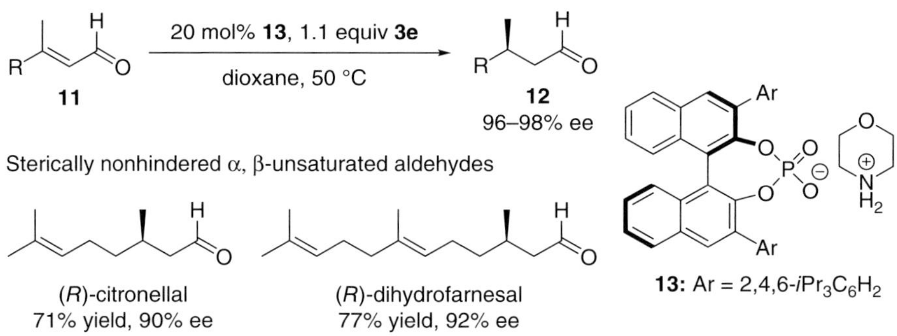

Scheme 9.3 Asymmetric transfer hydrogenation of enals by the ACDC strategy.

In 2006, the List group extended the ACDC strategy to asymmetric transfer hydrogenation of cyclic and linear enones **14** in high activity and enantioselectivity using a catalytic amount of the ammonium salt of valine *t*-butyl ester with the TRIP anion **16** [15]. Interestingly, when using the opposite enantiomer of the TRIP anion, the same enantiomer of the product was formed but with low ee value, illustrating a dramatic case of a matched/mismatched catalyst–ion pair combination. Almost at the same time, MacMillan applied furyl imidazolidinone **17** and TCA as catalyst combination in the asymmetric reduction of cyclic enones **18**. The sense of asymmetric induction observed in all cases was consistent with selective engagement of the HEH reductant

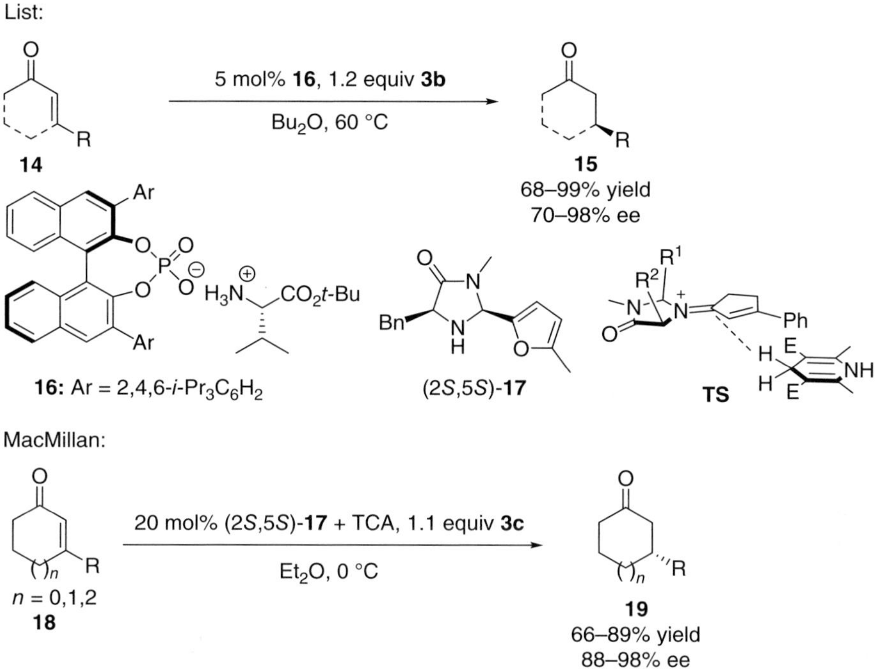

Scheme 9.4 Asymmetric transfer hydrogenation of linear and cyclic enones.

with the *Si* face of *cis*-iminium isomer (Scheme 9.4) [16]. In addition, the Houk group employed hybrid density functional (B3LYP/6-31G(d)) theory to investigate the transition state (**TS**) of the organocatalytic transfer hydrogenation of 3-phenyl-2-cyclopentenone with imidazolidinone catalysts and suggested that HEH (bottom-*anti*) attack on the *E*-iminium was favored by 1.1 kcal/mol over the attack on the *Z*-iminium, resulting in the high enantioselectivity [17]. Based on this work, Lear and Ramachary groups applied chiral amine catalyzed transfer hydrogenation of cyclic enones in the synthesis of the key intermediate of (−)-platensimycin and biologically active 5β-dihydrosteroids, respectively [18].

As the bifunctional organocatalyst, chiral thioureas could also promote the transfer hydrogenation of C=C bonds. In 2007, the List group reported that β,β-disubstituted nitroolefins **20** and β-nitroacrylates **23** could be asymmetrically reduced by the chiral thiourea with excellent enantioselectivities (Scheme 9.5) [19]. Contrary to the enantioconvergent process mentioned in the reduction of enals, the enantioselectivity of the nitroolefin reduction strongly depended on the nitroolefin geometry. However, stereoconvergence could be established upon adding a catalytic amount of triphenylphosphine, creating a rapid equilibrium between *E*- and *Z*-nitroolefins *via* a conjugate addition/elimination pathway.

Later, the Paradies group developed a readily accessible thiourea catalyst derived from amino alcohols and realized an asymmetric reduction of nitroolefins, showing that the hydroxyl group was essential to form a ternary complex with both the substrate and the hydride donor through hydrogen

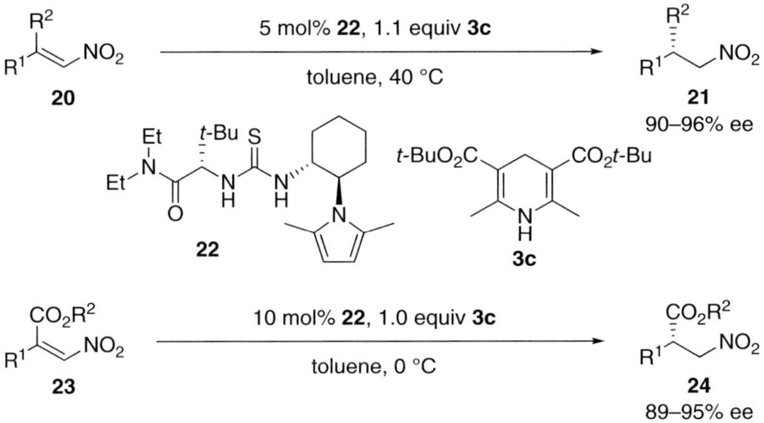

Scheme 9.5 Asymmetric transfer hydrogenation of nitroolefins by List.

bonding with the nitro group and the N–H moiety in the HEH [20]. In 2005, the groups of Fochi, Bernardi, and Benaglia exploited chiral thiourea catalysts in asymmetric transfer hydrogenation of trisubstituted β-trifluoromethyl nitroalkenes with up to 98% ee and 97% ee, respectively [21]. However, as for the α,β-disubstituted β-trifluoromethyl nitroalkenes, the substrates were not configurationally stable (less than 2 kcal mol between *E* and *Z* isomer by computational analysis), which resulted in low ee values (<70% ee). In addition, Fochi found that high activities (40–97% yield) and enantioselectivities (93 –> 99% ee) would be obtained when replacing a trifluoromethyl with an amide group [22].

Recently, asymmetric transfer hydrogenation of electron-rich C=C bonds was developed by Zhu, Lin, Sun, and coworkers using CPA **5b** [23], giving the corresponding chiral 2-(1-arylethyl)phenols **26** with 75–99% ee. The starting materials, vinyl phenols **25**, were initially protonated by the catalyst to form the carbocation intermediate, which interacted with the chiral phosphate anion through hydrogen bonding (IM-R1). The latter isomerized to form the *o*-quinone methide (*o*-QM) intermediate, and subsequent enantioselective hydride addition afforded the final reduction products. The stereochemistry was controlled by the chiral anion or the chiral acid (Scheme 9.6). This strategy was also suitable for the corresponding indole-substituted 1,1-diaryl ethylenes.

9.3 Asymmetric Transfer Hydrogenation of C=N Bonds

The first report of asymmetric transfer hydrogenation of C=N bonds to provide the chiral amines in the absence of metals dates back to 2005 [24]. The Rueping group developed the first Brønsted acid catalyzed reduction of ketimines with HEH as hydrogen source under mild conditions. The substrates bearing phenyl (Ph) and 4-methoxyphenyl (PMP) were evaluated successfully with up to 84% ee using CPA **5c**. Soon after, using the more sterically hindered CPA **5a** [25], the List group disclosed a highly enantioselective transfer hydrogenation of

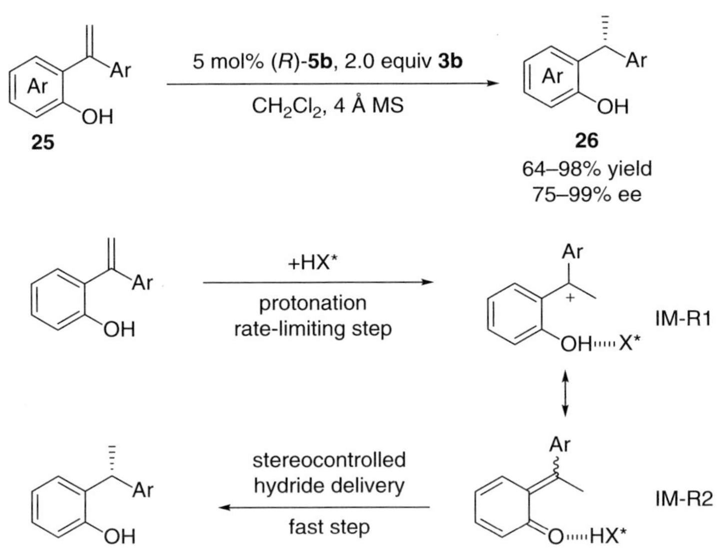

Scheme 9.6 Asymmetric transfer hydrogenation of electron-rich C=C bonds.

PMP-protected ketimines with up to 93% ee. Furthermore, both aliphatic and aromatic imines were transformed successfully with remarkably low catalyst loading (1 mol%) (Scheme 9.7).

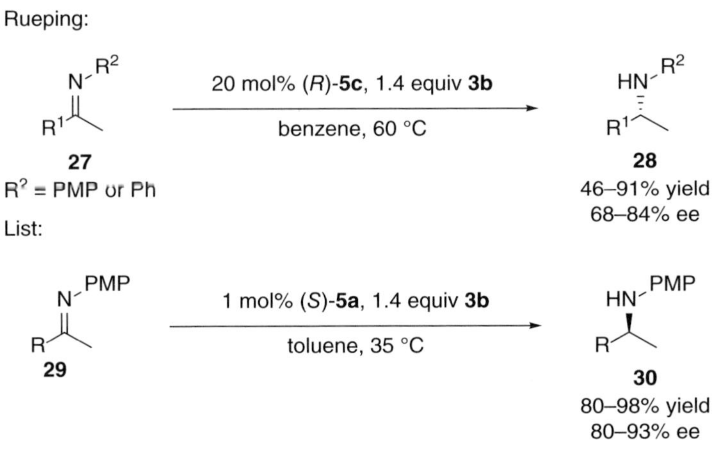

Scheme 9.7 Asymmetric transfer hydrogenation of imines by Rueping and List.

It was proposed that the catalytic cycle commences with the protonation of the ketimine by the CPA [24]. The resulting chiral iminium ion pair was stabilized by hydrogen bonding and reacted with HEH, giving a chiral amine and pyridine (Scheme 9.8). Additionally, Goodman and Himo groups carried out studies on the mechanism of the CPA-catalyzed asymmetric transfer hydrogenation of imine using density functional theory (DFT) [26]. The results showed that CPA acted as a bifunctional catalyst that activated the imine group while at the same time

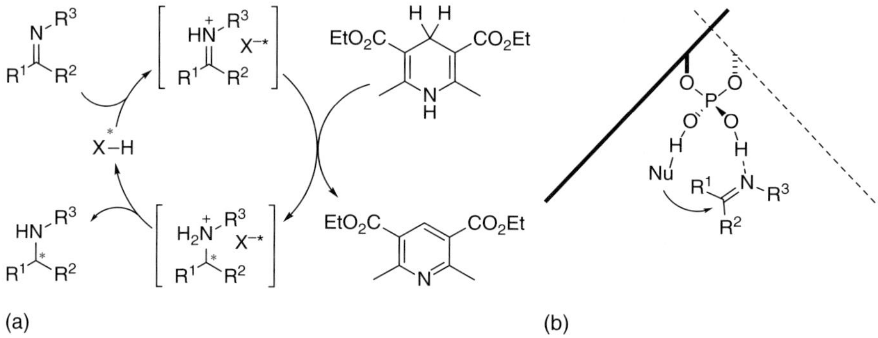

Scheme 9.8 Mechanism of asymmetric transfer hydrogenation of imines catalyzed by CPA.

it interacted with HEH through hydrogen bonding. These interactions satisfied the "three-point interaction model," well explaining the high enantioselectivity of the above reaction.

Benzothiazoline, another widely used hydrogen source in the asymmetric transfer hydrogenation, exhibits reducing ability by releasing the molecular hydrogen to form more stable aromatic benzothiazole [27]. In 2009, the Akiyama group reported the first example of benzothiazoline as hydrogen source in a CPA-catalyzed asymmetric transfer hydrogenation of ketimines **31** [28]. Even the reduction of aliphatic ketimines proceeded smoothly without any loss of enantioselectivity (Scheme 9.9). In addition, the Akiyama group found that 2-arylindoline could also be an efficient hydrogen donor to enable an asymmetric transfer hydrogenation of aromatic ketimines [29].

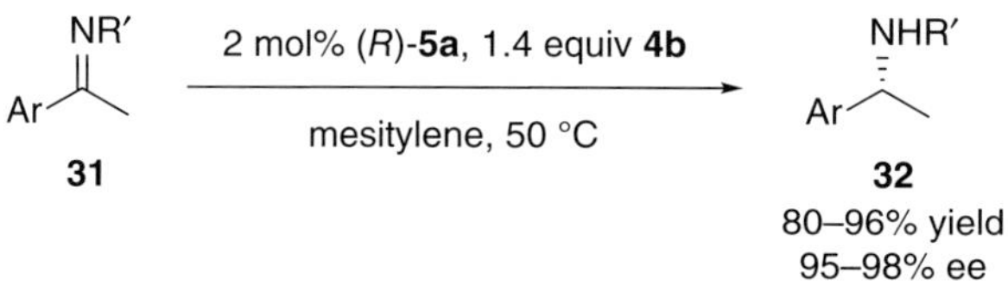

Scheme 9.9 Asymmetric transfer hydrogenation of imines using benzothiazolines by Akiyama.

Inspired by this work, a series of CPA-catalyzed asymmetric transfer hydrogenation of ketimines were reported using HEHs [30] or benzothiazolines [31] as reducing agents with excellent yields and enantioselectivities, with substrates such as linear imines (α-imino esters, β,γ-alkynyl-α-imino esters, and *ortho*-hydroxyaryl ketimines) and cyclic imines (benzoxazines, benzothiazines, benzoxazinones, benzodiazepinones, and 3*H*-indoles) among others (for an overview, see Figure 9.2).

Although CPA-catalyzed asymmetric transfer hydrogenation of imines has made great progress, generally the reductions were limited to *N*-aryl imines. In 2015, the List group reported the chiral disulfonimide-catalyzed HEH-mediated reduction of *N*-alkyl imines **33** [32]. The *in situ* protection of the basic *N*-alkyl amine product with the easily removable Boc group could solve the problem of product inhibition. In this reduction process, the chiral disulfonimide promoted

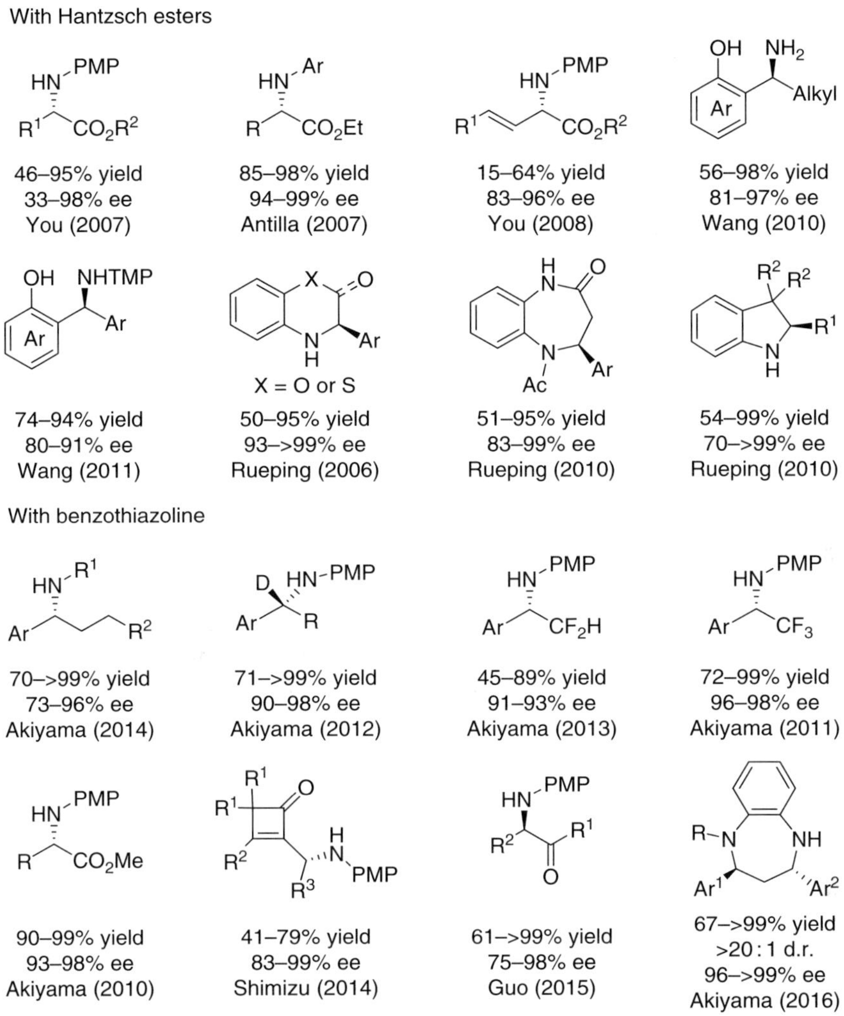

Figure 9.2 Asymmetric transfer hydrogenation of imines using Hantzsch esters and benzothiazolines as hydrogen sources.

the reaction of *N*-methyl imines with up to 97% ee while only disappointing conversion and enantioselectivity were obtained using the corresponding CPA. An aliphatic imine could also be employed but gave poor enantioselectivity (Scheme 9.10).

Reductive amination reactions remain a versatile coupling that enables the chemoselective union of diverse ketone- and amine-containing fragments, providing a rapid and general access to chiral amines with high efficiency and step economy. In 2006, the MacMillan group developed the first practical enantioselective organocatalytic reductive amination using a CPA [33]. Several bifunctional catalysts including chiral thioureas and TADDOLs did not promote the reductive amination while CPAs could provide the desired chiral amine

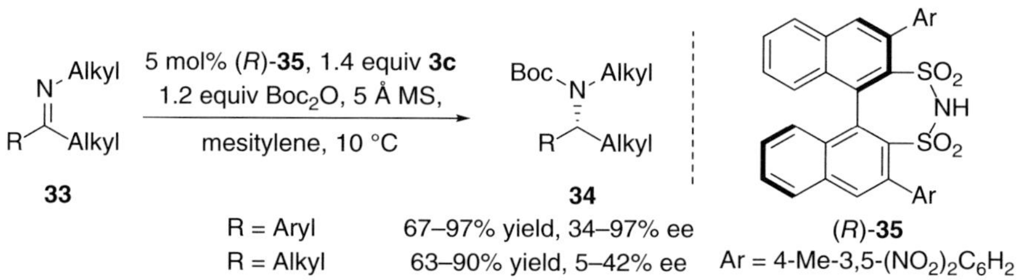

Scheme 9.10 Disulfonimide-catalyzed asymmetric transfer hydrogenation of *N*-alkyl imines.

products **38** with up to 97% ee. After a systematic optimization of the reaction conditions, the sterically hindered triphenylsilyl-substituted CPA **5b** could catalyze the reaction with satisfactory yields (49–92%) and enantioselectivities (83–97% ee) (Scheme 9.11).

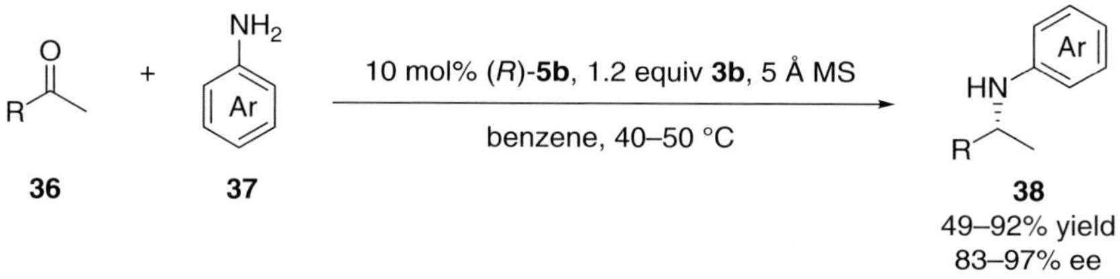

Scheme 9.11 Enantioselective organocatalytic reductive amination by MacMillan.

In the following, the List group disclosed the CPA-catalyzed asymmetric reductive amination of α-branched aldehydes and ketones through dynamic kinetic resolution (DKR) [34]. In this process, α-branched aldehydes **39** would undergo a fast racemization in the presence of amine **40** and acid catalyst **5a** *via* an acid-catalyzed imine/enamine tautomerization and subsequent enantioselective hydrogen transfer, giving chiral amines **41** with high enantioselectivities (up to 98% ee). Soon after, the List group developed a catalytic asymmetric reductive amination of cyclic ketones such as **42** with PMP-protected amines **43** through DKR, providing an efficient diastereoselective and enantioselective synthesis of the valuable chiral *cis*-2-substituted cyclohexylamines **44** (Scheme 9.12) [35]. In addition, significant contributions were also made by other groups in intermolecular organocatalytic reductive amination [36].

The Akiyama group performed an intramolecular asymmetric reductive amination and applied this process in the synthesis of fused piperidine and pyrrolidine heterocycles **46** with all-carbon stereogenic centers (Scheme 9.13) [37]. This transformation involved CPA-catalyzed symmetry breaking with good to excellent enantioselectivities (79–98% ee).

Recently, hemiaminals have been extensively involved in asymmetric transfer hydrogenations through the *in situ* formation of *N*-carbonyl iminium ions. Zhou group reported the enantioselective hydrogenolysis of 3-alkyl-3-hydroxyisoindolin-1-ones **47a** to produce cyclic *N*-carbonyl chiral amines in excellent enantioselectivities (65–93% ee) with **3c** as hydride source [38]. However, only moderate 61% ee was obtained when the scope extended

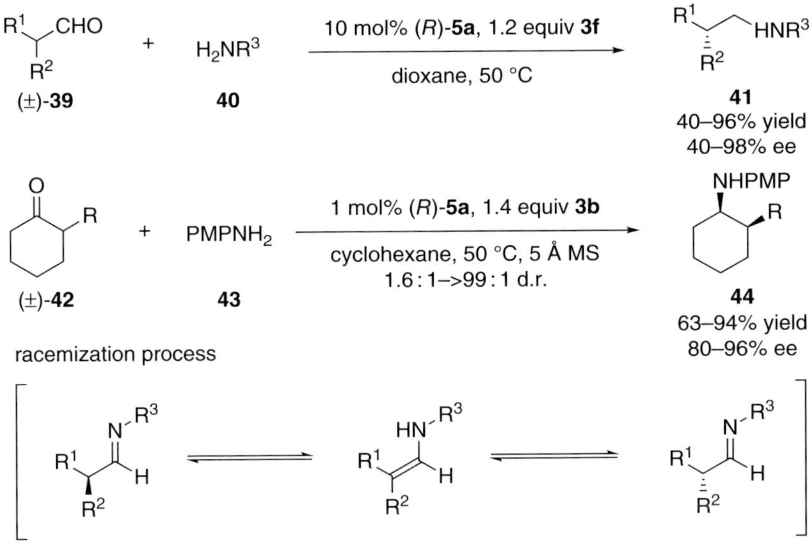

Scheme 9.12 Asymmetric reductive amination *via* DKR by List.

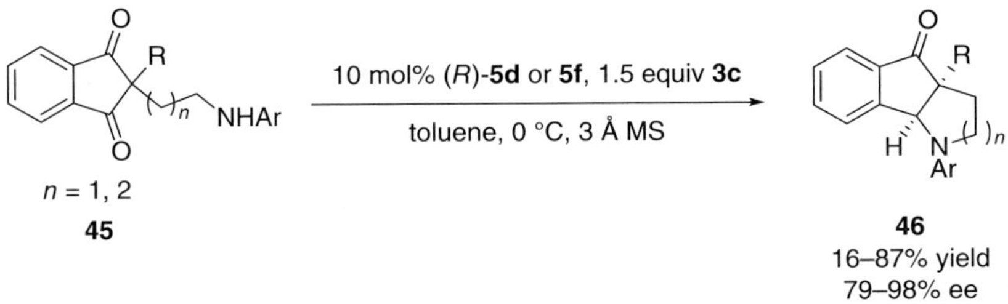

Scheme 9.13 Intramolecular asymmetric reductive amination.

to the 3-phenyl substituted substrate. Later, the Jia group realized highly enantioselective transfer hydrogenation of 3-alkyl-3-hydroxyisoindolin-1-ones **47b** with up to 91% ee while replacing HEH **3c** with benzothiazoline **4a** (Scheme 9.14) [39]. The You group used this strategy to synthesize a series of enantioenriched tetrahydro-β-carbolines *via* CPA-catalyzed enantioselective transfer hydrogenation of hydroxylactams with up to 94% yield and 90% ee [40].

The Antilla group disclosed a highly enantioselective transfer hydrogenation of enamides **50** catalyzed by dual acid catalyst system [41]. The loading of the CPA could be as low as 1 mol% but still excellent yields and enantioselectivities were obtained with a combination of CPA and achiral acetic acid as the catalyst. In this transformation, the iminium was the key intermediate and the achiral acetic acid could be used to generate a suitable concentration of iminium while being inactive in the transfer hydrogenation step (Scheme 9.15).

Tang and Li group developed a new method for asymmetric transfer hydrogenation of 1,2-dihydroquinolines [42]. In this reaction, 1,2-dihydroquinolines **52** could be transferred to the reactive *aza-o*-xylylene and *o*-quinone methide

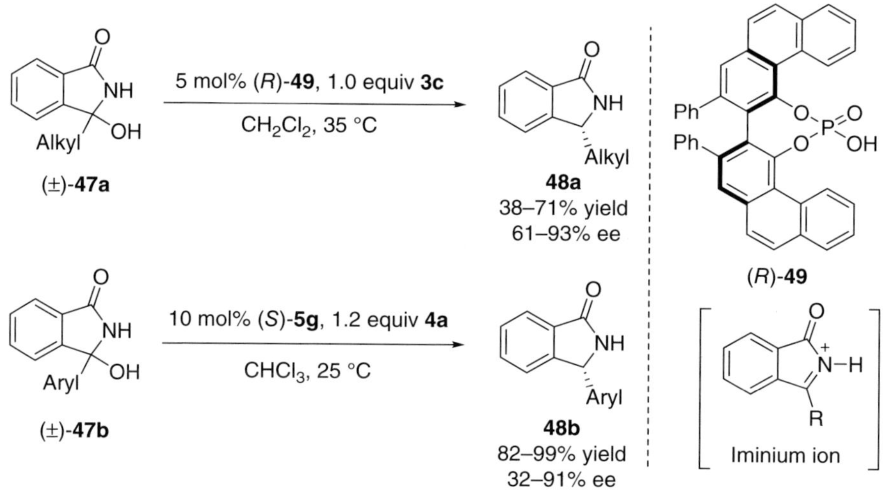

Scheme 9.14 Asymmetric hydrogenolysis of racemic 3-substituted 3-hydroxyisoindolin-1-ones.

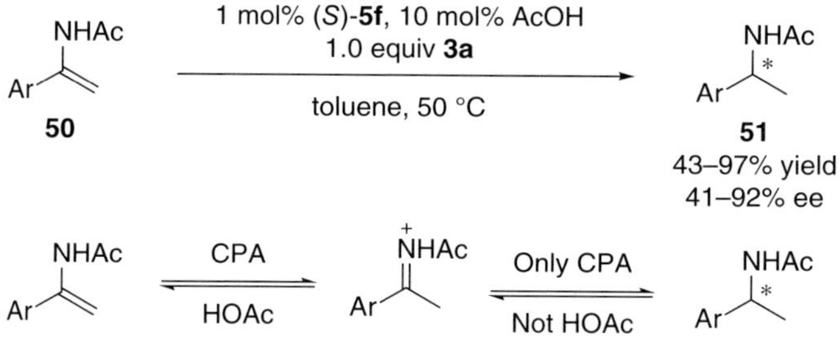

Scheme 9.15 Asymmetric transfer hydrogenation of enamides catalyzed by CPA.

imine (AOX) intermediate through dearomatization with a catalytic amount of CPA (Scheme 9.16). The resulting intermediate formed *in situ* could efficiently be transfer hydrogenated with HEH, giving good to excellent enantioselectivities (65–94% ee).

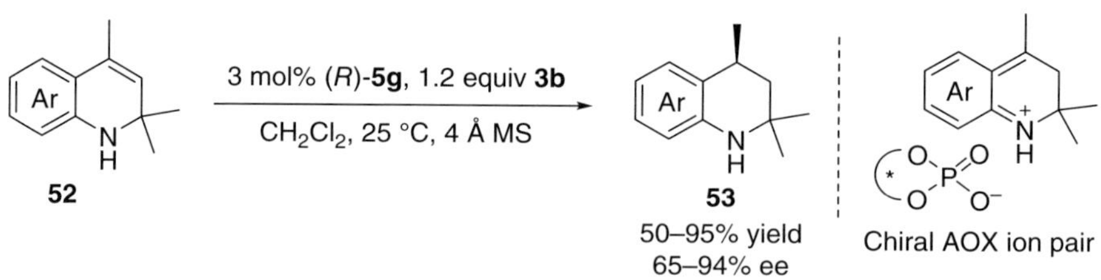

Scheme 9.16 Enantioselective organocatalytic transfer hydrogenation of dihydroquinolines through the formation of *aza-o*-xylylene.

An asymmetric transfer hydrogenation of C=N bonds was also involved in the construction of axially chiral biaryls, which are widely applied as chiral catalysts. The groups of Liu and Tan group applied CPA-catalyzed asymmetric reductive amination in the kinetic resolution of axially chiral 1,1′-binaphthyl-2,2′-diamine (BINAM) (±)-**54** with selectivities up to $s = 340$ (Scheme 9.17) [43]. Unsatisfactory results were obtained in the kinetic resolution of simple non-protected BINAM under optimized conditions, suggesting that the bulky protecting group was necessary for good interaction between the catalyst and the substrate.

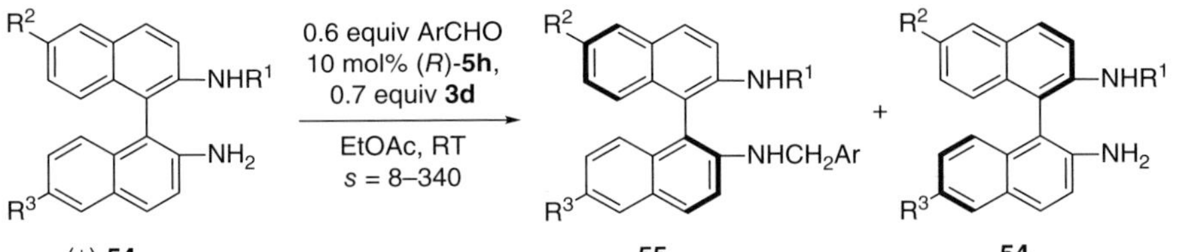

Scheme 9.17 Kinetic resolution of axially chiral BINAM *via* asymmetric reductive amination.

The Akiyama group exploited a CPA catalyzed DKR in the enantiodivergent atroposelective synthesis of the chiral biaryls through asymmetric transfer hydrogenation (Scheme 9.18) [44]. The ring opening/ring closing equilibrium between the biaryl *N,O*-acetal and biaryl imine and subsequent kinetic resolution type asymmetric transfer hydrogenation of imine proceeded to axially chiral compounds with excellent enantioselectivities (77–97% ee). Interestingly, the atroposelectivity of the products **58** was completely controlled by the choice of the hydroxyaniline **57**: whereas the use of *o*-hydroxyaniline derivatives favored the *R* enantiomer of **58**, the corresponding *m*-hydroxyaniline derivatives furnished the *S* enantiomer of the products **58**.

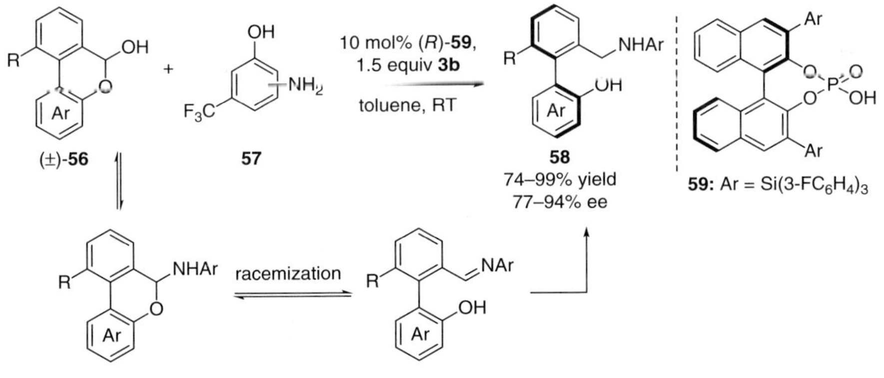

Scheme 9.18 Atroposelective synthesis of chiral biaryls through asymmetric transfer hydrogenation involving DKR.

9.4 Asymmetric Transfer Hydrogenation of C=O Bonds

Asymmetric transfer hydrogenation of C=O bonds is an efficient access to chiral alcohols and great progress has been made by transition metal catalysts [45].

In addition, great efforts have also been devoted to design a series of chiral organic hydride donors involved in the asymmetric transfer hydrogenation of C=O bonds, which were activated by stoichiometric amount of Lewis acids [46]. However, a fairly limited number of reports focused on the asymmetric organocatalytic transfer hydrogenation of C=O bonds have been disclosed.

In 2007, the Connon group developed the first class of thiourea-based bifunctional organo-catalyst **62** incorporating a chiral NADH analoge component that could both activate the carbonyl group and transfer a hydride to 1,2-diketone **60** without the use of stoichiometric amount of Lewis acids [47]. The active hydropyridine catalytic species could be generated and recycled *in situ* by sodium dithionite. Moderate enantioselectivity (15 hours, 29% conv., and 25% ee) could be obtained while the product underwent racemization (24 hours, 97% conv. and 2% ee) under the reaction conditions, hampering the synthetic potential of the methodology (Scheme 9.19).

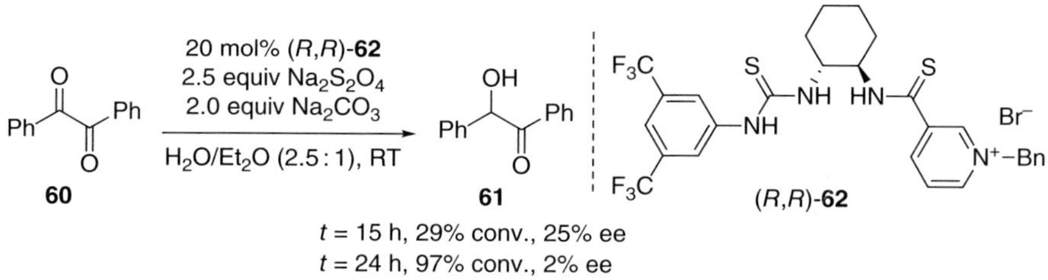

Scheme 9.19 Enantioselective organocatalytic transfer hydrogenation of C=O bond.

9.5 Asymmetric Transfer Hydrogenation of Heteroaromatics

The direct asymmetric transfer hydrogenation of heteroaromatics proved to be an efficient method to afford the chiral saturated or partially saturated heterocycles, widely prevalent in ubiquitous biologically active compounds [1]. However, some challenges remain due to the high resonance stabilization energy of the substrate, and the ability to deactivate the catalysts by coordination. In 2006, the Rueping group first realized a CPA-catalyzed asymmetric transfer hydrogenation of quinolines using HEH as hydrogen source (Scheme 9.20) [48]. The mild reaction conditions of metal-free reduction, the operational simplicity, as well as the low catalyst loading (2 mol%) led to an attractive method for the synthesis of biologically active tetrahydroquinoline alkaloids: (+)-cuspareine, (+)-galipinine, and (−)-angustureine. A proposed mechanism is depicted in Scheme 9.20. The activation of quinolines was achieved by a catalytic protonation through chiral Brønsted acid, which subsequently allowed a cascade hydrogenation involving a 1,4-hydride addition, proton transfer, and final 1,2-hydride addition.

Inspired by Rueping's work [48], the organocatalytic asymmetric hydrogenation of quinolines has attracted wide attention [49]. Among them, the Du group

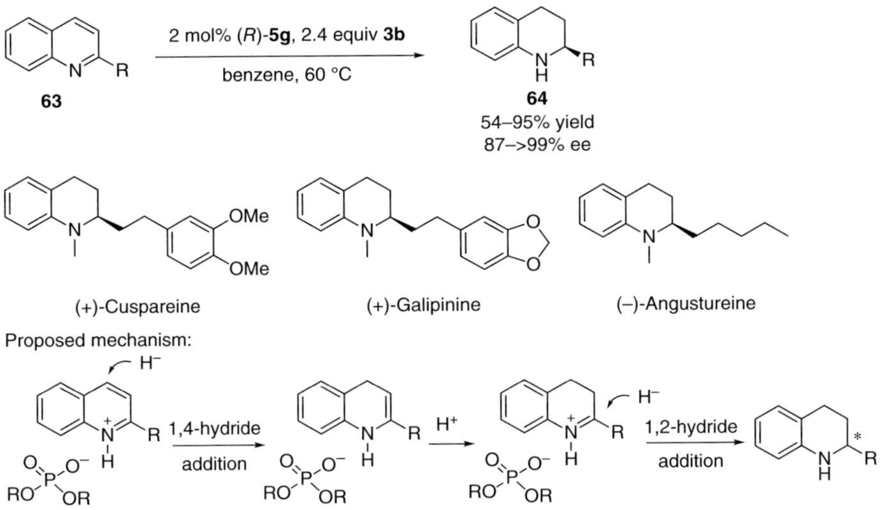

Scheme 9.20 Asymmetric transfer hydrogenation of quinolines and proposed mechanism.

designed a new double axially chiral phosphoric acid **65** as the chiral catalyst [50], achieving 2-substituted tetrahydroquinolines with excellent enantioselectivities (82–98% ee) and low-loading catalyst (0.2 mol%). Compared to CPA based on the 3,3-substituted binol scaffold, this catalyst showed higher efficiency in the asymmetric transfer hydrogenation of both the 2-aryl and 2-alkyl substituted quinolines. In 2013, the groups of Betzer and Marinetti developed planar CPA **66** based on the ferrocene-bridged paracyclophane structure and realized the asymmetric hydrogenation of 2-aryl quinolines (82–92% ee) [51]. Besides, Voituriez, Guinchard, and coworkers incorporated a thiophostone structure into CPA **67** and applied it in the asymmetric transfer hydrogenation of 2-phenylquinoline with moderate enantioselectivity [52]. The stereochemistry on the phosphorus atom played a crucial role for the enantioselectivity, as the catalysts bearing β-configured phosphorus performed better than α-configured phosphorus (Scheme 9.21).

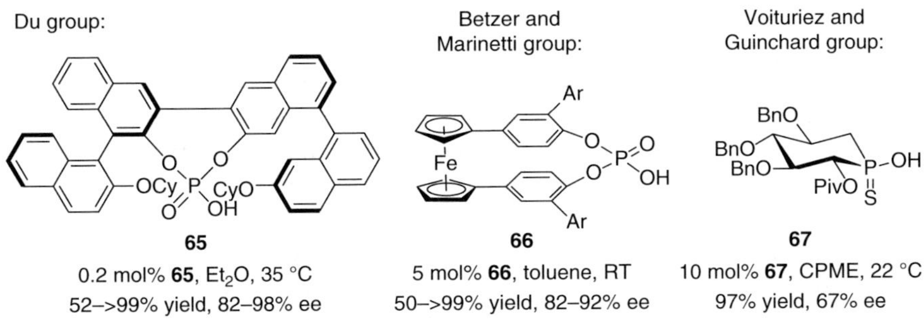

Scheme 9.21 CPAs developed by Du, Betzer and Marinetti, Voituriez and Guinchard and their application in the asymmetric transfer hydrogenation of 2-substituted quinolines.

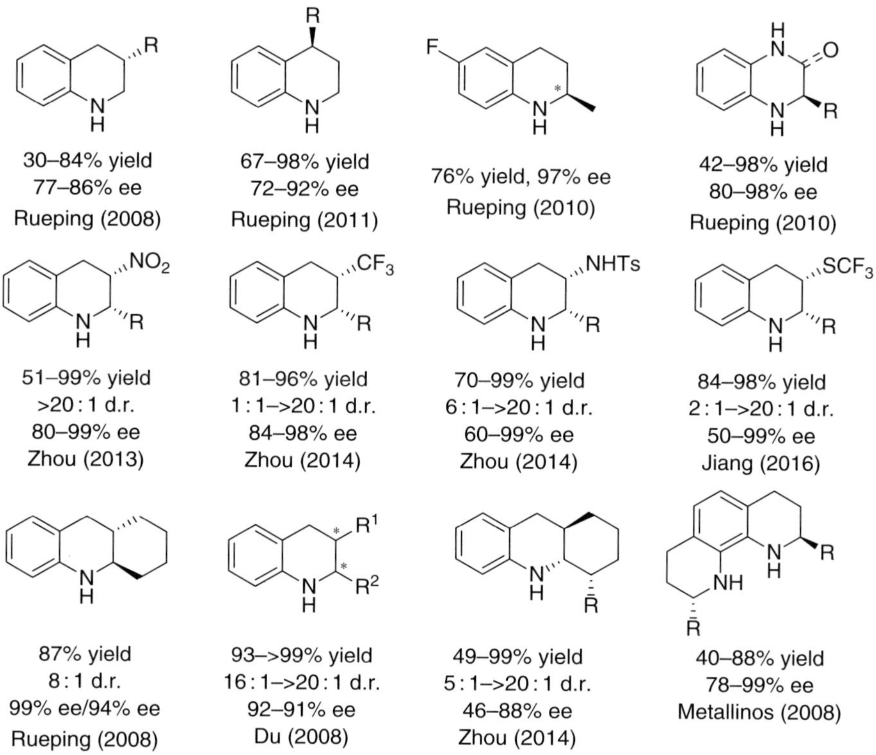

Figure 9.3 Asymmetric transfer hydrogenation of various substituted quinolines by other groups.

In addition, asymmetric hydrogenation of various substituted quinolines have been realized by the CPA/HEH system, including 3-substituted quinolines, 4-substituted quinolines, 3-trifluoromethyl substituted quinolines, 2-methyl-6-fluoroquinoline, 2-substituted quinoxalines, 2,3-disubstituted quinolines, 2,9-disubstituted 1,10-phenanthrolines as substrates (Figure 9.3) [53].

Contrary to the 2-substituted quinolines, the mechanism of asymmetric transfer hydrogenation of 2,3-disubstituted quinolines involved DKR in the rapid subsequent enamine–iminium isomerization after 1.4-hydride addition [53e]. In the following, a highly diastereoselective 1,2-hydride transfer provided solely the *cis* products. The high diastereo- and enantioselectivity achieved in the transfer hydrogenation was attributed to the fact that the rate of the tautomerization (k_1) was faster than the rate of 1,2-hydride addition and the reduction rate of 1,2-hydride addition (k_2) was faster than (k_3) (Scheme 9.22).

The Pélinski group disclosed the first partial transfer hydrogenation of lactone-fused quinolines **68** with achiral phosphoric acid to obtain *aza*-podophyllotoxin derivatives that exhibited potent anticancer activity with good yields (23–92%) and enantioselectivities (27–96% ee) [54]. The exclusive formation of stable 1,4-dihydroquinolines was rationalized by computational study (Scheme 9.23) [54a].

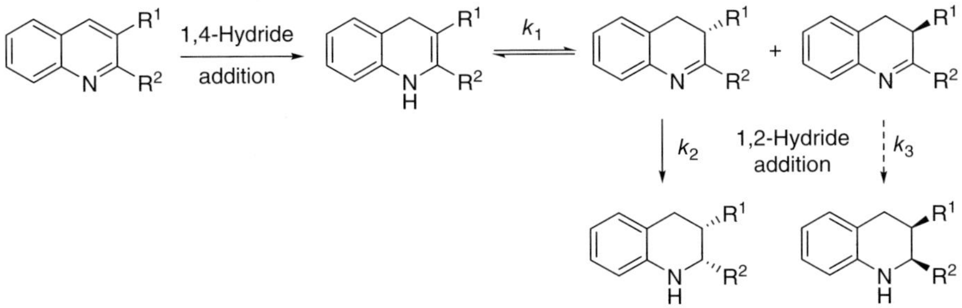

Scheme 9.22 Proposed mechanism for the asymmetric transfer hydrogenation of 2,3-disubstituted quinolines *via* DKR.

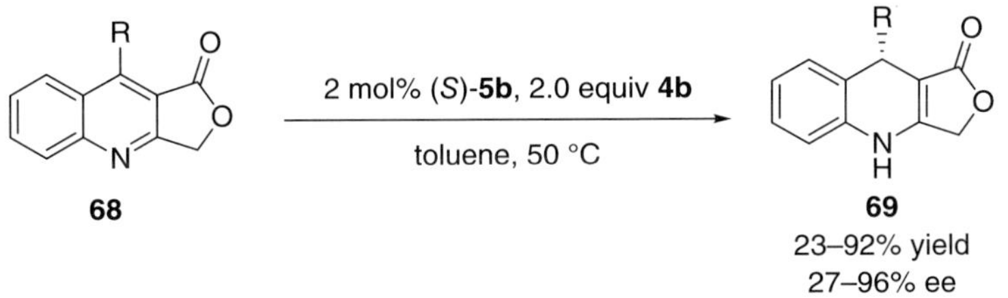

Scheme 9.23 Partial asymmetric transfer hydrogenation of lactone-fused quinolines.

The Rueping group disclosed the first enantioselective reduction of trisubstituted pyridines in the presence of Brønsted acids [55]. After optimization of the reaction conditions, anthryl-substituted CPA **5f** proved ideal, affording the chiral azadecalinones and tetrahydropyridines **71** with good yields and excellent enantioselectivities. The developed asymmetric transfer hydrogenation of pyridines could be used as a key step in the synthesis of decahydroquinolines from the pumiliotoxin family such as *diepi*-Pumiliotoxin C (Scheme 9.24).

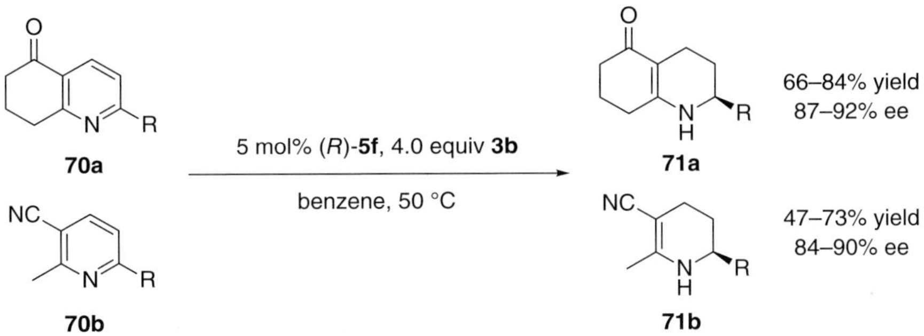

Scheme 9.24 Asymmetric transfer hydrogenation of electron-poor pyridines.

Recently, phosphothreonine (pThr) was found to constitute a new class of CPA catalyst upon insertion into peptides [56], which could be applied in the asymmetric transfer hydrogenation of 2-substituted quinolines containing a C8-amino substituent **72** with up to 88% ee. Strikingly, this catalyst framework is lacking the C2-symmetry of the better known CPA scaffolds, and yet was able to overcome the existence of both nonequivalent tautomeric states and the many rotatable bonds within the catalyst **74**. The selectivity was achieved through interactions with substrates in the context of the peptide secondary structure. Besides, NMR studies indicated that hydrogen bonding interactions promoted strong complexation between substrates and a rigid β-turn catalyst (Scheme 9.25).

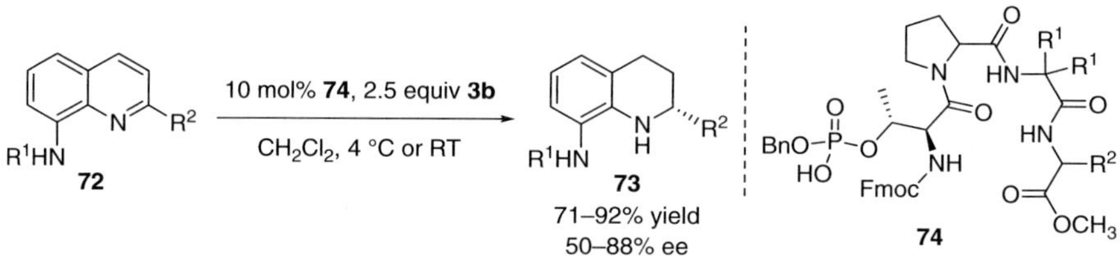

Scheme 9.25 Asymmetric transfer hydrogenation of 8-aminoquinolines catalyzed by a pThr-containing peptide.

Asymmetric transfer hydrogenation of isoquinolines provides an efficient approach to tetrahydroisoquinolines, which are the key structures in the biologically active compounds and naturally occurring alkaloids such as (−)-carnegine and (+)-solifenacin [57]. The Zhou and Shi groups developed an asymmetric transfer hydrogenation of isoquinolines **75** through a chiral anion metathesis strategy [58], affording chiral 1,2-dihydroisoquinolines with up to 79% ee. In this transformation, isoquinolines first reacted with 2,2,2-trichloroethylchloroformate (ClTroc) in the presence of the CPA, generating a chiral contact ion pair. The subsequent hydride transfer led to the formation of the chiral N-protected dihydroquinolines **76** (Scheme 9.26).

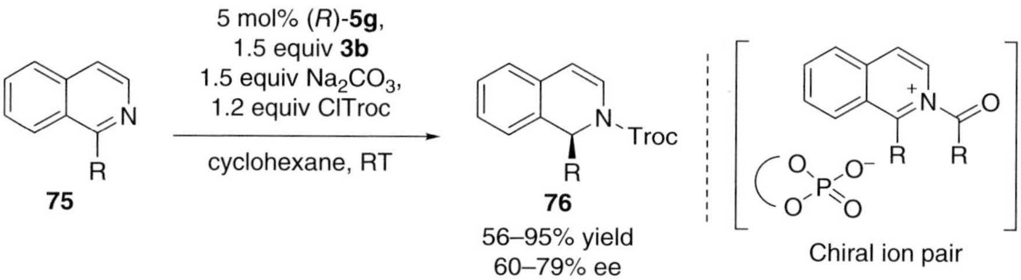

Scheme 9.26 Asymmetric hydrogenation of isoquinolines via chiral anion metathesis strategy.

Axially chiral biaryls scaffolds are prevalent in materials, natural products, chiral catalysts and ligands [59]. The Zhou group first successfully applied asymmetric transfer hydrogenation to the kinetic resolution of axially chiral 5- or 8-substituted quinolines, simultaneously obtaining two kinds of axially chiral skeletons with up to $s = 209$ [60]. As for the axially chiral substrates **77** recovered and the heterocyclic products **78** afforded, they could be transformed into each other by simple reduction or oxidation process in subsequent steps (Scheme 9.27).

The Rueping group applied a photocyclization–reduction strategy in the synthesis of chiral 4*H*-chromenes though benzopyrylium ion intermediates [61], starting with 2-hydroxylchalcones **79**. The reaction proceeded well under

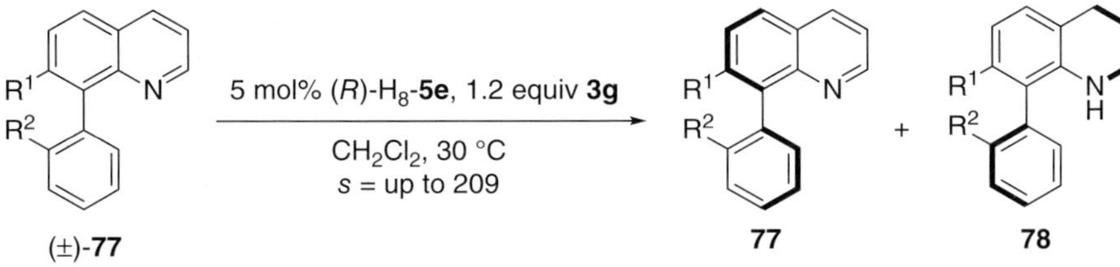

Scheme 9.27 Kinetic resolution of axially chiral biaryls *via* asymmetric transfer hydrogenation of quinolines.

optimized conditions and a wide scope of 4*H*-chromene derivatives **80** could be prepared in high enantioselectivities (80–94% ee) and high yields. In a different approach, the Terada group utilized the protonation of racemic 2*H*-chromenols (±)-**81** to form the same benzopyrylium ion intermediates in the presence of CPA $[H_8]$-**5a** and subsequent enantioselective 1,4-reduction of the 1-benzopyrylium ions to provide the same chiral 4*H*-chromene derivatives **80** [62].

The Liu group developed the first one-pot redox deracemization of a series of cyclic benzylic ethers(±)-**82**, including 6*H*-benzochromenes, isochromanes, and 1*H*-isochromenes with high enantioselectivities (83–99% ee) [63]. An "acetal pool" strategy was adopted to harmonize the complete oxidation of secondary ethers with imidodiphosphoric acid catalyzed asymmetric transfer hydrogenation. A suitable protic additive (R'OH, methanol) could react quickly and thermodynamically favorably with the intermediate oxocarbenium ion, enabling the two processes of ether oxidation and enantioselective reduction to be integrated successfully (Scheme 9.28).

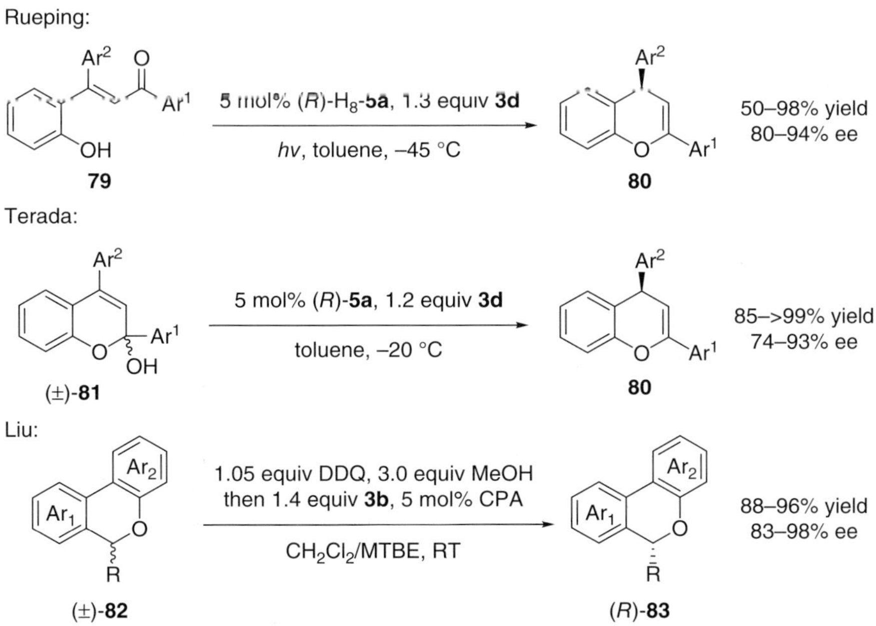

Scheme 9.28 Asymmetric transfer hydrogenation of oxocarbenium ion intermediates.

9.6 Summary and Outlook

Organocatalytic transfer hydrogenation has attracted wide attention and asymmetric transfer hydrogenations of a number of C=C, C=N, C=O bonds, and heteroaromatics under mild conditions have successfully been realized with high activities and enantioselectivities. These organocatalytic systems were developed on the basis of reductive transformation in biological systems, combining chiral organocatalysts (chiral phosphoric acid, chiral amine, chiral thiourea, and others) and hydrogen sources (Hantzsch ester, benzothiozoline, and others). In particular, this approach provides a potential application in the industry due to ease of operation, safety, and the absence of metal residues. Although significant progress has been made, there remain some limitations: a stoichiometric amount of hydrogen source is needed in most cases, resulting in low atom economy and difficulty in removing the by-products. The substrates bearing C=C bond are limited to functional alkenes; unactivated alkenes have not been realized as substrates. With respect to heteroaromatics, the results are not satisfactory for asymmetric transfer hydrogenation of isoquinolines and electron-rich pyridines. In addition, the current organocatalysts have some disadvantages in asymmetric transfer hydrogenation of C=O bonds, probably due to the too weak interaction with the carbonyl group. However, recent studies clearly demonstrate that new strategies and organocatalysts are emerging to overcome these obstacles. It is believed that organocatalytic transfer hydrogenation holds plenty of future opportunities for further applications.

Abbreviations

NADH	reduced nicotinamide adenine dinucleotide
NADPH	reduced nicotinamide adenine dinucleotide phosphate
$FADH_2$	reduced flavine adenine dinucleotide
TCA	trichloroacetic acid
TFA	trifluoroacetic acid
Piv	pivaloyl
TMP	3,4,5-trimethoxyphenyl
TRIP	2,4,6-triisopropylphenyl
PMP	4-methoxyphenyl
Troc	2,2,2-trichloroethoxycarbonyl
TADDOL	2,2-dimethyl-$\alpha,\alpha,\alpha^1,\alpha^1$-tetraaryl-1,3-dioxolane-4,5-dimethanol
DFT	density functional theory
Ar	aryl (substituted aromatic ring)
MTBE	methyl *tert*-butyl ether
CMPE	cyclopentyl methyl ether
DDQ	2,3-dichloro-5,6-dicyano-1,4-benzoquinone
MS	molecular sieves

References

1 (a) J.G. de Vries, C.J. Elsevier, *The Handbook of Homogeneous Hydrogenation*. Weinheim: Wiley-VCH, 2007, 1–1568; (b) W. Tang, X. Zhang, *Chem. Rev.* 2003, 103, 3029–3070; (c) J.-H. Xie, S.-F. Zhu, Q.-L. Zhou, *Chem. Rev.* 2011, 111, 1713–1760.
2 (a) W.S. Knowles, *Angew. Chem. Int. Ed.* 2002, 41, 1998–2007; (b) R. Noyori, *Angew. Chem. Int. Ed.* 2002, 41, 2008–2022.
3 (a) S.J. Roseblade, A. Pfaltz, *Acc. Chem. Res.* 2007, 40, 1402–1411; (b) R. Noyori, S. Hashiguchi, *Acc. Chem. Res.* 1997, 30, 97–102; (c) S. Gladiali, E. Alberico, *Chem. Soc. Rev.* 2006, 35, 226–236.
4 (a) J. Gebicki, A. Marcinek, J. Zielonka, *Acc. Chem. Res.* 2004, 37, 379–386; (b) R.H. Houtkooper, C. Cantó, R.J. Wanders, J. Auwerx, *Endocr. Rev.* 2010, 31, 194–223.
5 (a) M. Rueping, J. Dufour, F.R. Schoepke, *Green Chem.* 2011, 13, 1084–1105; (b) S.G. Ouellet, A.M. Walji, D.W.C. MacMillan, *Acc. Chem. Res.* 2007, 40, 1327–1339; (c) C. Zheng, S.-L. You, *Chem. Soc. Rev.* 2012, 41, 2498–2518; (d) A.M.F. Phillips, A.J.L. Pombeiro, *Org. Biomol. Chem.* 2017, 15, 2307–2340.
6 (a) C. Zhu, J.R. Falck, *ChemCatChem* 2011, 3, 1850–1851; (b) C. Zhu, K. Saito, M. Yamanaka, T. Akiyama, *Acc. Chem. Res.* 2015, 48, 388–398.
7 (a) J.W. Yang, M.T.H. Fonseca, B. List, *J. Am. Chem. Soc.* 2005, 127, 15036–15037; (b) Y. Huang, A.M. Walji, C.H. Larsen, D.W.C. MacMillan, *J. Am. Chem. Soc.* 2005, 127, 15051–15053; (c) S.-G. Wang, W. Zhang, S.-L. You, *Org. Lett.* 2013, 15, 1488–1491; (d) S.-G. Wang, S.-L. You, *Angew. Chem. Int. Ed.* 2014, 53, 2194–2197; (e) F. Shi, W. Tan, H.-H. Zhang, M. Li, Q. Ye, G.-H. Ma, S.-J. Tu, G. Li, *Adv. Synth. Catal.* 2013, 355, 3715–3726; (f) H.-H. Liao, C.-C. Hsiao, E. Sugiono, M. Rueping, *Chem. Commun.* 2013, 49, 7953–7955.
8 J.W. Yang, M.T. Hechavarria Fonseca, B. List, *Angew. Chem. Int. Ed.* 2004, 43, 6660–6662.
9 J.W. Yang, M.T. Hechavarria Fonseca, N. Vignola, B. List, *Angew. Chem. Int. Ed.* 2005, 44, 108–110.
10 S.G. Ouellet, J.B. Tuttle, D.W.C. MacMillan, *J. Am. Chem. Soc.* 2005, 127, 32–33.
11 (a) I. Paterson, N.A. Miller, *Chem. Commun.* 2008, 4708–4710; (b) H. Lee, Y.-K. Lee, D.-G. Kim, M.-S. Son, T.-G. Nam, B.-S. Jeong, *Tetrahedron Lett.* 2014, 55, 5895–5897; (c) A. Michrowska, B. List, *Nat. Chem.* 2009, 1, 225–228.
12 T.J. Hoffman, J. Dash, J.H. Rigby, S. Arseniyadis, J. Cossy, *Org. Lett.* 2009, 11, 2756–2759.
13 (a) K. Akagawa, H. Akabane, S. Sakamoto, K. Kudo, *Org. Lett.* 2008, 10, 2035–2037; (b) K. Akagawa, H. Akabane, S. Sakamoto, K. Kudo, *Tetrahedron: Asymmetry* 2009, 20, 461–466.
14 S. Mayer, B. List, *Angew. Chem. Int. Ed.* 2006, 45, 4193–4195.
15 N.J.A. Martin, B. List, *J. Am. Chem. Soc.* 2006, 128, 13368–13369.

16 J.B. Tuttle, S.G. Ouellet, D.W.C. MacMillan, *J. Am. Chem. Soc.* 2006, 128, 12662–12663.
17 O. Gutierrez, R.G. Iafe, K.N. Houk, *Org. Lett.* 2009, 11, 4298–4301.
18 (a) S.T.-C. Eey, M.J. Lear, *Org. Lett.* 2010, 12, 5510–5513; (b) D.B. Ramachary, R. Sakthidevl, P.S. Reddy, *RSC Adv.* 2013, 3, 13497–13506.
19 (a) N.J.A. Martin, L. Ozores, B. List, *J. Am. Chem. Soc.* 2007, 129, 8976–8977; (b) N.J.A. Martin, X. Cheng, B. List, *J. Am. Chem. Soc.* 2008, 130, 13862–13863.
20 J.F. Schneider, M.B. Lauber, V. Muhr, D. Kratzer, J. Paradies, *Org. Biomol. Chem.* 2011, 9, 4323–4327.
21 (a) E. Martinelli, A.C. Vicini, M. Mancinelli, A. Mazzanti, P. Zani, L. Bernardi, M. Fochi, *Chem. Commun.* 2015, 51, 658–660; (b) E. Massolo, M. Benaglia, M. Orlandi, S. Rossi, G. Celentano, *Chem. Eur. J.* 2015, 21, 3589–3595.
22 A. Ferraro, L. Bernardi, M. Fochi, *Adv. Synth. Catal.* 2016, 358, 1561–1565.
23 Z. Wang, F. Ai, Z. Wang, W. Zhao, G. Zhu, Z. Lin, J. Sun, *J. Am. Chem. Soc.* 2015, 137, 383–389.
24 M. Rueping, E. Sugiono, C. Azap, T. Theissmann, M. Bolte, *Org. Lett.* 2005, 7, 3781–3783.
25 S. Hoffmann, A.M. Seayad, B. List, *Angew. Chem. Int. Ed.* 2005, 44, 7424–7427.
26 (a) L. Simón, J.M. Goodman, *J. Am. Chem. Soc.* 2008, 130, 8741–8747; (b) T. Marcelli, P. Hammar, F. Himo, *Chem. Eur. J.* 2008, 14, 8562–8571.
27 (a) H. Chikashita, M. Miyazaki, K. Itoh, *Synthesis* 1984, 308–310; (b) H. Chikashita, M. Miyazaki, K. Itoh, *J. Chem. Soc. Perkin Trans.* 1, 1987, 699–706.
28 C. Zhu, T. Akiyama, *Org. Lett.* 2009, 11, 4180–4183.
29 K. Saito, H. Miyashita, T. Akiyama, *Org. Lett.* 2014, 16, 5312–5315.
30 (a) M. Rueping, A.P. Antonchick, T. Theissmann, *Angew. Chem. Int. Ed.* 2006, 45, 6751–6755; (b) G. Li, Y. Liang, J.C. Antilla, *J. Am. Chem. Soc.* 2007, 129, 5830–5831; (c) Q. Kang, Z.-A. Zhao, S.-L. You, *Adv. Synth. Catal.* 2007, 349, 1657–1660; Corrigendum: *Adv. Synth. Catal.* 2007, 349, 2075; (d) Q. Kang, Z.-A. Zhao, S.-L. You, *Org. Lett.* 2008, 10, 2031–2034; (e) T.B. Nguyen, H. Bousserouel, Q. Wang, F. Guéritte, *Org. Lett.* 2010, 12, 4705–4707; (f) M. Rueping, C. Brinkmann, A.P. Antonchick, I. Atodiresei, *Org. Lett.* 2010, 12, 4604–4607; (g) M. Rueping, E. Merino, R.M. Koenigs, *Adv. Synth. Catal.* 2010, 352, 2629–2634; (h) T.B. Nguyen, H. Bousserouel, Q. Wang, F. Guéritte, *Adv. Synth. Catal.* 2011, 353, 257–262; (i) A. Aillerie, C. Gosset, C. Dumont, V. Skrzypczak, P. Champetter, S. Pellegrini, T. Bousquet, L. Pélinski, *RSC Adv.* 2016, 6, 54185–54188; (j) Z.-Y. Han, H. Xiao, L.-Z. Gong, *Bioorg. Med. Chem. Lett.* 2009, 19, 3729–3732.
31 (a) C. Zhu, T. Akiyama, *Adv. Synth. Catal.* 2010, 352, 1846–1850; (b) A. Henseler, M. Kato, K. Mori, T. Akiyama, *Angew. Chem. Int. Ed.* 2011, 50, 8180–8183; (c) T. Sakamoto, K. Mori, T. Akiyama, *Org. Lett.* 2012, 14, 3312–3315; (d) T. Sakamoto, K. Horiguchi, K. Saito, K. Mori, T. Akiyama, *Asian J. Org. Chem.* 2013, 2, 943–946; (e) K. Saito, K. Horiguchi, Y. Shibata, M. Yamanaka, T. Akiyama, *Chem. Eur. J.* 2014, 20, 7616–7620; (f) I. Hachiya, A. Ito, M. Shimizu, *Asian J. Org. Chem.* 2014, 3, 614–618; (g) W. Wen,

Y. Zeng, L.-Y. Peng, L.-N. Fu, Q.-X. Guo, *Org. Lett.* 2015, 17, 3922; (h) K. Horiguchi, E. Yamamoto, K. Saito, M. Yamanak, T. Akiyama, *Chem. Eur. J.* 2016, 22, 8078–8083.

32 V.N. Wakchaure, P.S.J. Kaib, M. Leutzsch, B. List, *Angew. Chem. Int. Ed.* 2015, 54, 11852–11856.

33 R.I. Storer, D.E. Carrera, Y. Ni, D.W.C. MacMillan, *J. Am. Chem. Soc.* 2006, 128, 84–86.

34 S. Hoffmann, M. Nicoletti, B. List, *J. Am. Chem. Soc.* 2006, 128, 13074–13075.

35 V.N. Wakchaure, J. Zhou, S. Hoffmann, B. List, *Angew. Chem. Int. Ed.* 2010, 49, 4612–4614.

36 (a) A. Kumar, S. Sharma, R.A. Maurya, *Adv. Synth. Catal.* 2010, 352, 2227–2232; (b) V.N. Wakchaure, M. Nicoletti, L. Ratjen, B. List, *Synlett* 2010, 2708–2710; (c) D. Enders, J.X. Liebich, G. Raabe, *Chem. Eur. J.* 2010, 16, 9763–9766; (d) K.-H. Kim, C.-Y. Lee, C.-H. Cheon, *J. Org. Chem.* 2015, 80, 6367–6374; (e) K. Saito, T. Akiyama, *Chem. Commun.* 2012, 48, 4573–4575; (f) Y. Wang, G. Li, H. Liu, Z. Tang, Y. Cao, G. Zhao, *Tetrahedron Lett.* 2017, 58, 2993–2996.

37 K. Mori, A. Miyake, T. Akiyama, *Chem. Commun.* 2015, 51, 16107–16110.

38 M.-W. Chen, Q.-A. Chen, Y. Duan, Z.-S. Ye, Y.-G. Zhou, *Chem. Commun.* 2012, 48, 1698–1700.

39 J.-Q. Zhou, W.-J. Sheng, J.-H. Jia, Q. Ye, J.-R. Gao, Y.-X. Jia, *Tetrahedron Lett.* 2013, 54, 3082–3084.

40 Q. Yin, S.-G. Wang, S.-L. You, *Org. Lett.* 2013, 15, 2688–2691.

41 G. Li, J.C. Antilla, *Org. Lett.* 2009, 11, 1075–1078.

42 G. Li, H. Liu, G. Lv, Y. Wang, Q. Fu, Z. Tang, *Org. Lett.* 2015, 17, 4125–4127.

43 D.-J. Cheng, L. Yan, S.-K. Tian, M.-Y. Wu, L.-X. Wang, Z.-L. Fan, S.-C. Zheng, X.-Y. Liu, B. Tan, *Angew. Chem. Int. Ed.* 2014, 53, 3684–3687.

44 K. Mori, T. Itakura, T. Akuyama, *Angew. Chem. Int. Ed.* 2016, 55, 11642–11646.

45 (a) Y.-Y. Li, S.-L. Yu, W.-Y. Shen, J.-X. Gao, *Acc. Chem. Res.* 2015, 48, 2587–2598; (b) T. Ikariya, J. Blacker, *Acc. Chem. Res.* 2007, 40, 1300–1308; (c) A. Bartoszewicz, N. Ahlsten, B. Martín-Matute, *Chem. Eur. J.* 2013, 19, 7274–7302.

46 (a) N.-X. Wang, J. Zhao, *Synlett* 2007, 18, 2785–2791; (b) C.-B. Bai, N.-X. Wang, Y. Xing, X.-W. Lan, *Synlett* 2017, 28, 402–414.

47 B. Procuranti, S.J. Connon, *Chem. Commun.* 2007, 1421–1423.

48 M. Rueping, A.P. Antonchick, T. Theissmann, *Angew. Chem. Int. Ed.* 2006, 45, 3683–3686.

49 (a) K. Isaac, J. Stemper, V. Servajean, P. Retailleau, J. Pastor, G. Frison, K. Kaupmees, I. Leito, J.-F. Betzer, A. Marinetti, *J. Org. Chem.* 2014, 79, 9639–9646; (b) Y. Wang, W. Liu, W. Ren, Y. Shi, *Org. Lett.* 2015, 17, 4976–4979; (c) Y. Zhang, R. Zhao, R.L.-Y. Bao, L. Shi, *Eur. J. Org. Chem.* 2015, 3344–3351; (d) X. Qiao, Z. Zhang, Z. Bao, B. Su, H. Xing, Q. Yang, Q. Ren, *RSC Adv.* 2014, 4, 42566–42568.

50 Q.-S. Guo, D.-M. Du, J. Xu, *Angew. Chem. Int. Ed.* 2008, 47, 759–762; Corrigendum: Angew. Chem. Int. Ed. 2008, 47, 1541–1541.

51 (a) J. Stemper, K. Isaac, J. Pastor, G. Frison, P. Retailleau, A. Voituriez, J.-F. Betzer, A. Marinetti, *Adv. Synth. Catal.* 2013, 355, 3613–3624; (b) J. Stemper, K. Isaac, V. Duret, P. Retailleau, A. Voituriez, J.-F. Betzer, A. Marinetti, *Chem. Commun.* 2013, 49, 6084–6086.

52 A. Ferry, J. Stemper, A. Marinetti, A. Voituriez, X. Guinchard, *Eur. J. Org. Chem.* 2014, 2014, 188–193.

53 (a) M. Rueping, T. Theissmann, M. Stoeckel, A.P. Antochick, *Org. Biomol. Chem.* 2011, 9, 6844–6850; (b) M. Rueping, T. Theissmann, S. Raja, J.W. Bats, *Adv. Synth. Catal.* 2008, 350, 1001–1006; (c) M. Rueping, M. Stoeckel, E. Sugiono, T. Theissmann, *Tetrahedron* 2010, 66, 6565–6568; (d) M. Rueping, F. Tato, F.R. Schoepke, *Chem. Eur. J.* 2010, 16, 2688–2691; (e) X.-F. Cai, M.-W. Chen, Z.-S. Ye, R.-N. Guo, L. Shi, Y.-Q. Li, Y.-G. Zhou, *Chem.-Asian J.* 2013, 8, 1381–1385; (f) R.-N. Guo, Z.-P. Chen, X.-F. Cai, Y.-G. Zhou, *Synthesis* 2014, 46, 2751–2756; (g) X.-F. Cai, R.-N. Guo, G.-S. Feng, B. Wu, Y.-G. Zhou, *Org. Lett.* 2014, 16, 2680–2683; (h) J. Zhou, Q.-F. Zhang, W.-H. Zhao, G.-F. Jiang, *Org. Biomol. Chem.* 2016, 14, 6937–6941; (i) M.-W. Chen, X.-F. Cai, Z.-P. Chen, L. Shi, Y.-G. Zhou, *Chem. Commun.* 2014, 50, 12526–12529; (j) C. Metallinos, F.B. Barrett, S. Xu, *Synlett* 2008, 2008, 720–724; (k) G.V. More, B.M. Bhanage, *Tetrahedron: Asymmetry* 2015, 26, 1174–1179.

54 (a) A. Aillerie, V.L. de Talancé, A. Moncomble, T. Bousquet, L. Pélinski, *Org. Lett.* 2014, 16, 2982–2985; (b) Y. Hitotsuyanagi, M. Fukuyo, K. Tsuda, M. Kobayashi, A. Ozeki, H. Itokawa, K. Takeya, *Bioorg. Med. Chem. Lett.* 2000, 10, 315–317.

55 M. Rueping, A.P. Antonchick, *Angew. Chem. Int. Ed.* 2007, 46, 4562–4565.

56 C.R. Shugrue, S.J. Miller, *Angew. Chem. Int. Ed.* 2015, 54, 11173–11176.

57 (a) R. Paul, J.A. Coppola, E. Cohen, *J. Med. Chem.* 1972, 15, 720–726; (b) J.D. Scott, R.M. Williams, *Chem. Rev.* 2002, 102, 1669–1730.

58 L. Shi, Y. Ji, W. Huang, Y. Zhou, *Acta Chim. Sin.* 2014, 72, 820–824.

59 (a) G. Bringmann, T. Gulder, T.A.M. Gulder, M. Breuning, *Chem. Rev.* 2011, 111, 563–639; (b) M. Terada, *Synthesis* 2010, 12, 1929–1982.

60 J. Wang, M.-W. Chen, Y. Ji, S.-B. Hu, Y.-G. Zhou, *J. Am. Chem. Soc.* 2016, 138, 10413–10416.

61 C.-C. Hsiao, H.-H. Liao, E. Sugiono, I. Atodiresei, M. Rueping, *Chem. Eur. J.* 2013, 19, 9775–9779.

62 (a) M. Terada, T. Yamanaka, Y. Toda, *Chem. Eur. J.* 2013, 19, 13658–13662; (b) Y. Li, M. Wan, S. Sun, Z. Fu, H. Huang, L. Liu, *Org. Chem. Front.* 2018, 5, 1280–1283.

63 M. Wan, S. Sun, Y. Li, L. Liu, *Angew. Chem. Int. Ed.* 2017, 56, 5116–5120.

Index

a

acceptorless dehydrogenation 180–181
acetal pool 279
aceto-and benzophenones 81
acetophenone 122
  derivatives 80
acrylates 78, 179, 208, 210
active monomeric species 122
acyclic enamines 118
adenosine monophosphate (AMP) 249
adenosine triphosphate (ATP) 249
Ae amide catalyst 152
Ae catalyzed hydrogenation 153
AeN''_2 catalysts 154, 161, 162
air-stable FLPs 184–186
Al-based complex 71
alcohol dehydrogenases
  in multicatalytic processes 232–236
  in non-conventional media 229–230
aldehydes and ketones 29–31, 186, 199, 270
aldimines 29, 32, 33, 153, 154, 161, 170, 187–193, 235
aliphatic imine 269
aliphatic PNP-pincer ligand-based cationic and neutral nickel(II) complexes 68
alkenes 7, 9, 23, 26, 30, 40, 45, 47, 48, 52–54, 57, 65–78, 82, 83, 89, 95–98, 100–102, 105, 112, 114, 115, 118–120, 134, 135, 137, 143, 145, 147–150, 152, 153, 160, 161, 215
alkoxide intermediate 121
alkoxide species 122
alkyl amines 181
alkyl-and aryl-substituted alkenes 65–76
3-alkyl-3-hydroxyisoindolin-1-ones 270, 271
alkylidene malonates 207–211
alkyl-substituted ketones 81
alkynes 6, 7, 10, 13, 23, 24, 26–28, 52, 54, 55, 78–79, 98–106, 114, 119, 120, 137, 184, 215, 216
allylbenzene 68, 69, 71
allylic substitutions with a hydride nucleophile generated from H_2 95–98
allyl *N*-(1-methyl-4-oxocyclohexyl)-carbamate 239
α-alkyl β-keto esters 231
α-alkyl esters 130
α-amino acid esters 130
α-amino-β-ketoester hydrochlorides 80
α-amino-γ-butyrolactones 235
α-and/or β-substituted aldehydes, ketones or esters 76
α-branched aldehydes 270
α-diimine ligand 74
α-fluoroketones 236
α-halogenated-α,β-unsaturated chiral alcohol key intermediate 237
α-halohydrins 234, 236
α-haloketones 234
α-imino esters 268
α-methyl styrene 152

Homogeneous Hydrogenation with Non-Precious Catalysts, First Edition. Edited by Johannes F. Teichert.

α-mono-or β-disubstituted α,β-unsaturated carbonyl compounds 76
α-tocopherol 263
α,β-disubstituted β-trifluoromethyl nitroalkenes 266
α-or β-substituted hydrocinnamates 76
α,β-unsaturated acids 134
α,β-unsaturated aldehydes 262, 264
α,β-unsaturated carbonyl and carboxyl compounds 105
  α,β-unsaturated ketones and aldehydes 91–92
  conjugate reduction 88–91
  simple (non-conjugated) ketones and aldehydes 92–93
α,β-unsaturated carboxylic acids 242
α,β-unsaturated (di)esters 242
α,β-unsaturated ester precursors 242
α,β-unsaturated esters 77, 249
α,β-unsaturated ketones and aldehydes 91–92
α,β-unsaturated lactones 242
α,β-unsaturated nitriles 242
Ambrox® 130
amides 32, 132, 149–154, 156, 157, 161, 266
amines 30, 32, 33, 45, 46, 48, 50, 51, 100, 153, 154, 170, 176–181, 187, 188, 193, 200, 228, 245, 248–250, 261, 266, 269, 270
amine-tethered cyclopentadienyl ligand 118
4-amino-1-phenylpentan-2-ol 234
ammonia borane 104–105, 191, 205
(–)-angustureine 274
aniline 105, 180, 251
aniline-derived aminoboranes 182
ansa-aminoborane 192
ansa-ammonium borates 191
ansa (bridged) cyclopentadienyl ligands 113
anthracene 55
anti-aminoalcohols 80
anti-cancer agent 10, 51, 52
arene cobalt complexes 53
aromatic benzothiazole 261, 268
$[Ar_3P{-}H][H{-}B(C_6F_5)_3]$ 213
3-arylalkan-2-ones 232
aryl-bridged phosphonium-borate 187
aryl ketones 92, 93
aryl-linked phosphonium-borate FLP system 187
(*S*)-2-arylpropanols 230
2-arylpropionic acids 242
3-arylquinoxalines 205
aryl substituted *Z*-allylic chlorides 96
asymmetric counteranion directed catalysis (ACDC) 264
asymmetric FLP-catalyzed hydrogenation 189, 205
asymmetric transfer hydrogenation
  C=C bonds 262–266
  C=N bonds 266–273
  C=O bonds 273–274
  heteroaromatics 274–279
atorvastatin 238, 240
axially chiral biaryls scaffolds 278
axially chiral 1,1′-binaphthyl-2,2′-diamine (BINAM) 273
axially chiral 5-/8-substituted quinolines 278
aza-o-xylylene and *o*-quinone methide imine (AOX) 272
azepanes 248

b

$B(2,4,6\text{-}F_3{-}C_6H_2)_3$ 175, 181, 208
$B(2,6\text{-}F_2{-}C_6H_3)_3$ 175, 185, 200, 209, 210
$B(C_6F_5)_3$ 144, 170, 172, 173, 185, 193–195, 198
$B(C_6F_5)_3$-activated nitriles 187
Baeyer–Villiger monooxygenase (BVMO) 246
base and bulk chemicals 1
base-catalyzed hydrogenation of ketones 160
6*H*-benzochromenes 279
benzodiazepinones 268
(benzo)quinolines 205
1,4-benzoquinone 87
benzothiazines 248, 268

benzothiazoline 261, 268, 269, 271
benzoxazines 268
benzoxazinones 268
benzyl complexes $(DMAT)_2Ca\text{-}(THF)_2$ 145
benzyl cyclohexane carboxylate 44
β-activated vinylphosphonates 242
β-acylamino-substituted nitroolefins 78
β,β-disubstituted nitroolefins 265
β-cyanoacrylate esters 243
β-cyano-α,β-unsaturated esters 242
β,γ-alkynyl-α-imino esters 268
β-hydroxy acids 233
β-hydroxy nitriles 233
β-hydroxytriazoles 234
β-keto nitriles 233
β-nitroacrylates 265
β-tetrahydrocarbolines 248
bidentate bis(silylene)xanthene ligand 72
bidentate phosphine ligand 48, 65, 77, 79
binaphthophosphepine (*S*) 92
1,1′-binaphthyl system 191
((*S*)-Binapine) 78
biocatalysis 3, 227, 236, 251
biocatalysts 4, 227, 228, 232, 234, 248, 249, 252
biocatalytic (hydrogen transfer) reductions
  carboxylic acid reductases 249–250
  ene-reductases 241–247
  imine reductases 248–249
  ketoreductases 228–241
  nicotinamide cofactor 227
biologically active 5β-dihydrosteroids 265
biphenyl-bridged titanocene catalyst 118
bisiminopyridine complex 28
bisligated iron(0) complexes 24
1,1′-bisnaphthyl-derived aminoborane 191
bis(N-heterocyclic silylene) ligand 73
bis(pentafluorophenyl)boron chloride $((C_6F_5)_2BCl)$ 182
bis(pentafluorophenyl)-pentachlorophenyl borane $(C_6F_5)_2B(C_6Cl_5)$ 185
bisphosphines 170, 195, 200
bis(4-toloyl)silyl amine 217
1-(3,5-bis(trifluoromethyl)phenyl) ethanol 239
BoPhoz ligand (*R*,*S*) 92
borane-based FLPs 172, 185
borane-based strong Lewis acid 167
boranes 55, 81, 82, 170, 172, 184–187, 190–192, 198, 199, 202, 213, 215, 217, 236
borohydrides 167, 170, 176, 179, 200, 213
2-bromoacetophenone 49
3-bromoacetophenone 50
(*S*)-bromo-2-cyclohexen-1-ol 237
bromo substituted styrene 101
Brønsted acid 121, 170, 172, 179, 181, 266, 274
(*R*)-butane-1,3-diol 238
2-butanone 122

c

camphor-derived borane structures 189
Candida krusei ZJB-09162 238
Candida parapsilosis carbonyl reductase (CPCR) 230
carbapenem antibiotics 231
carbon dioxide 20, 22, 25, 40, 41, 63, 81–83, 93, 132–134
carboxylic acid reductases 228, 249–250
carboxylic acids 40–46, 48, 57, 228, 242, 243, 246, 249–250
carboxylic esters 40–48
  and acids 44
(−)-carnegine 278
(*S*)-carvone 207, 210
cascade reactions 54, 234, 262
catalyst choice 4
catalyst productivity 4
catalytically active $[CuI(dbu)_2]$ 21, 22, 24, 66, 94, 95, 113, 115, 116, 142, 155, 185

catalytically active hydride species 128
catalytically inactive complex
  [(κ^4-PNNP)Fe(CO)] 21
catalytic hydrogenation 1, 18, 20, 28, 32, 69, 72, 111, 112, 127, 141, 142, 146, 148, 151, 153, 154, 156–160, 172, 176, 195, 217
cationic Me_3TACD-stabilized calcium hydride complex 147
C=C double bonds 43, 47, 92, 97, 144–151
chemoenzymatic oxidative system 246
chemoselective hydrogenation 1, 5, 47
chiral alcohol (*S*) 241
chiral alkylboranes 54
chiral amine 33, 188, 244, 265–267, 269, 270
chiral amino acids 49
chiral 1,3-amino alcohols 234
chiral anti-cancer agent 52
chiral 2-(1-arylethyl)phenols 266
chiral azadecalinones 277
chiral benzyl-substituted silanes 54
chiral binaphthol-derived phosphoric acids 264
chiral 1,1′-binaphthyl derivative 183
chiral bis-boranes 191
chiral catalyst 3, 83, 125, 191, 272, 278
chiral 3,4-dialkyl-3,4-dihydroiso-coumarins 231
chiral diphosphine ligand
  (*R*)-DTBM-Segphos 89
chiral 4*H*-chromene derivatives 278, 279
chiral indolines 191
chiral intramolecular frustrated Lewis pairs 184, 190, 191
chiral (1*R*,2*R*)-*N*,*N*′-dimethyl-1,2-diphenylethane-1,2-diamine 127
chiral N-protected dihydroquinolines 278
chiral partially oxidised tetradentate PNNO ligand 49
chiral phosphoric acids (CPA) 261, 262, 266, 275, 276, 280
chiral pincer-type ligand 51
chiral *tert*-butylsulfinimide 191
chiral thiourea catalysts 266
chiral thioureas 265, 269
2-chloro-1-(3,4-diflurophenyl)acetone 240
chloroiminium chlorides 200
2-chloropyridines 188
4*H*-chromene derivatives 279
$-CH_2PPh_2$ ligand 114
chromium 121–122
cinnamyl alcohol 249
(1*S*,2*S*)-*cis*-alcohol 233
cis-allyl *N*-(4-hydroxy-1-methyl-cyclohexyl)-carbamate 240
cis-and *trans*-4-substituted-4-amino-cyclohexanols 239
cis-cyclooctene 71
cis-2,6-substituted piperidines 204
cis-4-*tert*-butylcyclohexanol 238
(*R*)-citronellal 264
C=N double bonds 47, 143, 153–158
cobalt-catalyzed Diels–Alder reactions 45
cobalt-catalyzed hydrogenations
  carbon dioxide 40
  carboxylic acids 40–46
  carboxylic esters 40–46
  C=C 46–57
  C≡C 46
  C=N 46–57
  C=O 46–57
  dihydrogen-complexes *vs.* dihydride complexes 39–40
  (hetero)arenes 46–57
  nitriles 40–46
cobalt complex 53
  $Co(dppe)(CH_2SiMe_3)_2$ 48
  $CoH(N_2)(PPh_3)_3$ 46
cobalt dihydride species 41, 42
cobalt-hydride complexes 40, 52
cobalt monohydride species 41
cobalt-phosphine type complexes 42
cobalt(I) species 44
Co catalyzed hydroboration/hydrogenation sequence 10, 13
Codexis 5, 239
C=O double bonds 106, 157–160

cofactors/coenzymes 227
commercially available diphosphine (*S,S*) 92
commercially available perfluorated borane $B(C_6F_5)_3$ 172
conjugate Brønsted acids 172, 176
cooperative ligands 18, 23, 25, 26, 29, 34
copper(I) acetate (CuOAc) 87
copper-catalyzed hydrogenations
  allylic substitutions with a hydride nucleophile generated from H_2 95–98
  α,β-unsaturated ketones and aldehydes 91–92
  ammonia borane 104–105
  conjugate reduction 88–91
  dihydrogen mediated cross-coupling of internal alkynes and aryl iodides 105–106
  early studies on 87–88
  reduction of CO_2 to formate 93–95
  simple (non-conjugated) ketones and aldehydes 92–93
  Z-selective alkyne semihydrogenation 98–104
copper hydride monomer triphos/CuH 95
copper(II) salts 87, 88, 93
$Co(BF_4)_2$/triphos system 43
CPA-catalyzed asymmetric transfer hydrogenation 267, 268, 274
$[Cp_2Mo(\mu\text{-}OH)_2MoCp_2](OTs)_2$ 122
$[Cp_2ZrH(CH_2PPh_2)]_n$ 114, 119
Cu carbene catalyst 9
Cu catalyzed deuteride transfer 11
Cu/NHC complex [SIMesCuCl] 91
$[Cu(NO_3)(PPh_3)_2]$ 89
(+)-cuspareine 274
7-cyano-7-deazaguanine 250
cyclic and linear enones 264
cyclic benzylic ethers(±) 279
cyclic enones 89, 242, 264, 265
cyclic imines 115–118, 248, 250, 268
cyclohexane 44, 120, 132, 152, 233
cyclohexenyl phosphines 182
cyclohexyl and cycloheptyl keto nitriles 233
cyclohexyl-substituted ligands 66
cyclooctene 54, 66, 71, 73, 113
cyclopentadienyl cobalt complexes 42

d

deep-eutectic solvents (DES) 229
dehydro amino acid derivatives 8
dehydrogenative coupling 179–180
δ-keto acids 249
δ-keto aldehydes 249
(*S,S*)-3-dialkyl-3,4-dihydroisocoumarins 232
2,3-dialkylquinoxalines 206
dialkyl-substituted 1,3-diynes 98, 102
diamagnetic and trigonal bipyramidal complex 19
di-and trisubstituted *E*-α,β-unsaturated esters 105
di-and trisubstituted quinolones 205
diaryl-substituted 1,3-diynes 102
dibenz[c,e]azepines 248
dibenzylamine by-products 46
dibenzyl amine-type side-products 46
dicationic dimeric complex $[(Me_4TACD)_2Ca_2H_2]^{2+}$ 161
dichloride complex 44
dicopper hydride complex 95
dideuterated *Z*-allylic chloride 97
diethyl ether 170, 213
2,3-dihydrocarvone 47
(*R*)-dihydrofarnesal 264
dihydrogen mediated cross-coupling of internal alkynes and aryl iodides 105–106
3,4-dihydro-2*H*-1,4-benzothiazines 248
dihydronaphthalenes 51
1,4-dihydropyridine structure 261
1,2-dihydroquinoline 203, 271
1,4-dihydroquinoline 203, 276
2,6-di-*iso*-propylphenyl-substituted bis(imino)pyridine (^{iPr}PDI) 23
diketone 5-phenylpentan-2,4-dione 234
(dimeric) copper(I) complex 87

dimethoxymethane (DMM) 42, 43
2,3-dimethyl-but-2-ene 214
3,3-dimethyl-1-butene 68, 71
(3*R*,5*R*)-diol 240
1,4-dioxane 79, 100, 125, 126, 128, 198
1,2-diphenylethane 54, 127
1,1-diphenylethylene 146–148, 152
1,2-diphenylethylene 152
diphosphane-borane ligand supported nickel complex 70
diphosphine Xantphos 73
$[(DippNacNac)CaH\cdot(THF)]_2$ 145
$[(DippNacNac)CaH\cdot(THF)]_2$ (DIPPnacnac = HC[(CMe)$N(C_6H_3$-2,6-*i*|*i*Pr)$]_2$ 144
direct hydrogenation 26, 39
1,1-disubstituted alkenes 118
1,1-disubstituted α-methylstyrene 68
3,4-3,5-, and 4,5-disubstituted γ-butyrolactones 245
1,2-disubstituted indenes 75
2,9-disubstituted 1,10-phenanthrolines 276
2,3-disubstituted quinolines 205, 276, 277
2,4-disubstituted quinolines 205
2,3-disubstituted quinoxalines 205
1,4-disusbstituted cyclohexenes 75
1,3-diyne 98, 102–104
$(DMAT)_2Ca\cdot THF)_2$ 145, 152, 154, 161
$(DMAT)_2Sr\cdot(THF)_2$ 145
1-dodecene 73
double axially chiral phosphoric acid 275
Dowex MWA-1 231
dynamic kinetic resolution (DKR) 80, 81, 230, 270
dynamic reductive kinetic resolutions (DYRKRs) 230

e

E-and *Z*-α,β-unsaturated iminium ion intermediates 263
E-and *Z*-isomers 27
Ebalzotan 245
E,*E*-1,3-dienes 102
electron deficient alkenes 76–78
electron deficient allenes 208, 211
electron-donating bisphosphine ligands 66
electron-donating fluorinated phosphines 172
electron-donating Lewis bases 172, 213
electronically modified moisture-tolerant triaryl boranes 186
electrophilic phosphonium cations (EPC) 169, 170, 215–217
elemental magnesium 112
enamides 49, 271
enamines 118, 137, 182, 183, 191, 193–198, 201
enantioenriched tetrahydro-β-carbolines 271
enantiomerically pure mono-and disubstituted piperidines and pyrrolidines 249
enantiomerically pure (*R*)-profen derivatives 242
enantiopure (*R*)-2-bromo-1-(4-fluorophenyl)ethanol 236
enantiopure bromohydrin 237
enantiopure *cis*-(3*S*,5*R*)-amino lactone 235
enantiopure (*S*)-pentane-1,4-diol 235
enantioselective catalysts 4
enantioselective hydrogenation 5, 6, 8–10, 29–31, 33, 47, 51, 52, 78, 83, 92, 124, 191, 193, 197, 207
encounter complex 169
ene-reductases
  multicatalytic processes 244–247
  substrates 242–244
enoates 91
enones 89–91, 197, 207–211, 242, 264, 265
enzymes 5, 15, 64, 142, 227–229, 232, 234, 235, 237, 239–241, 247–252, 261
E-selective hydrogenation of alkynes 6, 7
E-selective semihydrogenation of alkynes 79, 80

esters 6, 11, 31, 55, 80, 91, 98, 99, 101, 105, 113, 127, 128, 130, 132, 242, 245, 280
  hydrogenation 6
  and lactone hydrogenation 130
(*R,R*)-EtDuphos 49
ethylarenes 73
ethyl β-methylcinnamate 76
ethyl (*S*)-4-chloro-3-hydroxybutanoate [(*S*)-CHBE] 238
ethyl 4-chloro-3-oxobutanoate (COBE) 238
ethylene-bridged ligand complex 66
ethylene-bridged phosphino borane 182, 193
(*S*)-ethyl-4-hydroxypentanoate 232
ethyl levulinate 232
$Et_2O/B(C_6F_5)_3$ 214
extracellular-regulated kinase inhibitor GDC-0994 240
E/Z-stilbene mixture 54

f

ferrate complex 24
ferrocenophane-derived dienamine 194
flexible ethylidene-bridged phosphinoborane 167
FLP-catalyzed hydrogenations 170, 171, 185, 194, 196, 198, 201, 204, 207
fluorinated phosphines 172, 213
2-fluorobenzyl zinc chloride 49
(*S*)-5-fluoro-3-methylisobenzofuran-1(3*H*)-one 240
(*R*)-flurbiprofen 243
formamides 32, 56, 133
frustrated Lewis pair (FLP) 2, 143
  acceptorless dehydrogenation 180–181
  air-stable FLPs 184–186
  alkylidene malonates 207
  choice of Lewis acids 171–173
  dehydrogenative coupling 179–180
  electrophilic phosphonium cations (EPC) 215–217
  hydrogenation of
    aldimines and ketimines 187–193
    enamines 193–198
    enones 207–211
    heterocycles 201–207
    ketones 198
    olefins 178–179
    polycyclic aromatic hydrocarbons 211–215
    silylenol ethers 193–197
    unpolarized olefins 211–215
  intramolecular 181–184
  Lewis acid and Lewis base 171–173
  Lewis acidity *vs.* Lewis basicity 173–186
  mechanistic considerations 168–171
  nitroolefins 207–201
  reductive deoxygenations 198–201
  Stephan's intramolecular 167
furyl imidazolidinone 264
fused piperidine 270

g

(+)-galipinine 274
γ-aminobutyric acid analogs 243
γ-deuterated terminal alkenes 96
γ-valerolactones 235
geminal FLPs 183
genetically engineer enzymes 5
greenhouse gas and waste product CO_2 81
Grignard reagents 49, 144, 186
Guerbet alcohols 246, 247

h

hafnium 112–119
4-halobut-2-enals 247
Hammett substituent constant 176
Hantzsch esters (HEHs) 3, 11, 261, 268, 269, 280
Helional^{TM} 242
hemiaminals 270
(hetero)arenes 46–57
heteroaromatic compounds 6, 8, 10
heteroaromatics 99, 261, 274–280
heteroatom-based ether 88
heterobimetallic complexes 162

heterocycles 48, 54–56, 170, 201–206, 212, 270, 274
heterogeneous catalysts 1–3, 5–8, 26, 45, 47, 55, 68, 141
heterogeneous Pd, Pt, Rh, Ru and Ni catalysts 5
heteroleptic boranes 186
heterolytic dihydrogen activation 88, 89, 95
hexamethylphosphoramide (HMPA) 146
hexan-1-ol 246, 247
1-hexene 113, 134, 147–149, 152, 153, 161
2-hexene 113, 150, 152
H_2-mediated *formal* alkyne semihydrogenation 98
homogeneous nickel catalyzed hydrogenations
  alkenes 65–76
  alkynes 78–79
  carbon dioxide 81–83
  ketones 79–81
homogeneous non-noble catalysts 6
homogeneous transition metal hydrogenation catalysts 142
homogenous catalyzed olefin hydrogenation 113
homoleptic dibenzyl calcium complex $(DMAT)_2Ca{\cdot}(THF)_2$ 146
HOMO–LUMO combinations 16, 141, 142
homolytic dihydrogen activation 88
horse liver alcohol dehydrogenase (HLDH) 230
hydricity 20, 28, 82
hydrogenases 15, 142, 228
hydrogenated quinolines 11
hydrogenation
  aldehydes and ketones 29–31
  aldimines and ketimines 187–193
  alkylidene malonates 207–211
  alkynes 27–28
  amides 32
  C=C double bonds 144–153
  chemoselective 6–8
  C=N double bonds 153–157
  C=O double bonds 157–160
  of enamines 193–196
  enantioselective 8–9
  enones 207–211
  esters 31–32
  heterocycles 201–206
  ketones 198
  nitroolefins 207–211
  olefins 26–27, 178–179
  polycyclic aromatic hydrocarbons 211–215
  reductive deoxygenations 198–201
  silylenol ethers 193–198
  unpolarized olefins 211–215
hydrogen–dihydrogen complexes *vs.* dihydride complexes 39–40
hydrogenolysis 5, 66, 88, 114, 270, 272
hydrogen source 19, 53, 55, 104, 135, 191, 261, 262, 266, 268, 269, 274
hydrometalation 39, 78, 142
hydrosilanes 81, 82, 87, 88, 95, 106, 112, 122, 180, 198
1,4-hydrosilylation 195
hydroxyaniline 273
2-hydroxycyclopentanecarbonitrile 233
hydroxyethylamine moieties 237
hydroxyethylene 237
2-hydroxylchalcones 278
(*R*)-3-hydroxy-2-methylpropanoate 242
4-hydroxy-5-phenylpentan-2-one 234
hydroxy-substituted quinoline 56

i

imido complex 128
imine-based iron pincer complex 30
imine reductases 228, 248–249
imines 8, 29, 30, 33, 34, 48, 57, 115–117, 122, 125, 135, 136, 144, 153, 156, 170, 171, 176, 183, 187, 189, 193, 201, 212, 228, 248–249, 267–270
indenes 51, 52, 75
3*H*-indoles 268
indole-substituted 1,1-diaryl ethylenes 266

indolines 191, 248
1-*H*-indolinium 181
in situ product removal (ISPR) approach 238
intermediate copper-*O*-enolate 89
intermolecular “frustrated” Lewis pairs 167
internal alkynes and aryl iodides 12, 105–106
intramolecular frustrated Lewis pair (FLP) 168, 181–184, 191
2′-iodo-[1,1′-binaphthalen]-2-amine 183
[IPrCuOH] 101, 102, 104
iron-catalyzed homogeneous hydrogenation of C—C multiple bonds
  alkynes 27–28
  olefins 26–27
iron-catalyzed homogeneous hydrogenation reactions
  carbon dioxide 25
  HOMO-LUMO-combinations 16
  nitriles 26
  noble *vs.* 3d metal complexes 17–22
  non-polar substrates 22–24
  polar substrates 24–25
iron-catalyzed hydrogenation of C—N-multiple bonds 32–34
iron-catalyzed hydrogenation of C—O-multiple bonds
  aldehydes and ketones 29–31
  amides 32
  esters 31–32
iron(II) complex $[(PP_3)Fe(H)(H_2)](BPh_4)$ 23
iron pincer catalyst 26, 28
(*S*,*S*)-isochroman 232
isochromanes 279
1*H*-isochromenes 279
isolable cationic triphos/copper(I) complex 95
isolable copper hydride hexamer $[PPh_3CuH]_6$ 88
isolated cationic dicopper hydride complex 95
isolated copper(I) formate complex 95
isophorone 89–91
isopropanol (IPA) 3, 19, 43, 53, 122, 126–128, 132, 135, 136, 229, 261
isoquinolines 278
isosolenopsin A 204, 206

j

Josiphos-type ligand 80

k

Kempe catalyst 48
ketimines 33, 81, 154, 170, 187–193, 266–268
ketones 6–8, 19, 26, 29–31, 44, 45, 47–50, 79–81, 89, 91–93, 121–127, 136, 137, 159, 162, 198, 229, 230, 245, 246, 270
ketoreductases (KREDs)
  alcohol dehydrogenases
    in multicatalytic processes 232–236
    in non-conventional media 229–230
  coupled enzyme approach 229
  coupled substrate approach 229
  dynamic processes 230–232
  valuable molecules 236
Knölker’s catalyst 30
Knölker’s complex 29

l

lactone-fused quinolines 276, 277
lanthanide catalyzed alkene hydrogenation 143
levulinic acid 232, 233, 235
Lewis acidic ligands 64
Lewis acidity *vs.* Lewis basicity 173–186
Lewis acids 172–173
  $B(C_6F_5)_3$ 144
  $(HSiR_3/B(C_6F_5)_3)$ 134
  and Lewis base 171–173
Lewis adduct 167–169, 172, 182

Lewis base 143, 144, 167, 169–174, 178–182, 187, 191, 198, 200, 201, 209, 213, 217
Lewis pair's heteroatoms 169
Lilial^{TM} 242
linear imines 268
lorlatinib 240
low-valent titanocene species 112
2,6-lutidine 168, 200, 201, 209

m
macrocyclic alkyne 79
malonates 179, 185, 207–211
manganese 121–136
marine macrolide (+)-neopeltolide 263
Me-DuPhos 76
Meerwein–Ponndorf–Verley reduction mechanism 122
(+)-menthol 115
$MesB(C_6F_4H)_2$ 202
mesitylcopper(I) and imidazolinium salt 99
Mes_3P 170
$(Me_4TACD)_2Ca_2H_3^+$ 148
Me_4TACD-stabilized cationic calcium hydride complex 148
Me_4TACD-stabilized dicationic calcium hydride complex 148
metal
  abundance 2
  catalyst separation and trace metal removal 3
  cost 2
  different reactivity 3
  toxicity 2
metal-free imine hydrogenation 143
metal-free *syn*-selective hydrogenation of alkynes 182
metalloenzymes 15, 34
methyl aryl ketimines 189
methyl (2*S*,3*R*)-2-benzamidomethyl-3-hydroxybutyrate 231
methyl 2-benzamidomethyl-3-oxobutanonate 231
methyl (*S*)-2-bromobutanoate 243
2-methylcyclohexanone 47
methylcyclohexene 74, 214
3-methylcyclohex-2-enone 241
2-methyl-6-fluoroquinoline 276
4-methylheptan-3-ol 245
2-methyl-3-substituted tetrahydrofurans 245
Mg(I) complex 162, 163
Michael-initiated ring closure (MIRC) reaction 247
moisture-tolerant $(C_6F_5)_2BC_6Cl_5)$/THF system 186
moisture tolerant FLP-catalyst 185
molecular hydrogen 39–40, 153, 167–217
molybdenum 121–122, 137
mono-and di-substituted olefins 26
Montelukast sodium (Merck) 238
Muguesia® 245

n
NADH-dependent KRED 230
N-alkyl imines 268
(*S*)-*N*-Boc-3-hydroxypiperidine 239
N-Boc-pyrrolidinone 241
n-butyllithium 112
(−)-neomenthol 115
neutral tetradentate ligand Me_4TACD 148
NHC/copper(I) complex 95
N-heterocyclic carbene 68
N-heterocyclic silylenes 71
$[Ni(dppe)_2]$ 81
nickel ammonia complex 67
nickel-bis-silylene complex 74
nickel catalyzed olefin oligomerizations 63
nickel, properties of 63
$Ni(CO)_4$-promoted cyclooligomerizations 63
nicotinamide biomimetics (mNADHs) 242
nicotinamide cofactor 227–229, 241, 245, 249, 251
niobium 119–121
nitriles 40
  reductases 250–251
nitroalkenes 242, 266

nitrobenzene 251
nitroolefins 78, 79, 179, 185, 207–210, 212, 265, 266
4-(4-nitrophenyl)butan-2-one 246
nitroreductases 250–251
N-methylhexahydrocarbazole 181–182
noble *vs.* 3d metal complexes
  catalytically inactive complex [(κ^4-PNNP)Fe(CO)] 21
  diamagnetic and trigonal bipyramidal complex 19
  hydricity 20
  penta-coordinated iron(0) complexes 21
  pincer-type complexes 21
  potentially paramagnetic, penta-coordinated intermediates 20
  ruthenium(II) complexes 18
  Shannon radii 17
  square planar and diamagnetic Vaska's complex 18
  square pyramidal amido iron complex 19
  square pyramidal complex 19
  two-electron-redox steps 18
non-functionalized olefins 8
non-innocent/redox-active ligands 23
non-noble metal catalysts 2–3
non-polar substrates 20, 22–24, 29, 34
non-traditional hydrogenation catalysts 2
norbornene 71, 72, 152, 153, 161
Noyori's Ru-catalyzed asymmetric hydrogenation of ketones 159
N-protected chloroketones 237
N-*t*butylimines 174, 176
N-terminal prolyl peptide 264

o

1-octene 53, 65–68, 71, 112, 149
2-O*i*|*i*Pr-5-*t*Bu-C_6H_3-derivative 205
olefins 26–27
  hydrogenation 23, 24, 26–27, 113, 142, 153, 179–180, 215
oligomeric products 146
optically active α-alkyl β-hydroxy esters 231
optically active 3-arylpropan-2-ols 232
optically active β-hydroxy ketone 236
o-quinone methide (*o*-QM) intermediate 266
organocatalytic metal-free hydrogenation 143
ortho hydroxyaryl ketimines 268
ortho-substituted aromatic ketones 125
ortho-substituted aryl ketones 92
Osborn complex 88
o-Tol_3P 173
oxalyl chloride 198–201
oxime ethers 195
2-(3-oxobutan-2-yl)benzonitrile 231
oxygen-trapping reagents 198

p

[2.2]paracyclophane bisphosphine-derived FLP 195
para-hydrogen induced polarization of NMR (PHIP) signals 40
p-chlorobenzaldehyde 236
pentacarbonyl metal acetate 121
pentacarbonyl metal hydride 121
penta-coordinated iron(0) complexes 21
pent-4-yn-1-ol 235
1-phenyl-but-1-yne 55
1-phenyl-1-cyclohexene 75
S-1-phenylethanol 30, 31, 49
1-phenylethylene-bridged ammonium borate 194
$[PhMe_2NH][B(C_6F_5)_4]$ 119
phosphine-derived Lewis base 167
phosphine-free imine hydrogenation 201
phosphine oxide 199–200
phosphinimines 170
phosphinoborane 167, 182
phosphothreonine (pThr) 277
phthalic acid 43
Piers' borane ($HB(C_6F_5)_2$) 182, 191, 204, 205, 214

pincer-type complexes 21, 26, 32, 44, 45
pincer-type ligand design 42
piperidines 204, 248, 250, 270
planar chiral phosphoric acid 275
(−)-platensimycin 265
platinum black 141
p-methoxy styrene 152
polar silanes (R_3SiH) 162
polar substrates 22, 24–25, 34
polycyclic aromatic hydrocarbons 211–215
polyleucine tether 264
potentially paramagnetic, *penta*-coordinated intermediates 20
$[PPh_3CuH]_6$ 87, 88
pregabalin [(*S*)-3-aminomethyl-5-methylhexanoic acid] 243
prochiral aryl-alkyl ketones 49
prochiral 1,1-diarylethene derivatives 51
prochiral ketimines 188
  yields chiral secondary amines 33
prochiral ketone 30, 237–240
prodrug (*S*)-licarbazepine 239
prokaryotic [NiFe]-hydrogenases 64
propane-1,2-diol 249, 250
propargylic silyl ether 97
propylene-bridged ligand 66
protodemetalation 39, 103, 104
pyrene 55
pyridine-based iron pincer complex 29, 32
pyridines 3, 101, 124, 168, 204–206, 277
2-and 4-pyridyl substituted alkynes 79
pyrophosphate (PPi) 249
pyrrolidine heterocycles 270
pyrrolidines 118, 248, 249, 270

q

quinazolines 193
quinoline-derivatives 202
quinolines 8, 87, 191, 203, 205, 209, 274–278
quinoxalines 205–207

r

rac-cuspareine 203
racemic 2*H*-chromenols 279
racemic 3-phenylbutan-2-one 232
Raney cobalt 46, 50
Raney Co 2724® 50
Raney nickel 50, 63, 141
reactive bisalkenyl derived bisborane 196
Red-Al ($[LiH_2Al(OCH_2CH_2OCH_3)_2]$) 115
redox active bis(imino)pyridine ligand 26
redox-active ligands 17, 23
reduction of CO_2 to formate 93–95
reductive aminases 249
reductive deoxygenations 198–201
$RhCl(PPh_3)_2$ 142
rhenium 122–136
rhodium complex $[Rh(nbd)PPh_3^+][PF_6^-]$ (nbd = norbornadiene) 142
(+)-ricciocarpin A 263
rigid tetrafluorophenylene bridged phosphonium fluoroborate 167
Robalzotan 245
Roche ester 242
Rotigotine 245
ruthenium(II) complexes 18, 20

s

saturated alcohols 89
scandium 111–112, 137
Schlenk equilibrium 144
Schwartz's reagent 114
selective C=O reduction of unsaturated carbonyl groups 6–8
semihydrogenation 7, 23, 27, 78–80, 97–105
Shannon radii 17
Shvo/Noyori catalysts 155
Shvo's ruthenium catalyst 135, 136
σ-bond metathesis 89
  mechanism 145
silylene-based ligands 73
silylenol ethers 193–196
simple enamines 182, 194

simple (non-conjugated) ketones and aldehydes 92
(+)-solifenacin 278
square pyramidal amido iron complex 19
square-shaped tetrametallic Ni(I) cluster 68
stable and isolable secondary ketimines 33
Stephan's intramolecular frustrated Lewis pair 167
stereodivergent catalyst system 52
sterically demanding β-diketiminate ligand 144
sterically encumbered mesityl-substituted borane 184
sterically hindered chiral phosphoric acid 266
sterically hindered TRIP anion and morpholinium ion 264
sterically hindered triphenylsilyl-substituted chiral phosphoric acid 270
sterically nonhindered α,β-unsaturated aldehydes 264
stilbene 105
stronger/weaker Lewis-acidic boranes 172
strongly Brønsted acidic 3-H-indolium 181
Stryker's reagent 87, 88
styrene-butadiene-styrene (SBS) copolymer 115
styrene-divinylbenzene co-polymers 113
styrenes 47, 70, 73
2-substituted acetophenone derivatives 49
3-and 4-substituted aryl derivatives 49
2-substituted piperidines 250
2-substituted quinolines 275–277
3-substituted quinolines 276
4-substituted quinolines 276
substituted quinolines and quinoxalines 10
2-substituted quinoxalines 276
2-substituted tetrahydroquinolines 275
synthetically valuable allylic alcohols 91

t

tantalum 119–121
(TBD)-protected propargylic alcohol 102
t-Bu_3P 170
terminal alkenes 9, 53, 95–98
tert-amyl alcohol 125
4-*tert*-butylcyclohexanone 238
tert-butyl-(5*R*)-6-cyano-5-hydroxy-3-oxohexanoate 240
tertiary borane C_6F_5–CH_2CH_2–$B(C_6F_5)_2$ 204
tethered NHC/copper(I) alkoxide complex 99
tetracarbonyl metal acetate 121
tetradentate monoanionic ligand Me_3TACD^- 147
tetradentate *N,N,O,O*-ligand 68
1,2,3,4-tetrahydroisoquinolines 248
tetrahydropyridines 277
tetrasubstituted acyclic alkenes 75
tetrasubstituted indene 75
tetrasubstituted silyl enol ethers 195
tetrasubstituted tetramethylethylene 73
3-thiazolidines 248
thiophene-substituted internal alkene 101
thiophenyl-or furyl-substituted nitroolefins 185
thiourea-based bifunctional organo-catalyst 274
3d metals 15, 17–22, 34
three point interaction model 268
threo-or *erythro*-N-protected chlorohydrins 237
Ticagrelor 240
titanium 112–119, 122
titanocene catalyst 115, 118, 143
transfer hydrogenations (TH) 3, 19, 28–30, 39, 52–56, 76, 79, 81, 104–106, 122, 123, 125–127, 135,

transfer hydrogenations (TH) (*contd.*) 161, 180, 191, 236, 248, 252, 261–279
trans-2-octene 71
triarylboranes 185, 186
trichloroacetic acid (TCA) 262
tridentate *N,N,O*-ligand 68
3-trifluoromethyl substituted quinolines 276
trimethylborane 168
trimethylcyclohexane 56
1,4,7-trimethyl-1,4,7,10-tetraaza-cyclododecane 147
triphenylphosphine oxide ($O{=}PPh_3$) 199
triphos/copper complex 94
triphos-type ligand 42, 57
trisubstituted alkenes 53, 74, 105, 118
tri-substituted β-trifluoromethyl nitroalkenes 266
trisubstituted methylcyclohexene 74
tungsten 121–122

u
unfunctionalized tetrasubstituted olefins 119
unfunctionalized trisubstituted alkenes 118
unpolarized olefins 211–215
unreactive dicarbonyl complex 44
unstabilized Meisenheimer anion ($C_6H_7^-$) 161
unsymmetrically substituted ketimines 188
unsymmetrical PNN-Pincer ligand 46

v
valuable chiral *cis*-2-substituted cyclohexylamines 270
valuable molecules 236
vanadium 119–121
Verkade's base 41
very strong Lewis acid $B(C_6F_5)_3$ 172, 175
vicinal B/N-based *intramolecular* frustrated Lewis pairs 194
vicinal Lewis pairs 183
vinyl phenols 266

w
water-tolerant borane-derived FLP 194

y
ynones 207, 211
yttrium 111–112, 137

z
Z-acrylate 98, 102
Z-allyl amine 101
Z-allyl silylether 102
Z-enyne 103
zirconium 112–119, 168
Z-selective alkyne semihydrogenation 98–104
Z,Z-1,3-dienes 102